Chemisorption and Reactions

on

Metallic Films

VOLUME 2

Physical Chemistry

A Series of Monographs

Edited by ERNEST M. LOEBL

Department of Chemistry, Polytechnic Institute of Brooklyn, New York

Chemisorption and Reactions

on Metallic Films

Edited by

J. R. ANDERSON

CSIRO, Division of Tribophysics,
University of Melbourne,
Australia

VOLUME 2

1971

ACADEMIC PRESS
London and New York

ACADEMIC PRESS INC. (LONDON) LTD
Berkeley Square House
Berkeley Square
London, W1X 6BA

U.S. Edition published by

ACADEMIC PRESS INC.
111 Fifth Avenue
New York, New York 10003

Library of Congress Catalog Card Number: 78–109033

SBN: 0 12 058002 0

PRINTED IN GREAT BRITAIN BY
W. S. COWELL LTD
IPSWICH, SUFFOLK

List of Contributors

J. R. ANDERSON, *CSIRO, Division of Tribophysics, University of Melbourne, Australia.*

B. G. BAKER, *School of Physical Sciences, Flinders University, Adelaide, Australia.*

*J. W. GEUS, *Central Laboratory, Staatsmijnen/DSM, Geleen, The Netherlands.*

*D. O. HAYWARD, *Department of Chemistry, University College, London, England.*

*D. F. KLEMPERER, *Department of Physical Chemistry, University of Bristol, England.*

*L. H. LITTLE, *Department of Physical Chemistry, University of Western Australia, Nedlands, Western Australia.*

I. M. RITCHIE, *Department of Physical Chemistry, University of Melbourne, Melbourne, Australia.*

D. R. ROSSINGTON, *State University of New York, College of Ceramics at Alfred University, Alfred, New York, U.S.A.*

*J. V. SANDERS, *CSIRO Division of Tribophysics, Melbourne, Australia.*

---

* Authors in Volume 1

Preface

The purpose of these two volumes is essentially to describe the contributions which evaporated metal films have made towards our understanding of the chemistry of the metal/gas interface. Metal films have been so extensively used for this purpose, and have contributed so much towards our understanding of adsorption and catalysis at metal surfaces, that little further justification is needed, save to note that, to the best of my knowledge, other books of similar intent to these do not exist. I hope that these volumes will be useful for all those who, working at the graduate student or research worker level, are interested in the chemistry of the metal/gas interface, particularly the fundamentals of adsorption and catalysis.

In a field as broad as this, one cannot expect to cover all aspects of it within a limited space. Some selection of material is inevitable. A choice was made with three principles in mind: that the work should be restricted mainly to *chemical* reactions; that this chemistry needs to be seen in relation to a detailed account of the structure of metal films and of their surfaces; that it was desirable to try, by the selection of material, to generate some sense of coherence and continuity. Some consequences of this policy will be immediately apparent. Rather more than three chapters deal with film structure, methods of studying it, and with theoretical and experimental matters concerning film growth. Some obviously important areas in metal/gas interactions such as physical adsorption, have only been dealt with in passing. In the main, we have avoided tying individual chapters to specific techniques. The detailed discussion of film structure is included at least as much with an expectation for the future as with an appreciation of the past, since it seems clear that the control of surface structure as an experimental variable will be of rapidly increasing importance for future work.

It is obviously impossible to write *exclusively* about metal film surfaces. However, all of the authors have made this their central theme, but all have also made use of ancillary data derived from other types of systems when this was useful for illustration and comparison.

There is, of course now a major technology based upon thin films, much of it relating to electronic devices. It has not been intended that the present work should concentrate at length upon the physics of thin films or upon the associated technology. Nevertheless, phenomena at the metal/gas interface are important to various aspects of electronic device technology, including that part which uses thin films, and these volumes should be useful to those who work in this area.

Technical processes based on adsorption and catalysis are of immense economic importance, and it is not surprising that, over the years, much effort has been devoted to their study. However, technical catalysis remains a subject of great complexity, and large areas of it are only understood at a relatively empirical level. In this situation, studies using model absorbents and catalysts assume great importance. The wish to work with well characterized surfaces has a long history in this field, and evaporated metal films have in this way played a major role, and there is no doubt that they will continue to do so, particularly as epitaxed films and modern methods for the characterization of their surface fine structure become more widely used.

My own contributions, both as editor and as an author, were made at three separate institutions. Most of the work was done at the General Electric Research and Development Center, Schenectady, and I must take this opportunity to thank my many friends there who made my stay so enjoyable and who contributed materially by way of discussions and other forms of help. In particular I must thank Drs. R. W. Roberts, W. Vedder, H. W. Schadler, C. B. Duke and Bruce McCarroll. The remaining work was done at Flinders University, Adelaide and more recently at the C.S.I.R.O. Division of Tribophysics, Melbourne, where a number of colleagues gave valuable assistance. Finally, I must thank Professor F. G. Gault of the University of Caen who generously gave me much help and advice concerning hydrocarbon reactions over metal film catalysts, particularly in relation to the results of the unpublished work from his own research group. The opportunity for discussions with him at Caen, and under more salubrious circumstances at Val d'Isère, are very much appreciated.

March, 1971

J. R. ANDERSON,
CSIRO
Division of Tribophysics,
University of Melbourne

CONTENTS

VOLUME ONE

Chapter 1

Structure of Evaporated Metal Films

J. V. SANDERS

Chapter 2

Experimental Techniques

DEREK F. KLEMPERER

Chapter 3

Fundamental Concepts in Film Formation

J. W. GEUS

Chapter 4

Gas Adsorption

D. O. HAYWARD

Chapter 5

The Influence of Adsorption on Electrical and Magnetic Properties of Thin Metal Films

J. W. GEUS

Chapter 6

Infrared Spectra of Surface Species

L. H. LITTLE

VOLUME TWO

Chapter 7

Adsorption, Kinetics and Surface Structure in Catalysis

J. R. ANDERSON and B. G. BAKER

Chapter 8

Catalytic Reactions on Metal Films

J. R. ANDERSON and B. G. BAKER

Chapter 8—*contd.*

Chapter 9

Properties and Reactions of Alloy Films

D. R. ROSSINGTON

Chapter 10

The Oxidation of Evaporated Metal Films

I. M. RITCHIE

Chapter 7

Adsorption, Kinetics and Surface Structure in Catalysis

J. R. ANDERSON* and B. G. BAKER

*School of Physical Sciences,
Flinders University, Adelaide, Australia*

I. INTRODUCTION

Widespread use of metal film catalysts dates from the pioneering work of Beeck *et al.* (1940), and since that time many studies have been carried out over catalysts of this type. Virtually all of these have been concerned with elucidating reaction mechanisms, and metal films have been adopted as a conveniently accessible form of metal surface of reasonable cleanliness and substantial surface area. The vast majority of these studies have used random polycrystalline metal films in which a variety of crystal planes are exposed to the gas, and in which various types of surface imperfection are also present: although these facts have undoubtedly been widely known, they have also been largely ignored on the general grounds that control and characterization of the surface details have been, at best, extremely difficult. This situation is gradually changing as better means become available for studying film and surface structure, and as the art of epitaxy becomes better developed and

*Present address: CSIRO Division of Tribophysics, University of Melbourne.

1

understood (cf. Chapters 1 and 3). Nevertheless, it is clear that for an appreciation of the bulk of the existing literature, some detailed comment is required concerning film structure and properties, particularly of random polycrystalline films. This chapter also presents some comments on general adsorption and kinetic phenomena, with particular emphasis upon concepts that are subsequently used in Chapter 8 where the chemistry of reactions over film catalysts is discussed in detail.

II. The Surface Structure of Metal Films

Films that are used for laboratory studies of adsorption and catalytic processes are typically a few hundred angstroms or more in thickness, and this section deals with such features as seem most important for their use in surface chemistry.

A. structure and sintering processes

It is often found that the ratio R of actual surface area accessible for gas adsorption to the apparent geometric film area exceeds unity, frequently by a substantial factor, and in these cases the porous nature of the film is the major factor. In general, R will depend on the nature of the metal, and on film deposition conditions such as substrate temperature and deposition rate.

The porosity is a consequence of the propagation of inter-crystal gaps which originate in the earlier stages of film growth, and this growth mechanism, particularly its early stages is fully discussed in Chapters 1 and 3. On a more macroscopic scale, porosity will also be enhanced by shadowing of film areas from the source by tall crystals. Of course, if the temperature is high enough the formation of a continuous film will be facilitated by a re-orientation of the crystals relative to one another and by recrystallization. Re-orientation may be of particular importance in the elimination of grain boundaries and the formation of an epitaxed deposit.

Implicit in this model is the assumption that the film is one crystal thick. This is in agreement with available evidence for films up to about 1000 Å thickness (Anderson *et al.*, 1962; Suhrmann *et al.*, 1963). It is possible that renucleation during growth may occur in very thick films, particularly when deposited at a low substrate temperature; however, the same general model for porosity will still be valid.

We may assess the degree of film porosity in terms of the parameter τ introduced by Sanders and equal to T/T_M, where T °K is the film temperature during deposition and T_M °K is the melting point of the metal. If τ is high enough the film will be produced with R approximately unity; such appears to be the case, for instance, for zinc (MP

419 °C), cadmium (321 °C), indium (156 °C), tin (231 °C) and lead (327 °C) deposited on a substrate at 0 °C (Trapnell, 1953; Anderson and Tare, 1964) when τ lies in the range 0.39–0.64. On the other hand, the evaporation of the transition metals which typically have $T_M > 1500$ °C on to a substrate at 0 °C produces porous films with R typically in the range 5–20, and for these $\tau < 0.15$, while the deposition of lead on a substrate at 90 °K with $\tau = 0.15$ gives R $\simeq$ 10 (Anderson and Tare, 1964). The deposition of copper (MP 1083 °C) on to a substrate at 0 °C with $\tau = 0.20$ gives R $= 1.5$–2 (Allen et al., 1950, 1959). Thus, the upper limit to τ for the retention of some degree of film porosity is probably in the region 0.15–0.20 as a general rule; that is, film porosity extends from range $1(\tau \leq \tau_{sd})$ into the lower part of range 2 ($\tau_{sd} < \tau < \tau_r$) on the classification adopted by Sanders.

The influence of film deposition rate on porosity arises because for a given temperature the rate of provision of metal by surface diffusion for the elimination of crystal gaps will be expected to increase with an increased flux of arriving atoms, and this agrees with the behaviour observed by Anderson et al. (1962) and by Thun (1963). In fact, it is usually observed that there is a lower limit to the deposition rate, below which the specific film area (cm²mg⁻¹) increases rapidly, but above which there is a region where the specific film area does not vary strongly with deposition rate. For deposition at 0 °C on baked glass from a source at a distance of 2–4 cm, this range of deposition rates is tungsten $> 2 \times 10^{-3}$ mg cm⁻² min⁻¹; nickel 2.5–4 $\times 10^{-3}$ mg cm⁻² min⁻¹; palladium 3–5 mg cm⁻² min⁻¹ (Anderson et al., 1962; McConkey, 1965).

The surface of a polycrystalline film has a degree of surface roughness that depends on the deposition conditions. For instance, for nickel deposited on glass at 0 °C to room temperature ($\tau \simeq 0.16$), replication of the surface shows a roughness due to the individual crystals, as shown in Figure 1 (a and b), while the surface of an individual crystal is also rough due to the presence of terraces and steps, as shown in Figure 1d. However, it should be emphasized that the surface roughness evident from such micrographs is far too little to explain the measured values of R by a factor that may be ten or more, and there is no doubt that the lower regions of the film must then be accessible to the gas phase via intercrystal gaps. These gaps are not seen in replication, but they have been resolved in some cases and their dimensions estimated by transmission electron microscopy. The possibility of direct observation diminishes with increasing thickness, and a comparison of the results of Suhrmann et al. (1963) with those of Anderson et al. (1962) shows that for nickel films deposited on glass at 0 °C to room temperature under

B

FIG. 1. (a) Nickel film deposited on glass substrate at 0 °C, specific film weight 0.05 mg cm^{-2}; (b) nickel film deposited on glass substrate at 20 °C, thickness about 3.1 μ (Reproduced with permission from Suhrmann, Gerdes and Wedler (1963). *Z. Naturf.* **18a**, 1212); (c) nickel film deposited on mica at 350 °C, specific film weight 0.05 mg cm^{-2}; (d) surface of (111) crystal in a polycrystalline silver film decorated with gold (courtesy J. V. Sanders).

ultrahigh vacuum conditions, the gaps are too uniformly narrow to make resolution possible, although they are visible with similar films deposited at pressures in the vicinity of 10^{-7} torr. As the substrate temperature increases the degree of surface roughness decreases, and Figure 1c shows the surface replica of a polycrystalline nickel film deposited on a cleaved mica substrate at 400 °C ($\tau = 0.39$); in this case the surface is almost featureless. Of course, such a replica does not imply complete surface perfection on a molecular scale. Furthermore, it should be remembered that the true surface area can only be measured with an absolute accuracy of about $\pm$ 20% so the measured value of R is only an index of a quasi-macroscopic porosity.

For metals deposited so that R > 1, for instance the transition metals

on to a substrate at 0 °C, the value of R increases with increasing film
thickness. Typical is the behaviour of nickel (Beeck *et al.*, 1940) and iron
(Porter and Tompkins, 1953) for which there is a linear increase in R
with increasing specific film weight (weight per unit area), indicating a
uniformly porous film structure in this range. In some cases this linear
relation is only valid above a lower limit to the specific film weight since
the film structure is not constant over the entire thickness range. A
great many, probably the majority of surface studies have been made
with specific film weights in the region 0.02–0.2 mg cm^{-2}. It is useful to
have an indication of the magnitude of R in relation to the specific film
weight (w) for various metals, and this is most readily expressed by
R/w(cm^{2}mg^{-1}). This ratio is itself subject to a number of uncertainties
such as evaporation rate and evaporation geometry, but the following
list culled from the literature gives the observed ranges of values as a
guide to the behaviour to be expected, in each case for a film deposited
on a substrate nominally at 0 °C: tungsten, 270–360; molybdenum,
550–680; chromium, 480–650; tantalum, 270; niobium, 350; iron, 200–
300; cobalt, 97; nickel, 86–130; rhodium, 200–330; platinum, 140;
palladium, 40–150; titanium, 200–240.

These values refer to films in the weight range 3–60 mg deposited
over an area in the region of 100–300 cm^2. For a given metal the ratio
R/w is strongly dependent on the substrate temperature during deposition, and this behaviour is exemplified in the following data (Crawford
et al., 1962; McConkey, 1965) where relative values of R/w are given
taking the value at 0 °C as unity: tungsten; 1.0, 0 °C; 1.0, 250 °C;
0.77, 350 °C; 0.32, 400 °C: molybdenum; 1.0, 0 °C; 0.64, 195 °C; 0.26,
317 °C: nickel; 5.3, −183 °C; 3.53, −78 °C; 1.0, 0 °C; 0.64, 100 °C;
0.17, 200 °C: palladium, 1.0, 0 °C; 0.29, 200 °C; titanium; 1.0, 0 °C;
0.10, 200 °C. At a given temperature there is a rough trend for metals
with high melting points to be associated with high values for R/w, and
there is a trend for R/w to decrease more slowly with increasing substrate temperature the higher the melting point. However, these trends
are not uniformly obeyed, and the factors governing film behaviour are
not understood well enough for an approach at other than an empirically
experimental level to be satisfactory.

The effect of gas pressure on film porosity may be discussed in terms
of two alternatives; nonadsorbable gas, and chemisorbed gas. Beeck
et al. (1940) have shown that the presence of an inert gas such as argon
during evaporation leads to an increase in R by a factor of about two
compared with corresponding vacuum deposited films. Such films are a
good deal rougher due to the presence of tall crystals (Sachtler *et al.*,
1954; McConkey, 1965). Argon pressures are in the region of 1 torr, that

is a factor of about 10^3 in excess of the pressure of metal vapour. There is
no doubt however that the metal atoms do not form a smoke in the gas
phase, and the effect of the argon occurs at the metal surface. At a film
temperature of 0 °C physical adsorption of argon will be negligible and
it seems likely that the argon acts by colliding with a condensing metal
atom so increasing its rate of thermal accommodation and reducing the
distance through which the metal atom may translate across the surface
in the condensation process. It is generally agreed that, except at very
high substrate temperatures, the overall condensation probability for a
metal atom on its own crystal is essentially unity (Henning, 1968).
However, there appears to be no independent experimental information
about the efficiency with which this process occurs. Although Lennard-
Jones' (1937) theory would suggest that this is rapid within a period of
a few lattice vibrations, the theoretical model is almost certainly in-
adequate for even an approximately accurate quantitative result. The
suggestion that the energy of the incident flux contributes significantly
to migration over the substrate surface has been made independently
by Melmed (1965). There seems no doubt that for a metal atom condensing on an unlike substrate as in the nucleation stage of film growth,
there may be a very high degree of mobility (Thun, 1963). It is generally
observed that film evaporation in a poor vacuum (say 10^{-5} torr) leads
to a more porous film with smaller crystals, caused presumably by a
reduction in the surface mobility due to chemisorbed atoms. This has
recently been demonstrated by Preece et al. (1967) in the deposition of
tin in a low oxygen pressure. However, the average diameter of nickel
crystals deposited at 0 °C to room temperature on glass is the same over
the range 10^{-7} torr to ultrahigh vacuum (Anderson et al., 1962; Suhrmann et al., 1963). The effect of adsorbed gas on surface mobility is
further discussed below when dealing with film sintering.

It is generally agreed that a decrease in substrate temperature will
result, other parameters constant, in a reduction in crystal size, and this
agrees with the electron microscopic results. This effect on crystal size
is probably the dominant reason for the increase in R with decreasing
substrate temperature at a given specific film weight.

Table 1 lists average crystal diameters for a number of metals,
together with the measured porosity R, and the table also includes for
comparison values of R_{calc}, computed on the assumption that each
crystal is a uniform cylindrical prism extending through the entire
thickness of the film. The reasonable agreement between R and R_{calc}
for films deposited at 0 °C lends support to the correctness of the model,
while the high values for R_{calc} for tungsten and molybdenum deposited
at 400 °C is probably a reflection of the failure of the pores in the corre-

TABLE 1

Values for average crystal diameter and film porosity*

Metal	Deposition Temperature (°C)	Specific Film Weight (mg cm^{-2})	MP of Metal (°C)	Average Crystal Width (Å)	R	R$_{calc}$
tungsten	0	0.067	3370	70	18.0	21.2
tungsten	400	0.069	3370	110	6.3	13.0
molybdenum	0	0.021	2620	75	14.3	13.4
molybdenum	400	0.033	2620	120	4.7	11.0
rhodium	0	0.034	1966	80	11.2	14.5
chromium	0	0.030	1890	130	14.4	15.1
palladium	0	0.022	1549	160	3.44	5.4
platinum	0	0.034	1774	180	4.66	4.4
iron	0	0.037	1535	240	10.9	8.8
nickel	0	0.065	1455	460	7.9	7.1

*from Anderson et al. (1962) and Anderson and McConkey (1967)

sponding actual film to extend the full depth of the film. It is seen that there is a clear trend for the average crystal diameter to increase with decreasing melting point of the metal.

Pore dimensions in evaporated films are only available in approximate estimates. Wheeler (1951) estimated from a guess at the actual film density that for a nickel film of about 1 mg cm^{-2} deposited at room temperature the average pore diameter (assumed cylindrical) was 100 Å. On the other hand, from an electron microscopic examination of a nickel film of about 0.1 mg cm^{-2} deposited at 0 °C, intercrystal gaps up to a few tens of Å at most were observed (Anderson et al., 1962). It may well be that the electron microscopic technique observed inter-crystal gaps at their minimum dimension, and furthermore, that a film of 1 mg cm^{-2}, which is about an order of magnitude thicker than those generally used for most surface studies, may have abnormally large pores. As an approximate working estimate we suggest that for films deposited with τ in the range 0.1–0.15, for instance many of the transition metals deposited at 0 °C, the average pore width is about 10–20 Å, and these pores probably extend through the depth of the film.

Porous films are inherently unstable and a reduction in R will occur if the temperature is raised above the deposition temperature. The rate at which this occurs depends on the metal, temperature and initial film structure. This instability is a manifestation of an excess surface free energy due both to the high surface area and the exposure of high index crystal planes, and Sachtler et al. (1968) have recently measured excess free energies between about 2–6 kcal mole^{-1} for random polycrystalline nickel films deposited at -192 °C. Film sintering is a complex process and preferably requires for its study a combination of measurements involving R vs time, electron microscopy and electrical conductivity.

At any temperature an ultimate (i.e. time independent) film area may be arrived at by sintering for a sufficient time at that temperature a film previously deposited on a substrate at a lower temperature. The ultimate area decreases with increasing sintering temperature, as shown in Figure 2 for nickel and molybdenum which are, respectively, examples

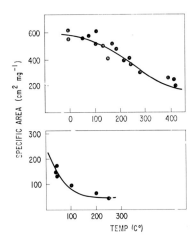

FIG. 2. (top): Dependence of ultimate specific area of sintered molybdenum films on sintering temperature. Films deposited at 0 °C. Points experimental, line calculated using activation energy parameters $E_0 = 40$ and $\sigma_E = 10$ kcal mole^{-1}. (bottom): Dependence of ultimate specific area of sintered nickel films on sintering temperature. Films deposited at -183 °C. Points experimental, line calculated using activation energy parameters $E_0 = 21$ and $\sigma_E = 5$ kcal mole^{-1}. Specific film weights in range 0.035 − 0.045 mg cm^{-2}. (Reproduced with permission from Anderson and McConkey (1967). *Proc. I.R.E.E. Australia* 132, April.)

of transition metals which sinter relatively easily and with difficulty. The form of these curves can be accounted for by the theory of rate processes with a distribution of activation energies as outlined below. The relation of the form log(area) $\propto T^{-1}$ suggested by Gundry and Tompkins (1957) appears to be without significance in terms of a physically realistic model.

Both the surface area and electrical resistance of the film decrease with increasing time during the sintering process. However, the film resistance decreases much more rapidly with time than does the surface area. It can be seen directly from the electron microscopic evidence in the upper temperature range for sintering that this is always accompanied by the partial or complete removal of gaps, and the resistance and area decrease are both associated with this, although there is no simple kinetic correlation between the processes. Indeed, at about the time that intercrystal gaps become no longer observable in the electron microscope there still exist gaps of sufficient size to permit ingress of gas between the crystals, since at this time the measured surface area is still much too large to be accounted for by the apparent surface roughness alone. Surface areas are preferable for use in a quantitative analysis of sintering kinetics, since the resistance change is confined to a relatively small part of the total observable sintering process.

Anderson and McConkey (1967) have used the temperature coefficient of resistance of the film to characterize structural changes during sintering. It should be noted that there are two distinct thickness ranges in which metal films may display a nonmetallic temperature coefficient of resistance. For ultrathin films where the individual crystals are each completely isolated, the resistance varies with temperature (in a range chosen so that no structural changes occur) according to $\log$(resistance) $\propto T^{-1}$, and the behaviour is well described by the theory of Neugebauer and Webb (1962). On the other hand, for much thicker films of the type being considered here (thicknesses typically > 500 Å), the temperature coefficient of resistance may have a nonmetallic sign, but the resistance varies linearly with T (again in a range chosen so that no structural changes occur), and the electron microscopic appearance of the films makes it highly unlikely that the crystals are each completely isolated in the sense used by Neugebauer and Webb. For these films the temperature coefficient of resistance moves towards that for the bulk metal as sintering progresses. This is illustrated in Figure 3. For relatively thick films of this sort where narrow intercrystal gaps are an important feature of their structure, the main component to the electrical resistance lies in the contribution from electron tunnelling across very narrow intercrystal gaps, including those surrounding small contact areas. When this is so, the most important contribution to the temperature coefficient of resistance probably lies in the temperature dependence of the gap width due to the differential thermal expansion between the metal and the substrate (cf. Anderson and McConkey, 1967). Clearly, the elimination of gaps and the provision of normal metallic contacts as sintering progresses provides an alternative conduction path which eventually becomes dominant.

Instances have been reported (Anderson et al., 1962; Anderson and Tare, 1964) where area/time data could be fitted to simple kinetic plots such as first order rate plots, while Roberts (1960) has obtained the activation energy for film sintering by the use of initial sintering rates. This has the severe disadvantage of focusing attention on only a very limited part of the sintering process, and it is clear furthermore that no

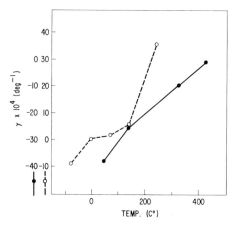

FIG. 3. Variation of temperature coefficient of resistance (γ) with the temperature at which the film was sintered to constant resistance; $-\bigcirc-$, nickel; $-\bullet-$, molybdenum. Nickel deposited at -183 °C, molybdenum at 0 °C. Specific film weights in range 0.035–0.045 mg cm^{-2}. (Reproduced with permission from Anderson and McConkey (1967). *Proc. I.R.E.E. Australia*, 132, April.)

simple kinetic plot of area/time data is of general applicability. These approaches are also unable to account for the temperature dependence of ultimate area. Anderson and McConkey (1967) showed that sintering data can be consistently interpreted in terms of the treatment due to Vand (1953) and Primak (1955) (cf. section IV) for kinetic processes occurring with a spectrum of activation energies. It was assumed that the activation energies were distributed about a most probable value E_0 according to a Gaussian distribution with a standard deviation σ_E, and that the frequency factor was constant. A typical example of the fit between experimental area/time points and a "best" curve calculated from a normal distribution of activation energies is shown in Figure 4 and observed and calculated ultimate areas for sintering are compared in Figure 2. Table 2 lists the parameters E_0 and σ_E for sintering of nickel and molybdenum films, of specific weights in the range 0.035–0.045 mg cm^{-2}.

Analysis of resistance/time data gave activation energy parameters consistent with, but less well defined than those obtained from area/time data. Anderson and McConkey (1967) compared these activation energy parameters with activation energies for metal transport by surface diffusion and by bulk diffusion by vacancy and interstitial mechanisms. This comparison alone does not permit a decision as to which of these processes is the likely mechanism for metal transport in sintering. However, it is known (Anderson et al., 1962; Anderson and Baker, 1962; Logan and Kemball, 1959) that the rate of film sintering

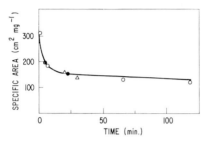

Fig. 4. Comparison of area/time data for sintering of nickel films (points) and curve calculated from activation energy parameters. The different symbols distinguish three different experiments, each sintered at 40 °C. The curve was calculated using $E_0 = 21$ and $\sigma_E = 5$ kcal mole^{-1}. Films deposited at -183 °C; specific film weights in range 0.035–0.045 mg cm^{-2}. (Reproduced with permission from Anderson and McConkey (1967). *Proc. I.R.E.E. Australia*, 132, April.)

TABLE 2
Activation energy parameters for film sintering

	E_0 (kcal mole^{-1})	σ_E (kcal mole^{-1})	Temperature Range (°C)
nickel	21	5	-185 to 250
molybdenum	40	10	0 to 400

is greatly retarded by the presence of chemisorbed hydrogen, and while this may be readily understood in terms of its influence on metal transport by surface diffusion, it is less clear how this effect may arise if sintering were to occur only by a process the rate of which was controlled by the migration of vacancies or interstitials.

Adsorbed gas may influence film sintering in two possible ways; the first is during the act of chemisorption, the second after adsorption equilibrium has been established. A number of authors (Dell et al., 1956; Klemperer and Stone, 1957; Brennan et al., 1960) have reported that chemisorption of a gas which has a high heat of chemisorption such as oxygen results in area reduction during the act of chemisorption. On the other hand Roberts (1960) reported an area increase as a result of hydrogen chemisorption on iron, and it was shown by Brennan et al. (1960) that, in the adsorption of oxygen on manganese, if the system was maintained in vacuum to regenerate further adsorption capacity for oxygen (probably by solution of adsorbed oxygen into the bulk), the apparent surface area of the specimen increased. In these experiments the film areas were measured by rare gas adsorption using the BET method, and the films were prepared and used for chemisorption in the range 0 °C to room temperature. In one case (Klemperer and Stone, 1957) it is known that the area measurement technique was inaccurate, and this together with the inconsistency with regard to the direction of the area change makes one suspect that the effect may be an artifact. The question has recently been reviewed by Campbell and Duthie (1965) who provide persuasive arguments that the effect is not due to any substantial change in actual film area, but that the effective area occupied by each adsorbed rare gas atom in the BET measurement depends on the presence or absence of chemisorbed gas; the suggested reasons for this are essentially electrostatic, and the effect is thus dependent on the nature of the chemisorbed gas.

There is some evidence from surface potential data that chemisorption of gases such as oxygen or chlorine can result in a local rearrangement of the metal atoms near the adsorption site. This has been particularly observed with films, but has also been reported with a tungsten field emitter tip by Duell *et al.* (1966). However, it appears that such rearrangements as may then occur are not sufficient to result in appreciable film sintering. It still remains feasible that chemisorption sintering might be found with films of very high porosity prepared at low substrate temperatures with metals of relatively low melting points.

Nickel films which have reached adsorption equilibrium with hydrogen are much more resistant to sintering than are the corresponding clean films (Logan and Kemball, 1959; Roberts, 1960; Anderson and Baker, 1962), and this conclusion has also been inferred from reaction kinetic data by Kemball (1951, 1952). It has been suggested by Anderson and Baker (1962) that the effect of chemisorbed hydrogen is to reduce the surface mobility of the nickel atoms. According to the models for chemisorbed hydrogen proposed by Takaishi (1958) and Toya (1960) at least part of the hydrogen is bound in an interstitial position on the surface (Toya's s-state), with a multicentered orbital. Irrespective of whether binding occurs by means of Coulson and Hume-Rothery's hybrid orbitals as originally suggested by Anderson and Baker, or whether Toya's proposals are more correct, the effect of this type of chemisorbed hydrogen will be to link together the surface metal atoms by bonds which are not formed in the absence of the chemisorbed atom. On this model not all chemisorbed species would be expected to reduce the sintering rate, for instance an adsorbed species bonded solely to a single surface metal atom. Moreover, the required binding mode may not be formed on all metals and, in fact, the effect appears to be less important on iron than nickel (Logan and Kemball, 1959). Furthermore, Clasing and Sauerwald (1952) have shown that the presence of adsorbed oxygen accelerated the sintering of copper and iron powders, although thick oxide layers had the opposite effect.

B. FILM POROSITY

Catalytic reaction kinetics in porous media have been analysed by Thiele (1939) and Wheeler (1951), and a detailed account has been presented by Satterfield and Sherwood (1963).

The behaviour of a porous catalyst may be characterized by an effectiveness factor η which is the ratio of the actual reaction rate to that which would occur if all the surface throughout the inside of the porous catalyst were exposed to reactant of the same concentration and temperature as that existing at the outside. For pore diameters <100 Å, diffusion through the pores will be in the Knudsen regime for pressures less than about 10 atmos. In the case of a porous film catalyst the geometry is that of a flat plate accessible to reactant only from one side. If we further assume isothermal behaviour and first order kinetics, then

$$\eta = \frac{\tanh \phi_{\mathrm{L}}}{\phi_{\mathrm{L}}} \qquad \text{(II)–(1)}$$

where ϕ_{L} is the Thiele diffusion modulus, and

$$\phi_{\mathrm{L}}(\tanh \phi_{\mathrm{L}}) = 1.43 \times 10^{-4} \left(\frac{\mathrm{L}}{\mathrm{r}}\right)^2 \frac{\beta \mathrm{V_R M^{\frac{1}{2}}}}{\mathrm{t_{\frac{1}{2}} S T^{\frac{1}{2}}}} \qquad \text{(II)–(2)}$$

where L(cm) is the depth of the porous film, $\mathrm{V_R}(\mathrm{cm}^3)$ the volume of the (static) reaction system, S (cm²) the *total* area of film surface, r(cm) the average pore radius, $\mathrm{t_{\frac{1}{2}}}$(sec) the half-period of the reaction, β an empirical factor to correct for pore tortuosity and for nonuniformity of pore cross section, M the molecular weight of gas, and T(°K) the temperature.

Satterfield and Sherwood (1963) have suggested from measurements on relatively well-characterized porous glass with r in the region 20–60 Å, that β is about 5.9. On the other hand, Wheeler (1951) suggested a value of 2. We suggest that a value of 4 is more reasonable by comparison with these and other values.

It is worth computing η for a relatively unfavourable case. If M = 140 (as for instance for a tetramethyl cyclohexane), T = 300 °K, and if reaction occurs on a typical nickel film with r = 12 Å, L = 1000 Å, S = 1000 cm², in a reaction volume with $\mathrm{V_R}$ = 1000 cm³ and $\mathrm{t_{\frac{1}{2}}}$ =

100 sec, we find from Eq. (II)–(2) that ϕ_L is 0.056, and thus η is 0.98; that is, the pores are without influence. With the above parameters it is possible for $t_{\frac{1}{2}}$ to fall to about 14 sec before η falls to 0.95. The vast majority of catalytic reactions studied over films have half-periods in the range 200–2000 sec, and under these conditions it is clear that the pore structure will have a negligible effect, even allowing for unfavourable variations in the parameters. This general conclusion is equally true for adsorption and desorption kinetics with half periods in the same range. However, it is also clear that *porous* films would not be satisfactory for kinetic studies in catalysis or in adsorption or desorption in a fast rate regime with half-periods, say 10 sec. If such fast reactions are intended, use should be made of continuous polycrystalline films produced by sintering, or of continuous single crystal films grown epitaxially. It is conceivable that because of the thermal history of a film, or because of the formation of surface reaction product, r could be so reduced that the reaction became pore dependent. There appear to be no reports of this occurring in catalytic reactions as such, at least for films deposited at 0 °C and above, and not of exceptional thickness. A specific search for such an effect with sintered nickel films was made by Crawford *et al.* (1962), while Dwyer *et al.* (1968) recently analysed the data of Kemball (1954) for the exchange of neopentane with deuterium over palladium films at 148 °C: in neither case was any diffusional influence found.

C. FILM SURFACE HETEROGENEITY

Surface heterogeneity arises in two different ways. Firstly, there is usually a structural heterogeneity resulting from a variety of exposed crystallographic planes and other surface features such as steps, kink sites and emergent dislocations. Secondly, there may be a chemical heterogeneity due to gas adsorption on the surface.

1. *Structural Heterogeneity*

If a film is polycrystalline, there are two limiting cases. The crystals may all have the same crystal axis normal to the substrate, but they may be randomly oriented about these axes, or the crystal axes may be randomly oriented with respect to the substrate. In the first case, approximately the same crystal face will be exposed by all crystals, although facets of other faces may well occur in the grain boundary regions, while in the second case a number of different crystal faces will be exposed on a gross scale. In all cases, the exposed faces will inevitably be somewhat imperfect. Under good vacuum conditions a single crystal substrate favors the first case, an amorphous substrate the second. Examples of the first type of behaviour are to be found in the

deposition of platinum and nickel at about 300 °C on a mica substrate
by Anderson and Avery (1966) and Anderson and Macdonald (1969), and
where the film structures were determined by transmission electron
diffraction and surface replication. Into the category of surface imperfection mentioned above we would class surface rearrangements such
as are known for covalent semi-conductors but which appear to be
absent for most metals, as well as surface imperfections such as those
arising from emergent screw dislocations, stacking faults, twin boundaries, surface vacancies, self-adsorbed metal atoms, and quasi-macroscopic features such as steps and terrace edges. The only well authenti-
cated cases of surface rearrangements of clean metals occur with the
(100) faces of platinum (Palmberg, 1969; Morgan and Somorjai, 1969)
and gold (Fedak and Gjostein, 1967; Palmberg and Rhodin, 1967;
Palmberg and Rhodin, 1968), and where Auger spectroscopy gives
reasonable evidence for believing the effects do not represent contamination artifacts. However, even here the rearranged surface structures
appear to be of low stability and at least with (100) platinum, adsorption of gases such as carbon monoxide or some olefins causes a reversion
back to the unreconstructed surface (Morgan and Somorjai, 1969).

When dealing with polycrystal films of random crystal orientation,
it has so far been impossible to obtain direct experimental information
about the proportions of the various crystal planes exposed to the gas.
The assumption is usually made that the planes of lowest surface
energy are the most important. On the reasonable model that the crystal
energy equals the sum over all pairwise interactions, the surface energy
may be computed from the interactions destroyed in creating the surface. The relative surface energies obtained by Nicholas (1968) are
given in Figure 5 as contours on a unit sterographic triangle. The results
depend to some extent on the pairwise interaction potential used, but
the data in Figure 5 are for a 5–7 Lennard-Jones potential, which is a
reasonable general compromise to the best potential function required
to reproduce experimental crystal shapes (Dreschsler and Nicholas, 1967).

Results from pairwise summations taken to third-nearest neighbours, and from an examination of crystal shapes (Stranski, 1949;
Wranglen, 1955) agree that the following planes stand in the indicated
order of increasing surface energy

$$\text{f.c.c.} \quad (111) \; < \; (100) \; < \; (110)$$
$$\text{b.c.c.} \quad (110) \; < \; (100) \; < \; (211) \text{ or } (111) \qquad \text{(II)--(3)}$$
$$\text{HCP} \quad (0001) \; < (10\bar{1}0) \; < (10\bar{1}1)$$

These data are in only approximate agreement with Nicholas (cf. the
position of (100) for the FCC and BCC systems). The assumption is
often made that the planes indicated in expression (II)–(3) are present

to equal extents in random polycrystalline films. This assumption would appear indefensible in the light of the surface energies in Figure 5. For instance, in the f.c.c. system there are a number of $(11l)$

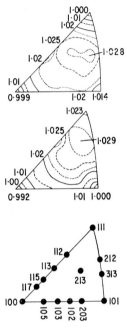

Fig. 5. Calculated relative surface energies; (top): f.c.c. crystal with surface energy for the (111) plane set equal to unity; (middle): b.c.c. crystal with surface energy for the (110) plane set equal to unity; (bottom): unit stereographic triangle. (Top and middle figure reproduced with permission from Nicholas (1968). *Aust. J. Phys.* **21**, 21.)

planes with surface energies lower than (110), and in the b.c.c. system there are a number of $(10l)$ planes with energies less than either (211) or (111). In so far as surface energies have any validity as a guide to the result of what is, after all, a kinetic rather than a thermodynamic problem, it would seem that an assumption of equal exposures of the planes listed in expression (II)–(3) may seriously overestimate the proportions of (110) (f.c.c.) and of (211) or (111) (b.c.c.) planes. This conclusion is in clear agreement with the conclusions reached by Suhrmann *et al.* (1963) who examined by electron microscopic replication the shapes of the tops of the crystals produced in very heavy nickel films deposited on glass at 20 °C. For films in the thickness region of $(1.5–3.0) \times 10^4$ Å, the crystals were large (average crystal width in the range 1900–2600 Å) and the shapes were uniformly well developed and (111) and (100) faces were clearly of dominant importance.

An assessment of the different planes exposed in a polycrystalline film should, in principle, be possible from the measurement of a physical property which is sensitive to the nature of the surface such as work function or heat of adsorption. If the results of such measurements are to be of much significance, however, they must identify a relatively small number of surface types. This has proven to be impossible with random polycrystalline films deposited on substrates at about room temperature. However, successful results have been achieved with nominally random polycrystalline nickel films deposited on glass at 250 °C under ultrahigh vacuum conditions. For such a film three distinct photoelectric work function thresholds can be extracted from the photoemission curve together with the proportions of each of the emitting surfaces, and the dominance of (111) and (100) surfaces is evident, accompanied by a small proportion ($\sim$ 5%) of low work function surface which is presumably atomically rough (B. G. Baker and B. B. Johnson, unpublished work from this laboratory). For a similar nickel film the heat of xenon adsorption has been measured and analysed in terms of three types of adsorbing areas (Baker and Fox, 1965).

Films deposited at high temperatures ($\tau >$ about 0.35) on relatively flat substrates such as cleaved crystal surfaces yield flat and featureless electron microscope replicas, and Figure 1c is one such example. Even if the film grows as an epitaxed crystal the surface may still be far from flat on a molecular scale, as well as having some grosser features as outlined by Sanders in Chapter 1. The actual surface structure must still be treated as requiring experimental characterization. Clearly such a single crystal surface is still energetically (and chemically) heterogeneous, but the spread of heterogeneity will be much smaller than for a random polycrystalline surface.

Attempts have been made to assess the heterogeneity of random polycrystalline metal films by means of adsorption studies using both stable and radio-isotopes. The essence of the technique is to determine what proportion of adsorbed gas equilibrates with an isotopically distinguished gas added subsequently. Gundry (1961) used H_2 and D_2 with nickel and tungsten films, while Boreskov (1961) used H_2, D_2 and T_2 on platinum films. The results show an increasing proportion of adsorbed gas able to be equilibrated with increasing temperature in the ranges -195 to 20 °C with nickel and tungsten, and -195 to 183 °C with platinum. The results clearly point to energetic surface heterogeneity for adsorption, and this conclusion is certainly still valid despite Toya's (1961) objections to the technique, since Suhrmann et al. (1962) have reached the same conclusion by working with carbon

monoxide adsorption on nickel under conditions where these objections appear to be essentially met.

Two groups of workers have demonstrated the chemical consequences
of the structural heterogeneity of random polycrystalline nickel films.
Crawford and Kemball (1962) showed that the pattern of catalytic
activity of nickel films towards the exchange of p-xylene with deuterium
was markedly different for a nickel film deposited at 0 °C compared to
the same film pre-sintered at 200 °C before the reaction. Although under
the vacuum conditions used, part of this effect may have come from the
adsorption of adventitious contaminant during sintering, the fact that
the pattern of exchange altered in a non-uniform way makes it clear
that the film has non-uniform catalytic properties with respect to this
reaction, a conclusion that is confirmed by the effect of pre-adsorbed
carbon monoxide as a poison towards this reaction (Phillips *et al.*, 1963).
It should be clear that heterogeneity is delineated in this way only
in relation to a specified surface reaction. For instance, the structural
changes which occur on the surface of a random polycrystalline nickel
film during sintering are without effect on the kinetic parameters of the
ethylene hydrogenation reaction, and in this case the rate only decreases by a factor roughly equal to that due to the decrease in the total
surface area (Crawford *et al.*, 1962). Anderson and Avery (1966) showed
that the (111) surface of platinum was preferred for the isomerization
of isobutane to n-butane, so that in this sense the surface of a random
polycrystalline platinum film is heterogeneous towards this reaction.
Recently, Anderson and Macdonald (1969) showed that in the exchange
of ethane and propane with deuterium over nickel films, the pattern of
exchange products is very dependent on whether the film is used after
deposition at 0 °C, or pre-sintered before the reaction, or whether the
film is deposited with the substrate at 300 °C. Again, heterogeneity of
the random polycrystalline film is clearly indicated, and the failure of
the effect to appear with methane exchange emphasizes again that it
must be considered in explicit relation to a specified chemical process.
The details of these catalytic reactions will be considered again in
Chapter 8; it remains to add, however, that there is a need for more
work with chemical reactions on single crystal surfaces, so as to explore
the fine structure in surface heterogeneity.

Ultra-thin metal films consist generally of extremely small isolated
crystallites dispersed over the support: often each crystallite amounts
to no more than a cluster of a few atoms which may be ordered if the
substrate is ordered (Allpress *et al.*, 1966). The surface and catalytic
properties of ultra-thin films have scarcely been explored at all, although one illustration is given in section (V) of Chapter 8.

2. Chemical Heterogeneity

The contamination of a film surface can arise in two ways. Impurity contained in the source material may evaporate and be condensed on the growing film, or residual gases in the vacuum system may adsorb on the final surface. Much attention has focused on the latter and we are frequently reminded of the short time needed for contamination under high vacuum conditions. Since much of the published data on adsorption and catalysis was obtained on films deposited at pressures no better than 10^{-6} to 10^{-7} torr, it is important to assess the likely conditions during evaporation and understand why such work has often shown reproducibility which is not indicative of gross contamination.

Firstly, the calculation of the maximum rate of contamination assumes that the ambient gas has unit sticking probability and an unlimited supply. It is apparent in the evaporation of many metals that the ambient gas measured by the pressure gauge is not readily chemisorbed by the film. Active gases have been gettered by the growing film during the evaporation leaving the gas phase rich in inactive gas. For a metal film of area $\sim$1000 cm^2 in a system at 10^{-6} torr it would require about 10^4 litres of gas to form a contaminating monolayer. Clearly the rate of outgassing of the vacuum system to produce this contaminant limits the rate of contamination.

Partial pressure analysis of the residual gases is obviously desirable and is frequently undertaken when films are evaporated under ultra-high vacuum conditions. However, the gas phase above the films contains little information about impurities from the source which are strongly adsorbed on the surface. The current practices of using very fast pumping or pulsed evaporation in order to maintain superior vacuum conditions do little to reduce contamination of this kind.

Adsorbed gas leading to chemical heterogeneity may be acquired as adventitious contaminant, and an important potential source of this is the impurity in the evaporating material and this is not necessarily measured adequately by the ambient pressure. The seriousness of this depends on the nature of the metal to be evaporated and on the adequacy of the outgassing process that can be used. In general we expect metals such as chromium which form highly stable oxides and which are thus difficult to purify, to be particularly bad from this point of view. Most workers have used spectrographically standardized materials of approximately the same specification but this has generally not included adequate information on the gas content.

When films have been used for catalytic studies another source of contaminant is the reaction mixture. This is frequently prepared in a system of poor vacuum in the presence of mercury vapour and is added

c

in such total quantity that the surface could be severely contaminated. Cold traps have usually been used to trap mercury but their effectiveness at high gas pressures may be questioned and their use may not be possible at all if the reaction mixture contains an easily condensable component.

It is concluded that while many studies have employed contaminated surfaces, the high surface area and the gettering action during evaporation have kept contamination levels lower than might be expected: nevertheless, ultra-high vacuum techniques should now be regarded as mandatory.

III. Adsorption Equilibria

There are two reasons why some knowledge of adsorption isotherms is useful in catalysis. First, the rate of a catalytic reaction is influenced by surface concentrations, and the latter must be related to bulk concentrations which are directly measurable. For this reason, it is clearly essential that for practical use the isotherms be expressed in the form $\theta = f(p)$. To obtain this form often requires an empirical approximation. Second, studies in adsorption offer information about the structure of the adsorbed layer, and this is also a determinant for catalytic behaviour. With the present state of knowledge it is only useful to employ relatively primitive adsorption isotherm equations in the construction of kinetic rate equations, but qualitative or semi-quantitative arguments concerning the influence of the structure of the adsorbed layer are nevertheless important.

It is therefore not the present purpose to give a completely detailed discussion of the forms taken by adsorption isotherms. However, it is worthwhile to recognize the main factors of which account should, in principle, be taken. The following factors are of general importance; (a) whether the adsorbed species are localized or mobile, (b) whether the surface is homogeneous or heterogeneous, (c) the magnitude of the interaction energy between adsorbed residues, (d) whether adsorption is dissociative and then yields identical or non-identical surface residues, (e) the number of effective surface sites occupied by an adsorbed residue. We shall use the term mobile if the adsorbed residues can translate freely across the surface so that the adsorbed phase can be considered as a two-dimensional gas. All other cases will be described as localized adsorption, where motion of an adsorbed residue across the surface occurs by a hop site mechanism, or by desorption/adsorption: immobile adsorption is an extreme limit where the lifetime of a residue at its adsorption site is infinite.

Catalysis is a chemically specific process with respect to adsorption at the surface: one can in general expect to find a chemisorbed surface species involved for which the localized model is likely to be the most appropriate description. However, some reactant species may be physically adsorbed, and only if the temperature is low enough will the periodic nature of the surface then lead to localized adsorption. From the analysis of Hill (1946) one may expect the transition from localized to mobile adsorption to be in progress at the temperature of $E_D/10R$, and the transition to have gone much of the way to completion at $E_D/3R$, where E_D is the heat of activation for surface diffusion. For adsorption on well-defined crystal surfaces where the adsorbed species is not so large that its energy of interaction with the surface becomes insensitive to translation across the surface, E_D is typically 20% of the heat of adsorption (Ehrlich, 1963). With this limitation on molecular size (for instance, N_2, CO, H_2O, CO_2, C_1–C_3 hydrocarbons, etc.) the heat of physical adsorption is unlikely to exceed about 8–10 kcal mole^{-1}, and mobile adsorption can certainly be expected above about 300 °K. With quite large adsorbing molecules (for instance, C_6 hydrocarbons and larger), although the heat of physical adsorption may reach 15–20 kcal mole^{-1}, the heat of activation for surface diffusion may be expected to be a smaller fraction of the heat of adsorption, and 300 °K may still be a useful estimate. On the whole, we conclude that above room temperature, and this refers to the majority of catalytic reactions, physically adsorbed molecules are likely to be present in a mobile layer, provided the surface coverage is not too high. This conclusion is in approximate agreement with mobility estimates made from entropy data for a number of systems (de Boer, 1953a). At higher surface coverages (θ greater than about 0.5), the low energy electron diffraction data for physical adsorption (often at about room temperature) suggests that the mobile model tends to break down because of the formation of ordered structures in the adsorbed layer, and that this is more important with larger adsorbed molecules (Lander and Morrison, 1967).

A. UNIFORM SURFACES

1. *Simple Adsorption Isotherms*

For mobile adsorption, the adsorption isotherm is related to the equation of state obeyed by the adsorbed two-dimensional (2D) gas (de Boer, 1953b). Neglecting the trivial case of an ideal 2D gas (which leads to direct proportionality between pressure and coverage), the first useful 2D equation of state is that of a hard-sphere model (molecules of finite size), but with no attractive interactions. The 2D equation of state then is

$$\phi(A-N\beta) = NRT \qquad \text{(III)–(1)}$$

where ϕ is the 2D pressure, A the area occupied by N adsorbed moles and β a constant related to molecular size in an analogous fashion to the relation between the van der Waal's constant b and molecular volume. Equation (III)–(1) corresponds to an adsorption isotherm of the form

$$p = B(T) \left(\frac{\theta}{1 - \theta} \right) \exp \left(\frac{\theta}{1 - \theta} \right) \qquad \text{(III)–(2)}$$

where B(T) is a temperature dependent constant and $\theta = N\beta/A$. In the approximation that the 2D gas is sufficiently dilute, equation (III)–(2) approximates to the form of the simple Langmuir adsorption isotherm.

Since $\exp\left(\dfrac{\theta}{1 - \theta} \right) \to \infty$ faster than $\dfrac{\theta}{1 - \theta} \to \infty$ as $\theta \to 1$, equation (III)–

(2) shows a stronger surface saturation character in a plot of θ vs p than does the Langmuir expression. The mobile model should not be extended to high coverages as it is possible for the adsorbate to pack in an orderly array characterized by the dimensions of the adsorbate rather than those of any surface lattice. Under these conditions there is a phase change and a breakdown of equation (III)–(2) (Stebbins and Halsey, 1964).

In the mobile model, attractive interactions between adsorbed species are most simply taken into account by using a 2D van der Waal's equation of state by which ϕ in equation (III)–(1) is replaced by $(\phi + \alpha N^2/A^2)$, where α is a constant dependent on the attractive potential. The corresponding isotherm equation is of the same form as equation (III)–(2) except that the argument of the exponential becomes

$$\left[\frac{\theta}{1 - \theta} - \frac{2 \alpha \theta}{RT \beta} \right]$$ and again approximates to the Langmuir form at

low θ. If dissociative adsorption occurs to give mobile species (an unlikely situation), the forms of the equations remain unaltered. However, if the 2D gas is a mixture, as might result from the simultaneous establishment of adsorption equilibrium with $A_{(g)}$ and $B_{(g)}$, or from dissociative adsorption of $A_{2(g)}$ and $B_{2(g)}$, the results are complex. The most useful result is due to Schay et al. (1956, 1957) who showed that for a hard-sphere model without attractive interactions, simultaneous mobile adsorption equilibrium of $A_{(g)}$ and $B_{(g)}$ gives adsorption isotherms of the same form as the Langmuir mixed isotherm equations, provided $(\theta_A + \theta_B)$ is not too large, that is, in the same approximation as equation (III)–(2) reduces to the Langmuir form. Equation (III)–(2) cannot be converted to the form $\theta = f(p)$ as a closed expression. However, at low coverage a reasonable approximation is to neglect the exponential term and obtain

$$\theta = \frac{ap}{1 + ap} \tag{III}-(3)$$

While this equation has the form of the Langmuir isotherm it has arisen here only as a low coverage approximation for the mobile adsorption. In practice, the use of equation (III)–(3) to represent equation (III)–(2) may not be too serious because, for physical adsorption under typical catalytic conditions, a low value for θ is frequently to be expected. For instance, if the heat of adsorption is 10 kcal mole^{-1}, θ at 500 °K and 1 atmosphere pressure would be less than 0.1.

The development of an adsorption isotherm for a localized model is a problem in lattice statistics, and should properly be solved by methods of statistical mechanics. Fowler and Guggenheim (1949) give isotherms in closed forms for a number of important cases.

If adsorption of the gas phase species $A_{n(g)}$ occurs dissociatively to give n identical species $A_{(s)}$ each occupying one adsorption site, if the adsorbed species are localized and there is zero interaction energy between them, and if the adsorption energy is the same for all sites, the Langmuir isotherm results

$$p^{1/n} = \frac{\theta_A}{1 - \theta_A} b_A(T)\, e^{-\epsilon_A/RT} \tag{III}-(4)$$

where $b_A(T)$ is a temperature dependent constant of well-known form, given by

$$[b_A(T)]^n = \frac{(2\pi m_{A_n}kT)^{3/2}(kT)(_g j_{A_n})}{h^3(_s j_A)^n} \tag{III}-(5)$$

with m_{A_n} the mass of the molecule A_n, $_g j_{A_n}$ and $_s j_A$ respectively the internal partition functions for $A_{n(g)}$ and $A_{(s)}$. The adsorption energy ϵ_A is the energy required to transfer the species A from the lowest energy level in the adsorbed state to the lowest energy level in the gas phase. Thus the change in internal energy for the adsorption of A_n (at 0 °K) is given by $(D(A_n) - n\epsilon_A)$, where $D(A_n)$ is the dissociation energy for the conversion of $A_{n(g)}$ to $nA_{(g)}$. By combining the last two factors of equation (III)–(4) this may be written alternatively as

$$\theta_A = \frac{a_{A_n}p_{A_n}^{1/n}}{1 + a_{A_n}p_{A_n}^{1/n}} \tag{III}-(6)$$

When $n = 1$, equations (III)–(4) and (6) take the familiar form for non-dissociative adsorption. Although the factor b(T) depends on temperature, it varies much more slowly with temperature than does $e^{-\epsilon/RT}$, and for most purposes the influence of temperature on the isotherm can be regarded as being confined to this latter term. Throughout Chapters 7 and 8 we follow the common practice of defining energies and heats of adsorption to be numerically positive quantities, so that for the adsorption process gas → surface they equal $-\Delta U$ and $-\Delta H$ respectively, where the latter follow the standard thermodynamic sign conventions.

When dissociative adsorption occurs to give non-identical adsorbed species, as for instance from the dissociative adsorption of $AB_{(g)}$ to $A_{(s)}$ and $B_{(s)}$, a single isotherm equation holds provided this is the only equilibrium established. For this example, the isotherm takes the form of equations (III)–(4) and (6) with $n = 2$,

$$p_{AB}^{1/2} = \frac{\theta_T}{1 - \theta_T} b_{AB}(T)\, e^{-\frac{1}{2}(\epsilon_A + \epsilon_B)/RT} \qquad \text{(III)–(7)}$$

where $\theta_T = \theta_A + \theta_B$, and

$$[b_{AB}(T)]^2 = \frac{(2\pi m_{AB}kT)^{3/2}(kT)(_g j_{AB})}{h^3(_s j_A)(_s j_B)} \qquad \text{(III)–(8)}$$

and

$$\theta_T = \frac{a_{AB} p_{AB}^{1/2}}{1 + a_{AB} p_{AB}^{1/2}} \qquad \text{(III)–(9)}$$

by stoichiometry $\theta_A = \theta_B$. Analogous expressions for related cases can clearly be written.

2. *Mixed Adsorption*

If there are a number of gas phase components competing for the surface sites while retaining the Langmuir assumptions, there will be an isotherm for each gas phase component. The isotherm equations are easily obtained from the basic statistical theory, provided the equilibria are unrelated. For instance, suppose there are two simultaneous equilibria described by

$$\left.\begin{array}{c} AB_{(g)} \rightleftharpoons A_{(s)} + B_{(s)} \\[2mm] C_{(g)} \rightleftharpoons C_{(s)} \end{array}\right\} \qquad \text{(III)–(10)}$$

The equations for θ_A, θ_B and θ_C are

$$\left.\begin{array}{l} \theta_A = \theta_B = \tfrac{1}{2}\,\dfrac{a_{AB}p_{AB}^{1/2}}{1 + a_{AB}p_{AB}^{1/2} + a_C p_C} \\[4mm] \theta_C = \dfrac{a_C p_C}{1 + a_{AB}p_{AB}^{1/2} + a_C p_C} \end{array}\right\} \qquad \text{(III)–(11)}$$

The extension to more extensive simultaneous equilibria leading to mixed adsorption is straightforward.

The problem needs closer examination, however, if the adsorption equilibria are inter-related, and this is a common situation. In practice, a given surface residue may be related to more than one gas phase species, but the position of equilibrium may be such that not all gas phase species are directly measurable. For instance, suppose the following equilibria are established

$$\left.\begin{array}{l} AB_{(g)} \rightleftarrows A_{(s)} + B_{(s)} \\[3mm] B_{2(g)} \rightleftarrows 2B_{(s)} \\[3mm] A_{(g)} \rightleftarrows A_{(s)} \end{array}\right\} \qquad \text{(III)–(12)}$$

We need isotherms for θ_A and θ_B. Although there are three gases present, their pressures are not independently variable since they are connected by the gas phase equilibrium

$$2AB_{(g)} \rightleftarrows B_{2(g)} + 2A_{(g)} \qquad \text{(III)–(13)}$$

Thus θ_A and θ_B can be expressed directly in terms of the two gas phase species to which they are directly related, giving

$$\left.\begin{array}{l} \theta_A = \dfrac{a_A p_A}{1 + a_A p_A + a_{B_2}p_{B_2}^{1/2}} \\[4mm] \theta_B = \dfrac{a_{B_2}p_{B_2}^{1/2}}{1 + a_A p_A + a_{B_2}p_{B_2}^{1/2}} \end{array}\right\} \qquad \text{(III)–(14)}$$

When it happens that one of the pressure terms required for equation (III)–(14), p_A or p_{B_2}, is not directly observable, use may be made of the equilibrium constant for equation (III)–(13). To take a specific example, suppose $AB_{(g)} \equiv CH_4$, $B_{2(g)} \equiv H_2$, $A_{(g)} \equiv CH_3$ etc., θ_{CH_3} is given by

$$\theta_{CH_3} = \dfrac{a_{CH_3}p_{CH_3}}{1 + a_{CH_3}p_{CH_3} + a_{H_2}p_{H_2}^{1/2}} \qquad \text{(III)–(15)}$$

but since p_{CH_3} is not observable, we may use the equilibrium constant K for

$$2 \, CH_{4(g)} \rightleftarrows 2 \, CH_{3(g)} + H_{2(g)} \left. \phantom{\begin{matrix}a\\b\end{matrix}} \right\}$$
$$K = p_{CH3}^2 p_{H2} p_{CH4}^{-2} \qquad\qquad\text{(III)–(16)}$$

to give

$$\theta_{CH_3} = \frac{(a_{CH_3}K^{\frac{1}{2}})p_{CH_4}p_{H_2}^{-\frac{1}{2}}}{1 + (a_{CH_3}K^{\frac{1}{2}})p_{CH_4}p_{H_2}^{-\frac{1}{2}} + a_{H_2}p_{H_2}^{\frac{1}{2}}} \qquad \text{(III)–(17)}$$

It should be noted that the constant $(a_{CH_3} K^{\frac{1}{2}})$ can have a temperature dependent part extracted from it in the form $e^{-E/RT}$, where E contains both the adsorption energy of CH_3 and the enthalpy change for the equilibrium reaction (III)–(6): here $E = -\epsilon_{CH_3} + \frac{1}{2}\Delta H^\circ$. The application of expressions of the type of (III)–(17) has been discussed by Kemball (1966).

In all of this discussion of localized adsorption, it has been assumed that the Langmuir assumptions are obeyed. Lifting the restrictions on zero interaction energy and on one adsorption site per adsorbed species leads to more complex isotherms.

3. *Adsorption with Interaction*

The simplest model that includes an interaction energy is that for a regular localized monolayer, which assumes that the entropy of the adsorbed layer is ideal (that is, the adsorbed species remain randomly distributed over the sites), and that the total interaction energy for the layer is summed from the number of nearest neighbour interactions. For non-dissociative adsorption, the isotherm (Fowler and Guggenheim, 1949) is

$$p = \frac{\theta}{1 - \theta} \, b(T) \, e^{ZV\theta} \, e^{-\epsilon/RT} \qquad \text{(III)–(18)}$$

where V is the nearest neighbour interaction (positive for repulsion) and Z is the number of nearest neighbour sites surrounding a given site; b(T) is similar to that term in equation (III)–(4). If adsorption is dissociative to give n identical adsorbed species, the term in V from equation (III)–(18) is unchanged, and the Langmuir equation (III)–(4) is only modified by the factor $e^{ZV\theta}$ on the right hand side. Dispersion interactions which contribute to V are always attractive, but if the adsorbed species have a surface dipole, interaction of like dipoles will lead to repulsion.

A better degree of approximation assumes that the interaction energy affects the ordering in the adsorbed layer. In the quasi-chemical approximation this is treated as an equilibrium between three types of nearest-neighbour pairs; filled-filled, vacant-vacant, and filled-vacant.

There appears at present to be no useful theoretical treatment for lateral interactions in mixed adsorbed layers. Such a treatment would clearly be complex since even in the regular monolayer approximation there would be three separate interaction energies due to A-A, A-B and B-B nearest neighbours. Nevertheless, it is important to recognize the possible importance to catalysis of ordering in adsorbed layers. For instance, if we have a mixed adsorbed layer consisting of species $A_{(s)}$ and $B_{(s)}$ which are reactant entities in a catalytic process, if ordering produces a change in the number of A-B nearest neighbour pairs, one could expect the reaction rate to be influenced. Furthermore, such ordering may also establish the orientation of $A_{(s)}$ and $B_{(s)}$ with respect to each other, thus introducing highly specific steric factors.

4. Adsorption with Site Exclusion

The relaxation of the Langmuir assumption that each adsorbed entity occupies only one adsorption site leads to quite complex adsorption isotherms.

Suppose the adsorption sites form a square lattice, and non-dissociative adsorption occurs and that the size of the adsorbing entity is sufficiently large that it blocks neighbouring sites. There are two cases for which solutions are available; (a) the four nearest neighbour sites are blocked, (b) the eight nearest and next-nearest neighbour sites are blocked. When only nearest neighbour sites are blocked, an isolated adsorbed atom "occupies" a total five sites; when both nearest and next-nearest neighbour sites are blocked, it "occupies" nine sites. It is convenient to define an occupation number $\bar{\nu}$ as the average number of blocked sites per adsorbed entity at a given coverage. Since, as the surface fills blocked sites are shared by more than one adsorbed entity, $\bar{\nu}$ will decrease as coverage increases.

It is readily shown (Baker, 1966) that the configurational differential entropy of adsorption is given by

$$\Delta \bar{S} = -R \ln \left(\frac{\theta}{\nu_M - \theta \bar{\nu}} \right) \qquad \text{(III)–(19)}$$

and there follows the isotherm equation

$$p = \frac{1}{\nu_M} \left(\frac{\theta}{1 - \theta(\bar{\nu}/\nu_M)} \right) b(T) \, e^{-\epsilon/RT} \qquad \text{(III)–(20)}$$

C*

where $b(T)$ and $e^{-\epsilon/RT}$ are the same as the corresponding terms in the simple Langmuir isotherm. Here ν_M is the number of blocked sites per adsorbed entity for a surface regularly packed to capacity, and θ is defined as the ratio of the number of adsorbed entities per unit area to the maximum allowed. Clearly θ is related to n the number of adsorbed entities per unit area, and to N the number of adsorption sites per unit area by $\theta = (n/N)\nu_M$. If in equation (III)–(20) $\nu_M = \bar{\nu} = 1$, the simple Langmuir isotherm is recovered.

Values of ν_M are simply obtained by geometry. The values for ν_M and ν_0, the latter the number of sites excluded by an isolated adsorbed entity, are: (a) square lattice with nearest neighbour exclusion 2, 5; (b) square lattice with nearest and next-nearest neighbour exclusion 4, 9; (c) triangular lattice with nearest neighbour exclusion 3, 7.

If dissociative adsorption of a molecule A_n occurred to give n identical surface species $A_{(s)}$, each of which behave with the above site exclusion criteria, the isotherm is

$$p^{1/n} = \frac{1}{\nu_M} \left(\frac{\theta}{1 - \theta(\bar{\nu}/\nu_M)} \right) b_A(T)\, e^{-\epsilon_A/RT} \qquad \text{(III)–(21)}$$

The form of the isotherm depends thus on the evaluation of $\bar{\nu}$ as a function of θ. Three methods have been employed for this purpose: the Kramers-Wannier matrix method, the cluster expansion method and the Monte Carlo technique. The results for a square lattice with nearest and next-nearest neighbour exclusion are given by the Monte Carlo method (Baker, 1966) and by van Craen et al. (1968). For the square lattice with nearest neighbour exclusion the result may be calculated from the cluster expansion coefficients given by Gaunt and Fisher (1965). However, the existence of a phase change near $\theta = 0.7$ renders the series unreliable in this region for use with equation (III)–(21). For exclusion on a triangular lattice, the result is available up to $\theta = 0.7$ from the Monte Carlo experiment (Baker, 1966). The $\bar{\nu}$ vs θ relations for the various cases are shown in Figure 6a, and the configurational parts of the corresponding isotherms are shown in Figure 6b plotted on a normalized pressure co-ordinate. The simple Langmuir case is provided for comparison, as is the isotherm function

$$\frac{\theta}{(1 - \theta)^{2.25}}$$

(Langmuir, 1940) which was suggested as an approximation for the case of nearest and next-nearest neighbour exclusion on a square lattice. The latter is a poor approximation, as also is the approximation due to Tonks (1940). The exclusion isotherms cannot be satisfactorily approximated by a Freundlich isotherm over a useful coverage range, and the

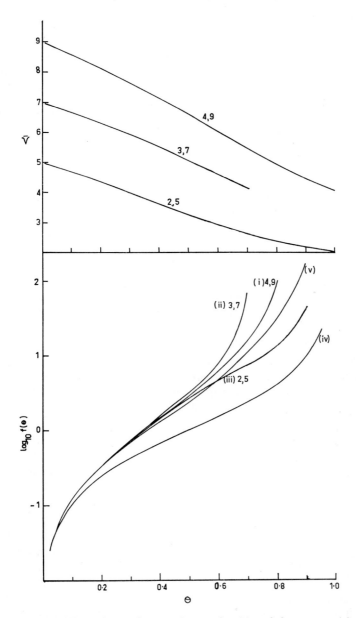

FIG. 6. (a) (upper): Dependence of occupation number ($\bar{\nu}$) on θ, for ν_M, $\nu_0 = 4,9$; $3,7$; $2,5$.

(b) (lower): Configurational parts of isotherms: for curves (i), (ii), (iii), $f(\theta) = \dfrac{\theta}{1 - \theta(\bar{\nu}/\nu_M)}$:

(iv), $f(\theta) = \dfrac{\theta}{1 - \theta}$: (v), $f(\theta) = \dfrac{\theta}{(1 - \theta)^{2,2}}$

Tempkin isotherm ($\theta \propto \log p$) is a useful approximation only over the restricted range $0.2 < \theta < 0.6$.

The site exclusion isotherms in Figure 6b approach surface saturation much more slowly with increasing pressure than does the simple Langmuir isotherm, so that in the region where the exclusion isotherms are relatively independent of pressure, there is still a significant proportion of free surface.

Mixed adsorption with one or more types of adsorbed entity dictating site exclusion has not been analysed. However, the data in Figure 6b suggest the large extent to which small adsorbed entities could be accommodated on the sites left unoccupied by the large ones. Consider, for instance, the adsorption of a large entity A of diameter $d_A = 1.6a$, where a is the lattice spacing of the surface lattice. This value of d_A lies in the range $a\sqrt{2} < d_A < 2a$ which leads to blocking of nearest and next-nearest neighbour sites on a square lattice to other A. However, of the sites which at any given θ_A are blocked for A adsorption, all are available for adsorption of another entity B, provided that $d_B < 0.4a$. Thus at $\theta_A = 0.8$, at which point the θ_A vs p_A isotherm is only very slowly dependent on p_A, the fraction of surface sites available to B equals $(N - n_A)/A$ and since $n_A = \theta_A N/\nu_M$ this fraction of sites available to B is 80%. Similarly, on a triangular lattice with $d_A = 1.3a$, $d_B < 0.43a$, and $\theta_A = 0.7$, the fraction of sites available to B is 77%. There are clearly a large number of possible geometric criteria for site blocking in mixed adsorption, and quantitative treatments are best contemplated by Monte Carlo methods. The application to catalytic situations is clear enough. For instance, on a (100) nickel surface for which $a = 2.49$ Å, d_A and d_B above would be 4 Å and 1 Å respectively, and the model could thus be applied to the adsorption of a methyl residue and a hydrogen atom, for which the van der Waal's radii are about 4 Å and 0.6 Å respectively.

It is important to recognize that even in the absence of an interaction potential V between adsorbed entities, component segregation can occur in mixed adsorption depending on the operating site exclusion criteria. These phenomena are best demonstrated by computer simulation of adsorption equilibria by Monte Carlo methods, and the following cases are offered as illustrative examples (Baker and Bruff, unpublished work from this laboratory): in all cases we retain $a\sqrt{2} < d_A < 2a$. When B is as small as the case mentioned in the preceding paragraph, no segregation occurs at any coverage. However, if B is such that $a < d_B < (a2\sqrt{2} - d_A)$, then B may not be adsorbed on nearest or next-nearest neighbour sites of A although B has no site exclusion properties toward B: segregation then becomes obvious when the coverage is high enough and Figure 7a shows a computer print-out of an equilibrium configuration. If B is slightly smaller than the previous case so that B may be adsorbed on a next-nearest neighbour site to A but not on a nearest neighbour site $(a2\sqrt{2} - d_A) < d_B < (2a - d_A)$, segregation is again apparent at a high enough coverage (Figure 7b), although here like species tend to occur in linear arrays.

These results have implications for surface kinetics which it is convenient to discuss at this juncture. Reaction between (say) the entities A and B which are both adsorbed on a catalyst surface may be visualized as requiring an A and a B to be closer than some critical distance so that they can "interact" for reaction to be possible. The reaction rate would then be proportional to the number (N_{int}) of such A–B "interactions". For instance, if both A and B were adsorbed with simple Langmuirian criteria, one might define an "interaction" as A and B on adjacent sites, and for this situation $N_{int} \propto \theta_A \theta_B$ accurately. This is the general argument which justifies a kinetic rate expression such as (IV)–(1) for localized adsorption and given subsequently in section (IV). When site exclusion criteria result in component segregation, it is clear that a simple proportionality between N_{int} and component coverage cannot be accurate. Again, results are only available from Monte Carlo simulations, but one example suffices to give a feel for the magnitude of the effect. For the blocking criteria $a\sqrt{2} < d_A < 2a$ and $a < d_B < (a2\sqrt{2} - d_A)$, an "interaction" was arbitrarily defined as a B occupying a third- or fourth-nearest neighbour site to A. Figure 7c shows a plot of N_{int} against the surface density of B (δ_B) for a constant δ_A of 0.12. Note that δ is defined to be the ratio of the number of entities to the total number of adsorption sites, and we use this parameter because of the ambiguity of defining θ scales in this type of situation. It is seen that N_{int} vs δ_B is approximately linear up to about $\delta_B = 0.07$. At this point the density of free sites for A ($_f\delta_A$) is 0.043, and $_f\delta_B$ is 0.104. It seems that, even in this situation where component segregation can occur, up to moderate coverages an assumption of direct proportionality between N_{int} and coverage is accurate enough for most kinetic purposes. However, we repeat again that one may expect N_{int} to be modified by the presence of an interaction potential between adsorbed entities, and a quantitative examination of this effect is badly needed.

B. HETEROGENEOUS SURFACES

The actual surface of a solid is generally not at equilibrium and variations in adsorption energy result from the different crystal planes, cracks, edges and lattice defects accessible to the adsorbing gas.

The problem of surface heterogeneity has been apparent in almost all

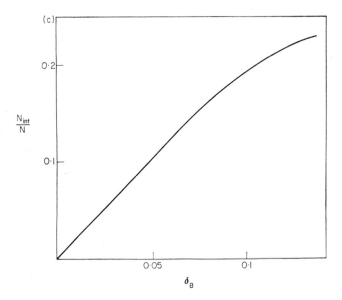

attempts to correlate adsorption data with theoretical models of adsorption. The variation of the heat of adsorption with coverage has often been regarded as a measure of heterogeneity but since interactions within the adsorbate can give similar effects, the surface energies cannot generally be deduced with much accuracy from an experimental adsorption isotherm.

A much clearer understanding of the influence of surface structure on adsorption has resulted from the careful studies of adsorption on single crystals (Rhodin, 1953), field emission microscopy and low energy electron diffraction (see Ehrlich, 1966) and it is clear that different crystal planes of the same metal will have different heats or energies of adsorption resulting in different equilibrium coverages, different rates of adsorption and in the case of chemisorption, possibly different adsorbed species.

The problem now is to decide if a metal film surface used for adsorption or catalysis can be treated as having a small number of discrete values for the heat of adsorption; values which hopefully, may be identified with those found in studies on single crystals or in the field emission microscope. The alternative is to treat the distribution of heats of adsorption as a continuous function to be determined from the adsorption data. This latter approach has been most often used as few surfaces have been sufficiently well-defined to allow discrete heat values to be deduced from adsorption data. While the physical unreality of such treatments of metal surfaces has been recognized (see Rideal, 1968) the identification of detailed surface structure has only recently become practicable.

The following discussion is given in terms of the isosteric differential heat of adsorption, q_{st}, which is defined by

$$q_{st} = RT^2(\partial \ln p/\partial T)_\theta \qquad\qquad \text{(III)–(22)}$$

and is thus conveniently extracted from experimental isotherm data:

FIG. 7. (a) top: Part of a computer print-out of an equilibrium configuration for the localized adsorption of a mixture of large and small entities. Key to symbols, S $\equiv$ large entity (called A in text); / $\equiv$ small entity (called B in text); * $\equiv$ vacancy for A; + $\equiv$ vacancy for B; totally blocked sites are unmarked. Blocking criteria, $a\sqrt{2} < d_A < 2a$ and $a > d_B > (a2\sqrt{2} - d_A)$. Occupation densities, $\delta_A = \delta_B = 0.125$.
(b) middle: As for (a) but with the blocking criteria, $a\sqrt{2} < d_A < 2a$ and $(a2\sqrt{2} - d_A) > d_B > (2a - d_A)$. Occupation densities, $\delta_A = \delta_B = 0.16$. Note that in both (a) and (b) the model is for a square lattice but the computer printer has a larger unit translation in the vertical than in the horizontal direction.
(c) bottom: Variation of the relative number of A–B "interactions" with δ_B. Blocking criteria as for (a) and $\delta_A = 0.12$. The "interaction" is defined in the text. N_{int}/N is the ratio of the number of interactions to the total number of adsorption sites.

the partial derivative is taken at constant composition of the adsorbed phase. The relation between q_{st} and other heats and energies of adsorption is summarized by Young and Crowell (1962).

For a heterogeneous surface, the total fraction covered (θ) is given by

$$\theta = \sum_i \chi_i \, \theta_i \qquad \text{(III)–(23)}$$

where the coverage on a homogeneous part of the surface

$$\theta_i = \phi(p, q_{st})_T \qquad \text{(III)–(24)}$$

is a function of the gas pressure and adsorption energy at constant temperature, and χ_i is the fraction of the total surface with the heat q_{st}.

There have been numerous attempts to represent the heterogeneity of the surface by a continuous function ($\chi = f(q_{st})$) so that the summation can be replaced by an integration.

This procedure can only succeed if the isotherm $\phi(p, q_{st})_T$ is an explicit expression for θ. The Henry's Law and Langmuir isotherms have this property and analytical solutions to equations of the form

$$\theta = \int f(q_{st}) \, \phi(p, q_{st}) \, dq_{st} \qquad \text{(III)–(25)}$$

have been limited to these isotherms. It is readily shown that Henry's law leads to a straight line isotherm regardless of the distribution function. The assumptions of the Langmuir model have been combined with continuous distribution functions, first by Zeldowitch (1934) and later by Roginsky (1944) and by Halsey and Taylor (1947). It has been shown that if the heat falls exponentially with coverage the overall adsorption is described by the Freundlich isotherm

$$\theta = kp^n$$

with $0 < n < 1$. This isotherm describes much experimental data with reasonable accuracy and has led to speculation on the physical reality of the exponential variation of the heat with coverage. However, it is now clear that other distributions lead to the same isotherm and that the treatment of the exponential variation of the heat requires that the entropy of adsorption be constant; an unjustifiable and physically unrealistic assumption (see Young and Crowell, 1962).

Although other isotherm equations for adsorption on a homogeneous surface can be used in conjunction with continuous distribution functions, the solution of the equation for a heterogeneous surface must be numerical. Ross and Olivier (1964) have used a Gaussian distribution of heats with a two-dimensional van der Waals equation to compute adsorption on graphite and other non-metallic adsorbents. The method does not appear useful for metals (Rideal, 1968). Harris (1968) has recently discussed the technique by which the experimental θ vs p functions may be used together with an assumed isotherm equation (Langmuir) to generate the corresponding (continuous) distribution of heats.

The loss of physical reality resulting from the use of a distribution function need not be tolerated for a non-analytic solution to the problem. It is unrealistic to have the discrete heats of adsorption of the different crystal faces, cracks, edges and surface defects of a metal conform to any continuous function. However, a patch treatment of the surface requires that the adsorption data shall result from a reasonably small number of different kinds of adsorption site.

A method of computing adsorption on a heterogeneous surface has been described by Baker and Head (1968) and applied to the adsorption of xenon on nickel films by Baker and Fox (1965) and Baker and Bruce (1968). The assumptions of the model are that only a small number of different types of gas adsorption sites exists on the surface and that sites of the same kind can be treated as patches where adsorption can be described by the established models for a uniform surface. The simultaneous filling of these several patches is then computed, subject to the condition that the adsorbed phases have the same chemical potential at any gas pressure.

The method employs a numerical integration routine to solve a set of simultaneous differential equations and in principle can use any isotherm equations on any number of patches. Each patch is characterized by its proportion of the total surface and heat and entropy constants for the isotherm equation. As an example, let us consider that the surface consists of patches, each of proportion χ_i which are in equilibrium with a gas, and that on each patch adsorption is described by a simple Langmuir isotherm

$$\theta_i = \frac{a_i\, p}{1 + a_i\, p} \qquad\qquad \text{(III)–(26)}$$

The constant a_i is a function of the heat and entropy of adsorption and has values which are different on each patch, so that different fractions, θ_i, of each patch are covered, subject to equation (III)–(23) for the total coverage θ.

While there need not be any direct functional relationship between the heat and entropy of adsorption, it is usually found that a high heat of adsorption is associated with a large loss of entropy. We may think of this as a result of the loss of degrees of freedom of translation and vibration as the energy of binding to the surface is increased. It follows that in many cases the influence of differences between heats of adsorption on coverage are to some extent compensated by an entropy effect.

The differential entropy of adsorption, $\Delta \bar{S}_i$, consistent with the Langmuir isotherm is

$$\Delta \bar{S}_i = -R \ln \frac{\theta_i}{1 - \theta_i} - C_i$$

The entropy of adsorption falls as the coverage increases, passing through an inflexion at $\theta_i = 0.5$ where it has a value $-C_i$ determined by the loss of translational and vibrational entropy accompanying the adsorption process. At any given total coverage θ, the average values $< q_{st} >$ and $< \Delta \bar{S} >$ are given by

$$<q_{st}> = \sum_i i q_{st} \, \chi_i (d\theta_i/d\theta) \qquad \text{(III)–(27)}$$

$$<\Delta \bar{S}> = \sum_i \Delta S_i \, \chi_i (d\theta_i/d\theta) \qquad \text{(III)–(28)}$$

For the purposes of an illustrative numerical example, suppose the surface consists of two patches with the parameters given in Table 3.

TABLE 3

Parameters for illustrative adsorption on two patch surface at 300 °K

Patch	χ	Isosteric heat of adsorption q_{st} (kcal mole^{-1})	entropy constant, C (cal mole^{-1} deg^{-1})
1	0.90	10	20
2	0.10	15	30

Ten per cent of the surface has the higher value of q_{st} of 15 kcal mole^{-1}, but this advantage is in part offset by the fact that a molecule bound with this energy undergoes a larger loss of entropy. Numerical solution of the problem shows that while patch 2 begins to fill first, patch 1 competes more and more favourably as the coverage on patch 2 increases.

The separate coverages on the two patches are plotted as a function of total coverage in Figure 8a. The overall heat and entropy of adsorption vary with total coverage as in Figures 8b and c. The adsorption isotherm that would be observed is in Figure 9 where logarithmic scales have been used for both axes.

In this example the isotherm at low coverage could be described by the Freundlich equation ($\theta \propto p^n$) with n = 0.65. Deviation from this equation occurs only when the coverage on patch 2 has become too high for it to play any further effective part in the filling of the surface.

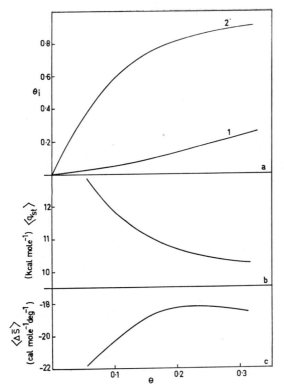

FIG. 8. Illustration of adsorption on a two-patch surface: for parameters see Table 3. (a) plots of θ_1 (curve 1) and θ_2 (curve 2) vs θ; (b), variation of average isosteric heat of adsorption ($<q_{st}>$) with θ; (c) variation of average differential entropy of adsorption ($<\Delta\bar{S}>$) with θ.

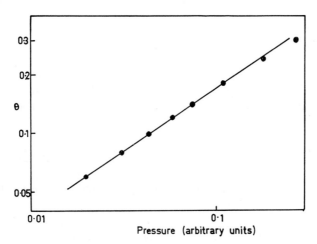

FIG. 9. Variation of θ with p, corresponding to the two-patch surface in Table 3 and Fig. 8.

It is seen in the present example that $< q_{st} >$ falls in a manner which is qualitatively reminiscent of an exponential decrease, and that the entropy rises through a maximum. However, this is simply the overall effect of the simultaneous filling of two patches each of which has a constant heat of adsorption, and has an entropy of adsorption which falls as the coverage is increased.

In practice, the required procedure is to evaluate the patch parameters which will fit a set of experimental adsorption isotherms. The physical reality of the model of a surface obtained in this way should be assessed by critical examination of the heat and entropy parameters returned by the computation; these should be reasonable, and internally consistent. Generally the limited range and accuracy of the experimental adsorption data will limit the resolution of the model. However, within the expected range of validity the same patch model should predict the adsorption of other gases with only a systematic displacement of the heat and entropy parameters.

A patch analysis has been used to analyse the adsorption data for xenon on nominally polycrystalline nickel films deposited on a pyrex substrate at 200 °C (Baker and Fox, 1965; Baker and Bruce, 1968). The general method of analysis followed that outlined above, except that adsorption within a patch was assumed to follow a two-dimensional van der Waals isotherm for which the differential entropy of adsorption is given by

$$\Delta \bar{S}_i = -R \ln \frac{\theta_i}{1 - \theta_i} - \frac{R\theta_i}{1 - \theta_i} - C_i \qquad \text{(III)-(29)}$$

and q_{st} was assumed to vary linearly with coverage

$$_iq_{st} = {}_iq^\circ_{st} + A_i\theta_i \qquad \text{(III)-(30)}$$

The analysis yielded the patch parameters which are recorded in Table 4 and the same χ_i values also fit krypton adsorption. Although an unequivocal identification of each patch in terms of surface structure

TABLE 4

Best-fit parameters for patch analysis of xenon adsorption on nickel films

	Patch	Fraction of total Surface (χ)	q°_{st} (kcal mole^{-1})	A (kcal mole^{-1})	C (cal mole^{-1} deg^{-1})
Ni film deposited	1	0.276	4.24	1.01	10.0
on pyrex glass at	2	0.661	5.26	0.18	19.8
200 °C	3	0.063	7.06	0	26.8

cannot yet be made, the following comments are possible. The nickel film deposited on pyrex at 200 °C, although nominally polycrystalline, can be satisfactorily characterized by only three patches, of which one is present to only 6.3% of the surface: thus only two patches make up most of the surface. In such a case, a model based on a continuous distribution of adsorption sites would clearly be grossly inapplicable. This relatively simple situation is due, at least in part, to the high temperature of the substrate during film deposition which prevents high energy surfaces from being frozen in. One would expect a nickel film surface deposited at (say) 0 °C to display a more extensive heterogeneity, and this should also be true for films of other metals which sinter less readily than nickel (e.g. tungsten). A patch model analysis has not yet been applied to rare gas adsorption data on 0 °C nickel films, but the isotherm given by Baker and Fox (1965) suggests that there is a greater proportion of high adsorption energy sites compared with films deposited at 200 °C.

The promise of the patch analysis method and the unequivocal identification of the patches with specific types of surface will be properly realized when sufficient data are available from single crystal studies independently to assign adsorption parameters to known crystal planes. A further possibility is to characterize the number and proportions of the surface patches by independent measurement of other surface properties (e.g. work function).

IV. KINETIC PROCESSES AT SURFACES

Two types of situation arise in surface reactions. In the first we are concerned with an adsorption or desorption process per se; that is, one in which the relative surface coverage changes with time. In the second we are concerned with a steady-state situation in terms of surface coverage, as in most catalytic reactions. It is in the latter situation that the approximation is usually made of applying an equilibrium adsorption isotherm to describe a steady-state surface concentration. It is the purpose of this section only to outline some of the main points which are relevant to the processes discussed in Chapter 8, and also to offer some comments about kinetics on heterogeneous surfaces. The subject has been frequently reviewed in detail for reactions on homogeneous surfaces, and the reader is referred to Laidler (1954) and Bond (1962).

A. UNIFORM SURFACES

1. *Steady-state Surface Coverage*

Rate expressions are constructed by analogy with simple kinetic theory in terms of the individual surface entities taking part, and it is the surface concentrations of the latter that are related to bulk pressures or concentrations via the adsorption isotherms. Thus, for instance, if the rate of a catalytic reaction between gases (A, B, ... X) is controlled by a surface reaction r between N types of surface species (1, . . .i, . . . N), we have

$$\text{rate} = A_r e^{-E_r/RT} \left[\prod_i^N \theta_i^{n_i} \right] \qquad \text{(IV)–(1)}$$

where n_i is the number of entities i entering the rate controlling reaction, and A_r is the frequency factor out of which the surface concentration terms have been explicitly extracted. The isotherms (IV)–(2) enable one

$$\theta_i = f(p_A, --- p_X) \qquad \text{(IV)–(2)}$$

to convert (IV)–(1) into a form containing measurable pressures. Some forms of the isotherms have been described in a previous section: for instance, if adsorption is random and localized a Langmuir isotherm is appropriate. Despite the obvious limitations, kinetics based on Langmuir isotherms are still the most extensively used method, mainly because the simplifications possible at the high coverage and low coverage limits enable equation (IV)–(1) to be easily integrated to forms useful with experimental data. The result of making assumptions about relative strengths of adsorption may be had from inspection of the isotherm functions. Two other factors are worth mentioning. All of the mixed adsorption isotherms discussed in section (III) assume that the gases compete for the same sites. If this is not so, the isotherm for each component can, of course, be written down independently. On the other hand, if a chemisorbed entity reacts with a gas phase molecule, the pressure of the latter will enter the kinetic expression directly, and will be indistinguishable from reaction with a physically adsorbed molecule the coverage by which is directly proportional to pressure. As an illustration of these ideas, we consider the following reactions which are relevant to exchange reactions discussed in Chapter 8:

$$NH_{3(sp)} + D_{(s)} \rightarrow NH_{2(s)} + HD_{(s)} \qquad \text{(IV)–(3a)}$$

$$NH_{2(s)} + D_{2(sp)} \rightarrow NH_2D_{(sp)} + D_{(s)} \qquad \text{(IV)–(3b)}$$

$$NH_{3(sp)} \rightarrow NH_{2(s)} + H_{(s)} \qquad \text{(IV)–(4a)}$$

$$NH_{2(s)} + D_{(s)} \rightarrow NH_2D_{(sp)} \qquad \text{(IV)–(4b)}$$

The subscripts (sp) denote non-dissociatively adsorbed species. Some pressure dependences of the reaction rates are listed in Table 5. These are obtained by considering the appropriate forms of the Langmuir isotherms (cf. section III). Laidler (1954) provides a summary which is useful for other situations.

TABLE 5
Pressure dependence of reactions (IV)–(3) and (IV)–(4) for
ammonia exchange with deuterium

Reaction	Assumed conditions					
	Low coverage		Ammonia strongly adsorbed compared to deuterium: reactants competing for sites		Ammonia strongly adsorbed compared to deuterium: reactants not competing for sites	
	exponent for		exponent for		exponent for	
	p_{NH_3}	p_{D_2}	p_{NH_3}	p_{D_2}	p_{NH_3}	p_{D_2}
(IV)–(3a)	1	0.5	−1	0.5	0	0.5
(IV)–(3b)	1	0.5	−1	1.5	0	1
(IV)–(4a)	1	0	−1	0.5	0	0
(IV)–(4b)	1	0	−1	1	0	0.5

Wei and Prater (1962) have described in detail the analysis by matrix methods of complex surface reactions, particularly those with interconnected pathways. The technique is only applicable in its present form if the adsorption of each component is weak enough for there to be a direct proportionality between pressure and coverage, and provided each step has overall first order kinetics. The application of computer techniques to the analysis of complex catalytic reaction schemes has recently been reviewed by Grover (1968).

An analysis based on a Langmuir isotherm will be in error if the adsorbed species are not randomly distributed, and probably grossly in error when interactions are strong enough to result in long range order.

The activation energy for a surface reaction may be defined in the usual way in terms of the energy of the transition state relative to that of the reactants. However, it should be remembered that if a surface reactant species in the rate controlling step is involved in a quasi-equilibrium prior to the rate controlling step, the observed activation energy will contain both the true activation energy for the surface reaction and the heat of reaction for the equilibrium reaction. A special case of this is when the equilibrium refers to reactant adsorption. The effect arises, of course, because the surface concentration of reactant becomes temperature dependent via the equilibrium reaction. An analogous situation arises if a poison to the surface reaction is in established adsorption equilibrium. As an example, equation (III)–(17) gives

$$\theta_{CH_3} = (a_{CH_3} K^{\frac{1}{2}}) p_{CH_4} p_{H_2}^{-\frac{1}{2}} \qquad \text{(low coverage)} \qquad \text{(IV)–(5)}$$

$$\theta_{CH_3} = p_{CH_4}^{0} p_{H_2}^{0} \qquad \text{(high coverage)} \qquad \text{(IV)–(6)}$$

Thus at low coverage the apparent activation energy for a rate controlling step in which $CH_{3(s)}$ is a reactant must contain the contribution $(\Delta H°/2 - \epsilon_{CH_3})$, where $\Delta H°$ is the enthalpy change for the equilibrium reaction (III)–(16) and ϵ_{CH_3} is the adsorption energy (approximately the heat of adsorption) of CH_3 defined in relation to equation (III)–(4). On the other hand, at high coverage θ_{CH_3} is approximately temperature independent and no such contributions occur. In an analogous way, the true frequency factor for the rate controlling step would be multiplied by the factor $a'_{CH_3} \exp(\Delta S°/R)$, where $\Delta S°$ is the entropy change for reaction (III)–(16), and a'_{CH_3} is the remainder from a_{CH_3} after having had $\exp(\epsilon_{CH_3}/RT)$ factored out. If the contributing entropy and heat terms are linearly related, a linear relation between E_{meas} and $\log A_{meas}$ results. This is one way in which a compensation effect may occur in surface kinetics.

2. Deuterium Exchange Kinetics

Since metal films have been so extensively used in catalytic exchange reactions, it is worthwhile to consider the appropriate kinetic representation.

(a) *Rate equations for a single group of exchangeable hydrogen atoms* (Kemball, 1953, Anderson and Kemball, 1954). Consider the exchange of the hydrogen atoms in the molecule $(R)H_m$. Here R represents any structural unit, and the only restriction is that the m hydrogen atoms shall all be equally susceptible to exchange. Let x_i be the percentage of the isotopic species $(R)H_{m-i}D_i$ present at time t. The extent of exchange at t may be defined by ϕ, where

$$\phi = x_1 + 2x_2 + \ldots\ldots + mx_m = \sum_{i=1}^{m} ix_i \qquad \text{(IV)–(7)}$$

ϕ merely represents the total amount of deuterium in the exchange products. If kinetic isotope effects are absent ϕ varies with t according to

$$-\log_{10}(\phi_\infty - \phi) = \frac{k_\phi t}{2.303\ \phi_\infty} - \log_{10}\phi_\infty \qquad \text{(IV)–(8)}$$

where ϕ_∞ is the value of ϕ at exchange equilibrium, and k_ϕ equals the number of deuterium atoms entering 100 molecules of reactant per unit time. Equation (IV)–(8) is equivalent to the general rate equation for exchange kinetics due to Harris (1951).

The disappearance of reactant may be represented by

$$-\log_{10}(x_0 - x_0^\infty) = \frac{k_0 t}{2.303(100 - x_0^\infty)} - \log_{10}(100 - x_0^\infty) \qquad \text{(IV)–(9)}$$

where x_0 is the percentage of reactant present at t, x_0^∞ is the value of x_0 at exchange equilibrium, and k_0 equals the initial rate of disappearance of reactant in percent per unit time. Equation (IV)–(9) is essentially an empirical rate expression, based on the assumption that exchange can be represented by opposing first order reactions. Its sole value is the evaluation of k_0 by a method that is additional to the evaluation of initial reaction rates by a graphical procedure.

The parameter M, defined by

$$M = k_\phi/k_0$$

equals the average number of deuterium atoms entering a reactant molecule under initial exchange conditions, and M may obviously also be evaluated from the initial product distributions themselves.

Values for ϕ_∞ and x_∞ can be easily calculated assuming that at equilibrium there is a random isotopic distribution between reactant and "hydrogen". For instance, with a mixture in which the initial molar ratio of $D_2/(R)H_m$ equalled y, ϕ_∞ would be given by

$$\left(\frac{100m \times 2y}{m + 2y}\right).$$

In fact, measured values of ϕ_∞ differ appreciably from those calculated by classical statistics, with the ratio $\phi_\infty^{obs}/\phi_\infty^{calc}$ ranging from 1.14 to 1.28 for some lower hydrocarbons (Gault and Kemball, 1961). For work of any accuracy, particularly for the use of equation (IV)–(8) at high conversions, ϕ_∞ should, if possible, be measured experimentally. It is usually desirable to work with reaction mixtures containing a substantial excess of deuterium, typically $D_2/(R)H_m > 5/1$, and when this is done the value of x_0^∞ is so small that a negligible error is introduced by using the statistically calculated rather than the experimental value: under these conditions x_0^∞ can often be set equal to zero without appreciable error.

(b) *Rate equations for more than a single group of exchangeable hydrogen atoms.* Not infrequently, the geometry or energetics of the reactant molecule is such that the hydrogen atoms exist as (say) two groups $(R)H_mH_n$, all atoms within the m-group being equally susceptible to exchange, but differing from the atoms in the n-group. If the exchange of such a molecule is treated as $(R)H_z$ with $z = m + n$ using the treatment outlined in the previous section, curved ϕ- and x-plots will obviously be obtained, since rate of deuterium entry must decrease as exchange is propagated into the difficult-exchange group. A typical example is given by Crawford and Kemball (1962) in their study of p-xylene exchange over nickel. The composition vs time plots clearly showed that six hydrogens were rapidly exchanged, and four exchanged more slowly: the geometry of the molecule invites the conclusion that the six easily exchanged hydrogens are from the two methyl groups (Me-group hydrogens) and the four slowly exchanged hydrogens are on the aromatic ring (R-group hydrogens). The standard application of equation (IV)–(8) to typical results is shown in Figure 10 using ϕ_∞ for

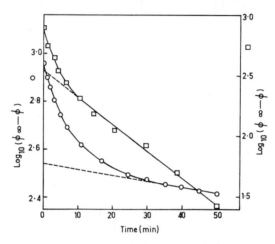

FIG. 10. Plots according to equations (IV)–(8) and (9) for the exchange of p-xylene with deuterium on nickel films: O, at -9.5 °C on 6.4 mg nickel; □, at 0 °C on 10.5 mg nickel (Reproduced with permission from Crawford and Kemball (1962). *Trans. Faraday Soc.* **58**, 2452.)

ten hydrogens. The initial and final sections of the plots are approximately linear and these parts correspond to the exchange of the two groups. The value of k_ϕ obtained from the slope of the first section of the plot is the sum of the initial exchange rates of the two groups,

$$k_\phi = k_\phi^{Me} + k_\phi^{R}$$

The final section of the plot may be represented by

$$-\log_{10}(\phi_\infty - \phi) = \frac{k_\phi^{R}t}{2.303\,\phi_\infty^{R}} - \log_{10}\phi_\infty^{R} \qquad \text{(IV)–(10)}$$

and this describes the exchange of the R-group on the assumption that the Me-group have attained exchange equilibrium. We have

$$\phi_\infty = \phi_\infty^{Me} + \phi_\infty^{R} \quad ; \quad \phi_\infty^{Me}/\phi_\infty^{R} = n_M/n_R$$

where n_{Me}/n_R equals the ratio of the numbers of hydrogens in the two groups, 6/4 in this case. The value of k_R may be determined by drawing the best line through the final section of the plot, with intercept $\log_{10}\phi_\infty^R$.

In general, if the easily exchanged group has m hydrogens, then the value of k_ϕ^{easy} may be estimated from

$$-\log_{10}(\phi_\infty^{easy} - \phi) = \frac{kt_\phi^{easy}}{2.303\phi_\infty^{easy}} - \log_{10}\phi_\infty^{easy} \qquad \text{(IV)–(11)}$$

where the value of ϕ_∞^{easy} is calculated on the basis that only m hydrogens are exchangeable: in the p-xylene case $\phi_\infty^{easy} \equiv \phi_\infty^{Me}$ with m = 6. Values of ϕ_∞ for various groups (as distinct from the entire molecule) must be calculated, and Crawford and Kemball outline how the statistically calculated values may be corrected by comparison with cases where both calculated and measured values are known. Finally, when the reactivities of the two groups differ by a very large factor, the result will be that the easily exchanged group will approach exchange equilibrium before a significant amount of the second group has exchanged. If a large excess of deuterium is being used, this will mean that at this stage the exchange product mostly consists of molecules in which all of the first group hydrogens have been exchanged: this may then be considered as the parent for exchange in the second group and the corresponding form of equation (IV)–(9) applied in the obvious way to give k_0^{diff}. Values for k_0^{easy} may be estimated from the first part of the plot from equation (IV)–(9).

More detailed analytical procedures for exchange kinetics have recently been published by Bolder et al. (1964). This treatment is formulated so as to be not necessarily confined to heterogeneous catalytic reactions; however terminology appropriate to the latter situation will be used in the following. The important feature of this treatment is that attention is focused on the time dependence of the proportion of each of the deuterated products and of the undeuterated reactant throughout the course of the reaction: they also treat a lumped parameter for the total deuterium content of the exchanged molecules (σ in their notation, equal to ϕ/m in the present notation), but this adds nothing further to the treatment given above. The algebra of Bolder's treatment is fairly lengthy in detail for any but trivial cases, so we confine ourselves here to providing some general comments. The detailed algebra makes use of matrix methods and some of the concepts discussed by Wei and Prater (1962). In principle, their treatment makes no assumptions about the deuterium distribution at the start of the reaction, however for reasonable mathematical tractability it is highly desirable to assume that the reactant $(R)H_m$ is initially completely undeuterated, and that the deuterium source is in such excess that isotopic dilution can be ignored.

A reaction which proceeds by stepwise exchange is characterized by

$M = 1$ using the procedures given previously. However, this conclusion can also be deduced by fitting the experimental to computed curves for the dependence of x_i (Bolder's d_i) on time (that is, fitting experimental and computed $X_i(t)$ functions). This includes the case of $i = 0$, that is the undeuterated parent. The computation of $x_i(t)$ is relatively straight-forward since stepwise exchange must lead to a binomial product distribution: Bolder *et al.* give an explicit expression (their equation 39) to compute $x_i(t)$ from $\sigma(t)$ (or $\phi(t)$). In stepwise exchange, kinetic isotope effects do not alter the nature of the product distribution, only the overall reaction rate. Furthermore, if the deuterium atoms are provided from the source not all with equal reactivity, the product distribution will remain binomial, although the $\sigma(t)$ (or $\phi(t)$) functions then become complicated, since one is dealing with parallel reaction paths. If the reactant molecule has (say) two groups of hydrogen atoms of unequal reactivity, $(R)H_mH_n$, stepwise exchange within each group gives a binomial distribution within each group, and the lumped parameter (ϕ) treatment described previously for this type of molecule is of course applicable.

The evaluation of $x_i(t)$ (including $i = 0$) is considerably more complex when multiple exchange occurs; that is, the introduction of more than a single deuterium atom in one total residence period of a reactant molecule on the catalyst surface. Multiple exchange involves inter-conversions between a sequence of surface intermediates. The virtue of the treatment due to Bolder *et al.* is that it provides a means of evalu-ating $x_i(t)$ for an arbitrary set of reaction paths for multiple exchange, these reaction paths being constructed in terms of specific assumed intermediates. One then seeks agreement between experimental and computed $x_i(t)$ as a test of the assumed model. The computation is limited to a reactant molecule with a single group of exchangeable hydrogens of equal reactivity. Under these assumptions, each function $x_i(t)$ contains two disposable parameters and, of course, each is inde-pendent of i for a given reaction scheme. In the most explicit of Bolder's formulations, these parameters are, k_1 the rate constant for "deactiva-tion" of the surface intermediate, that is by desorption to the gas phase, and β equal to k_2/k_1, where k_2 is the rate constant for transition between surface intermediates. This parameter β is, in a formal sense, the same as the parameter P introduced by Anderson and Kemball (1954), when discussing multiple exchange. Bolder *et al.* give alternative methods for the construction of the $x_i(t)$ functions, but the most straightforward and physically satisfying method is summarized in the recipe given on p. 333 of their paper, which can be applied without monumental difficulty. In any case, the fit between computed and observed $x_i(t)$ functions must be done numerically, and this of itself is certainly sufficiently tedious to require coding for a computer.

It should, of course, be pointed out that many of the main conclusions concerning the identity of the reaction intermediates deduced by Bolder's analysis would be arrived at in a qualitative fashion by an inspection of the initial exchange product distributions together with a lumped parameter analysis. However, there is also no doubt that the complete analysis is capable of returning a model which is more detailed and less subject to ambiguity.

Bolder *et al.* also provide an analysis for redistribution reactions, for instance the reaction between $(R)H_m$ and $(R)D_m$. The $x_i(t)$ functions are contained in the generalized treatment for exchange kinetics, since $(R)D_m$ is simply a type of deuterium source. Redistribution reactions avoid the requirement for having a large excess of deuterating agent, but in many heterogeneous catalytic reactions this can be a disadvantage because serious side reactions may occur (for instance catalyst poisoning) unless a large excess of (say) D_2 is present.

3. *Rates of Adsorption and Desorption*

On a uniform surface, the rate of adsorption will in general be given by an equation of the form

$$r_a = p\, A_a\, f_a(\theta)\, e^{-E_a/RT} \qquad (IV)-(12)$$

provided there is no weakly adsorbed precursor to the bound state. In the latter situation, the surface concentration of weakly adsorbed precursor may be less than directly proportional to the pressure, with the result that the rate of adsorption into the bound state may be proportional to pressure to an exponent less than one. In equation $(IV)-(12)$ E_a is the activation energy for adsorption, and $f_a(\theta)$ equals the chance that the adsorbing molecule collides with the necessary adsorption sites; that is, $f_a(\theta)$ is the fraction of surface sites available for adsorption. A_a is a parameter which varies only relatively slowly with temperature and represents the frequency factor for the adsorption process, thus A_a contains the entropy of activation.

A general expression for the rate of desorption is

$$r_d = A_d\, f_d(\theta) e^{-E_d/RT} \qquad (IV)-(13)$$

where A_d and E_d have analogous significance to those in the adsorption case, and $f_d(\theta)$ is the fraction of sites available for desorption. The parameters E and A may depend to some extent on θ via the influence of interactions between the adsorbed particles when θ is sufficiently high. The forms of $f_a(\theta)$ and $f_d(\theta)$ can in simple cases, be estimated from first principles. If adsorption occurs under simple nondissociative Langmuirian conditions, it is reasonable to write $f_a(\theta) = (1 - \theta)$ and $f_d(\theta) = \theta$; in confirmation, if E_a and E_d are independent of θ, equating r_a and r_d leads to the form of the simple Langmuir isotherm. For the corresponding case of binary dissociative adsorption, an expression for $f_a(\theta)$ may be written if we assume that adsorption requires a pair of vacant sites. If adsorption in the localized layer is random (zero interaction energy), the chance of obtaining each vacant site of the pair is $(1 - \theta)$, so $f_a(\theta) = (1 - \theta)^2$, and in an analogous way $f_d(\theta) = \theta^2$. In extending the discussion to n-fold dissociative adsorption, care needs to be taken to ensure that $f_a(\theta)$ and $f_d(\theta)$ are defined for the rate controlling step. Thus, if the overall adsorption reaction is

$$nS + A_n \rightarrow n\,A_{(s)} \qquad\qquad \text{(IV)–(14)}$$

the expression for r_a will contain a factor $(1 - \theta)^n$ provided equation (IV)–(14) is the actual rate controlling step: similarly r_d will contain θ^n provided the reverse of (IV)–(14) is rate controlling. In fact, a complex reaction is likely to occur via several consecutive elementary reactions, any one of which may be rate controlling, and the form of the expression for r_a or r_d must be written appropriately.

When the adsorbed species exclude neighbouring sites, simple expressions for $f_a(\theta)$ and $f_d(\theta)$ are still possible provided all the surface species have the same site exclusion properties and, as before, provided adsorption is random. The functions are written down in the same way as before, except that the chance of obtaining a vacant site is now $(1-\theta(\bar{\nu}/\nu_M))$ rather than $(1 - \theta)$, (cf. section III).

The functions $f_a(\theta)$ and $f_d(\theta)$ are much affected by interaction between adsorbed species and a solution is available for the particularly simple case of an interaction energy V limited to the four nearest neighbour positions of a square lattice. However, it has been shown (cf. Laidler, 1954) that the extent to which $f_a(\theta)$ or $f_d(\theta)$ is modified from its random value, is cancelled out by an opposite alteration in the value of A_a or A_d, so that the product $A_a f_a(\theta)$ or $A_d f_d(\theta)$ remains independent of the interaction energy. Thus, with this in mind we may usefully express $f_a(\theta)$ and $f_d(\theta)$ in terms of the simple random model.

The value of A_a depends in principle on whether the transition complex is mobile or localized; a transition from localized to mobile would, of itself, replace two vibrational by two translational degrees of freedom, and the entropy increment would be substantial. Within the framework of ordinary transition state theory this transition has been estimated to lead to an increase in A_a of one to two orders of magnitude. As pointed out in a previous section, it is unlikely that mobile adsorption (in the present sense) will be a valid model for the chemisorbed state, except at quite high temperatures, since chemisorption usually involves localized chemical binding. The extrapolation of this conclusion to the nature of the transition state is more uncertain, but we believe that the localized model is likely to be the more reasonable for most reactions of present interest.

B. HETEROGENEOUS SURFACES

1. *Discrete Distributions of Activation Energies*

In general terms, if on a heterogeneous surface the i'th area represents the fraction χ_i of the total surface, the reaction rate is

$$\text{rate} = \sum_i \chi_i A_i e^{-E_i/RT} \qquad \text{(IV)–(15)}$$

where A_i and E_i are, for the i'th type of surface, the total pre-exponential factor (containing pressure or concentration terms), and the activation energy. Fitting such an expression to experimental data is impossible without either assumptions or auxiliary information concerning the distributions in A_i and E_i. One simplification to equation (IV)–(15) (or its continuous distribution equivalent) is to lump the frequency factor A_i with the distribution parameter χ_i, so that an effective distribution is defined by $(\chi A)_i$: this will clearly be unsatisfactory if one wishes to use a distribution in χ_i available from independent data, except in the unlikely circumstance that A_i is a constant independent of i.

The main features of the problem can be realized by considering the simplest heterogeneous surface, one which has two kinds of structure with different activation energies E_1 and E_2 for the same reaction. If the pre-exponential factors A_1 and A_2 are equal then the reaction rate will be highest on the patch of lowest activation energy. Such a surface may not be readily recognized as heterogeneous in a catalytic experiment.

The real problems arise from experiments in which variations of the catalyst structure result in different observed activation energies for the same reaction. The attempts to correlate these changes in structure with changes in activation energy are usually frustrated by the fact that a change in the pre-exponential factor in part compensates for the change in activation energy.

Heterogeneity will be recognized experimentally if A_1 and A_2 have values which compensate for the difference between E_1 and E_2 at a temperature which is within or close to the range of the experiment. In a general way we can understand that a correlation may exist between E and A similar to that found for heat and entropy of adsorption. A transition state in the reaction requiring high energy would also normally be associated with a large entropy of formation.

The temperature dependence of reaction rate for a two patch surface near the temperature at which compensation occurs will be illustrated by a numerical example.

The overall rate on such a surface will be given by

$$\text{rate} = \chi_1 A_1 e^{-E_1/RT} + \chi_2 A_2 e^{-E_2/RT}$$

There will be a temperature (T') at which the pre-exponential factors (A_1, A_2) compensate for the difference between the activation energies E_1 and E_2. At this temperature the overall rate will be independent of any variations in χ_1 and χ_2 because the rates on each part of the surface are the same.

In Figure 11 the rates for $E_1 = 20$, $E_2 = 30$ kcal mole^{-1} have been plotted over the temperature range 300–400 °K, with the values of A_1 and A_2 chosen so that complete compensation occurs at $T' = 363$ °K.

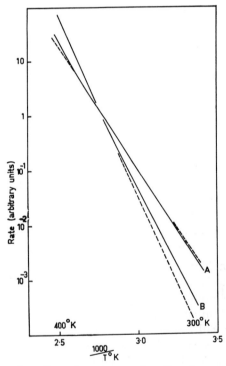

FIG. 11. Dependence of reaction rate on temperature for a two-patch model of a heterogeneous surface: $E_1 = 20$, $E_2 = 30$ kcal mole^{-1}. Curve A $\chi_1 = 0.9$, $\chi_2 = 0.1$; curve B $\chi_1 = 0.1$, $\chi_2 = 0.9$: frequency factors adjusted for complete compensation at 363 °K.

For $\chi_1 = 0.9$ and $\chi_2 = 0.1$ the overall rate (curve A) lies very close to the plot for patch 1. At temperatures between 300 and T' an almost linear plot with $E_{act} \sim 21$ is obtained. Above T' pronounced curvature occurs as the smaller patch with the higher activation energy plays an increasingly dominant role in the reaction. If we reverse the proportions so that now $\chi_2 = 0.9$, $\chi_1 = 0.1$ (curve B) then above T' the apparent activation energy is ~ 30 but at lower temperatures curvature away from this value gives successively lower apparent activation energies.

In this illustration rates have been plotted over six decades. It is unlikely that such a wide range would be investigated experimentally by the same technique. The experimental data are therefore likely to be plotted over a smaller range of temperature so that the slight curvatures would not be noticed. If the experimental range is not close to the compensation temperature (T') then the observed activation energy will be insensitive to variations of surface structure, as one part of the surface will be ineffective for the reaction.

A general quantitative treatment of the problem of a discrete distribution of two activation energies has been given by Nicholas (1959). The apparent activation energy E_{meas} and frequency factor A_{meas} are represented in terms of the parameters

$$\varepsilon = (E_2 - E_1)/RT$$

and
$$r = (A_2'/A_1') \exp(-\varepsilon)$$

by equations (IV)–(16) and (IV)–(17). In this treatment the distribution of the patches is incorporated in A_1' and A_2', that is, $A_1' = \chi_1 A_1$ and $A'_2 = \chi_2 A_2$.

$$E_{meas} = E_1 + r\varepsilon RT/(1 + r) \qquad \text{(IV)–(16)}$$

and

$$A_{meas} = A_1'(1 + r) \exp[r\varepsilon/(1 + r)] \qquad \text{(IV)–(17)}$$

The relationships are represented graphically in Figure 12. The parameter r represents the ratio of the effectiveness of mechanism (2) to the effectiveness of mechanism (1) at temperature T. It is clear from Figure 12 that for $r < 10^{-2}$ E_{meas} and A_{meas} are effectively equal to E_1 and A_1' while for $r > 10^2$ they are effectively equal to E_2 and A_2'. In the intermediate range E_{meas} and A_{meas} are obviously not the E and A values for any single mechanism.

The interpretation of experimental results according to this simple discrete model has been particularly successful for the decomposition of formic acid on silver (Sosnovsky, 1959; Jaeger, 1967).

D

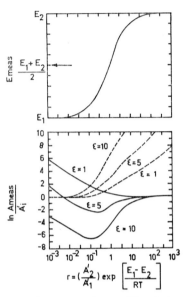

FIG. 12. Upper graph – plot of E_{meas} against r. The choice of scale makes the curve independent of ε. Lower graph – plots of $\ln(A_{meas}/A_1')$ (broken curve) and $\ln(A_{meas}/A_2')$ (full curve) against r for $\varepsilon = 5, 10$. Note that r, the ratio of effectiveness of mechanisms (2) to that of mechanisms (1), is plotted on a logarithmic scale. (Reproduced with permission from Nicholas (1959). *J. Chem. Phys.* **31**, 922.)

2. *Continuous Distributions of Activation Energies*

For reasons given in section III, a continuous distribution of adsorption energies is frequently assumed on a heterogeneous surface, and a similar assumption is also often made for activation energies.

Constable (1925) was the first of many authors to investigate an exponential distribution of activation energies and to show that such an assumption leads to a rate expression in which the activation energy is involved in the temperature independent factor. Nicholas (1959) has investigated the general case of an exponential distribution within a limited range of activation energies. The number of sites with activation energy E is assumed to have the form,

$$N(E) = \begin{cases} N_0 \exp(E/\alpha) \text{ for } E_1 \leq E \leq E_2 \\ \\ 0 \quad \text{otherwise} \end{cases} \qquad \text{(IV)–(18)}$$

The overall rate is given by,

$$\text{rate} = \int_0^\infty N(E)\, e^{-E/RT}\, dE$$

where, again, a lumped pre-exponential factor defines the distribution. The apparent activation energy E_{meas} and frequency factor A_{meas} as found from a logarithmic plot of rate against T^{-1}, are given by

$$E_{meas} = \frac{\int_0^\infty E\, N(E)\, exp(-E/RT)\, dE}{\int_0^\infty N(E)\, exp(-E/RT)\, dE} \qquad (IV)-(19)$$

and

$$A_{meas} = \int_0^\infty N(E)\, exp[(E_{meas} - E)/RT]\, dE \qquad (IV)-(20)$$

Direct substitution of (IV)–(18) in (IV)–(19) and (IV)–(20) yields

$$E_{meas} = E_1 + RT\left[\frac{1}{1-y} - \frac{\varepsilon}{exp[\varepsilon(1-y)]-1}\right] \qquad (IV)-(21)$$

and

$$\ln A_{meas} = \ln N + \frac{E_{meas} - E_1}{RT} + \ln \frac{y\{1 - exp[\varepsilon(y-1)]\}}{(1-y)(exp(\varepsilon y) - 1)} \qquad (IV)-(22)$$

where $\varepsilon = (E_2 - E_1)/RT$, $\qquad y = RT/\alpha$

and $N = $ total area under the distribution curve

$$= N_0\, \alpha\, exp(E_1/\alpha)\{exp(\varepsilon y) - 1\}$$

For various values of y and E, plots of E_{meas}, $\ln(A_{meas}/N)$ and $\ln(A_{meas}/N_1)$ are shown in Figure 13. $N_1 = N_0 RT\, exp(E_1/\alpha)$, is proportional to the height of the distribution function at its first non-zero value, $E = E_1$. It can be seen that E_{meas} changes from near E_1 to near E_2 as y increases, that is, as either the steepness of the distribution function or temperature increases.

The reverse problem of attempting to deduce a continuous distribution of activation energies from experimental rate data has also been considered. One approximation to the solution for concurrent processes occurring over a continuous range of activation energies has been given by Vand (1953) and Primak (1955). The assumption is made that only the activation energy is distributed, so that the frequency factor is constant.

The general form for the kinetics of any given process, that is, one for any given E is

$$-\frac{dc}{dt} = Ae^{-E/RT}c^n \qquad (IV)-(23)$$

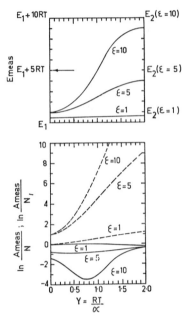

FIG. 13. Upper graph – plots of E_{meas} against y for $\varepsilon = 1,5,10$ for case 3. Note that by a suitable change of scales these curves can be made to coincide. Lower graph – plots of $\ln(A_{meas}/N)$ (full curve) and $\ln(A_{meas}/N_1)$ (broken curve) against y for $\varepsilon = 1,5,10$, where $N_1 = N_0 RT \exp(E_1/\alpha)$. (Reproduced with permission from Nicholas (1959). $J.$ $Chem.$ $Phys.$ **31**, 922.)

where c is some reactant concentration, and n is assumed independent of E. At time t, c will be the sum over all reaction processes, or

$$c = \int_0^\infty c_0 \phi(E,t) dE \qquad (IV)-(24)$$

where c_0 is the value of c at t = 0 for each process with a specified E, that is, c_0 is proportional to the number of processes with each specified E. Equation (IV)–(24) serves to define the function $\phi(E,t)$. Now the course of the reaction may be thought as one in which processes with the lowest activation energy "occur first", and those with the highest activation energy "occur last", so as the reaction proceeds we can imagine a suitable sweep function moving across the activation energy distribution, the position of the sweep function at a given time specifying the operating process with some specified E which we call E_*. The sweep function is $\phi(E,t)$, and since at some time t the value of c will be the sum from all processes yet to occur,

$$c = \int_{E_*}^{\infty} c_0 \, dE \qquad \text{(IV)–(25)}$$

It may be shown that

$$E_* = RT\log(Bt) \qquad \text{(IV)–(26)}$$

where $B = A \, c_0^{n-1}$. From IV–(18) and (19)

$$c_0 = \frac{1}{RT} \frac{dc}{d(\log t)} \qquad \text{(IV)–(27)}$$

The problem is usually to evaluate an E distribution given data for c vs t at various temperatures, and the theory may be applied in two related ways: (a) insertion of c vs t data into (IV)–(27) gives c_0 for various t at each T. Use of equation (IV)–(26) converts t to E_* at each temperature, and c_0 vs E_* is the required distribution. If no prior assumptions are made about the nature of the distribution function, this procedure requires data of high accuracy to obtain a distribution without high uncertainty. Alternatively, if some information is available to suggest the functional form of the distribution, one may proceed by generating c vs t for comparison with experiment, and the distribution parameters adjusted for best fit. In both cases a value for B must be assumed, and this is best treated as an adjustable variable for best fit. The distribution function is temperature independent, so the distribution parameters and B must fit the experimental data over the entire temperature range.

3. *Kinetics of Adsorption*

A surface which responds to the patch treatment for the analysis of adsorption isotherms can also be treated in similar terms for surface kinetics using equation (IV)–(15).

The reality of a heterogeneous surface in adsorption kinetics may be illustrated from the data in Figure 14 which shows the variation with uptake of the first order rate constant for adsorption of ethane and neopentane on nickel films deposited on pyrex at 250 °C. It is difficult to define an accurate θ scale for this type of system, since such a definition requires exact knowledge about the nature and number of adsorbed residues formed. Nevertheless, it is clear that the θ range in Figure 13 is very low: for instance, if $\theta = 1$ were arbitrarily set at a surface stoichiometry of one adsorbed hydrocarbon per surface nickel atom, an uptake of 10^{12} molecules cm^{-2} corresponds to $\theta < 0.001$. At such very low coverages as this, the fall in rate of adsorption with increasing coverage cannot possibly be explained within a homogeneous

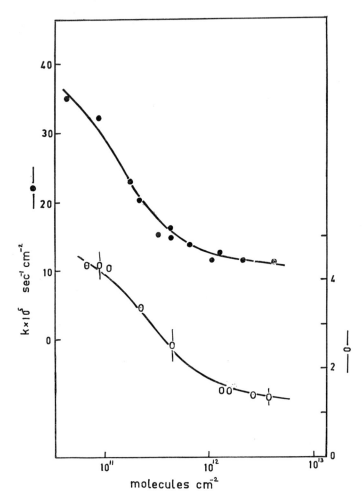

FIG. 14. Dependence on uptake of the first order rate constant (k) for adsorption of ethane (0) and neopentane (●) on nickel films deposited on pyrex at 250 °C.

surface; for instance, by site availability factors of the sort discussed in section IV.A.3, or by interactions between adsorbed residues. An explanation in terms of surface heterogeneity appears mandatory. A detailed analysis of these data in terms of a two-patch model has been given by Baker and Hoveling (1969) using the concepts given in this Chapter: however, it should be noted that the difficulty of defining an unambiguous θ scale makes a unique solution impossible.

For adsorption on a patch model surface, the rate may be expressed in terms of the rate of change of coverage with time by,

$$\frac{d\theta}{dt} = \sum_i \chi_i \, k_i \, f(\theta_i) \qquad \text{(IV)–(28)}$$

where k_i is the specific rate constant for the i'th patch, given by

$$k_i = A_i \, e^{-E_i/RT}$$

and $f(\theta_i)$ previously discussed for the homogeneous surface represents the chance of finding in the i'th patch the necessary adsorption sites. A numerical solution to the equation is readily computed. Since

$$\frac{d\theta_j}{dt} = \frac{d\theta_j}{d\theta} \cdot \frac{d\theta}{dt}$$

and

$$\frac{d\theta_j}{d\theta} = \frac{k_j \, f(\theta_j)}{\sum_i \chi_i \, k_i \, f(\theta_i)} \qquad \text{(IV)–(29)}$$

a numerical integration returns the overall rate as a function of coverage and the corresponding separate rates and coverages on the various patches.

In practice, the problem that is usually faced is that of obtaining patch parameters given the overall rate data. Adsorption studies themselves may provide information about the patch area distribution, thus enabling χ_i to be eliminated from equations (IV)–(15) or (IV)–(28) and (29). If this is so, a reasonably unique set of specific rate constants can be evaluated numerically (at a given temperature) provided the number of patches is small (2–3). However, the evaluation of E_i and A_i with any accuracy is difficult without additional information such as some specific values for E and A on single crystal surfaces.

A numerical example will illustrate the type of rate law that can result from a simple heterogeneous surface consisting of only two patches. For the simplest case, where each adsorbed species occupies just one site, $f(\theta_i) = 1 - \theta_i$. The overall rate of filling of the surface is then described by,

$$\frac{d\theta}{dt} = \chi_1 k_1(1 - \theta_1) + \chi_2 k_2(1 - \theta_2)$$

A calculation for $\chi_1 = 0.9$, $k_1 = 1$; and $\chi_2 = 0.1$, $k_2 = 10$ (i.e. the adsorption is 10 times as fast on the smaller patch) shows that the separate coverages plotted as a function of total coverage look very much the same as in Figure 7a. In this case, however, the coverage on patch 2 is the greater, not because of the equilibrium condition, but simply because it fills faster. The overall rate of adsorption which would be observed is in Figure 15. Since the plot of the logarithm rate

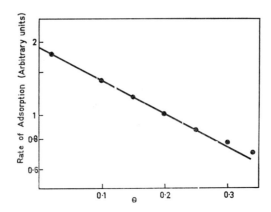

FIG. 15. Dependence of rate of adsorption on coverage (θ) for a two-patch model of a heterogeneous surface: $\chi_1 = 0.9$, $\chi_2 = 0.1$, $k_1 = 1$, $k_2 = 10$.

as a function of coverage is linear (for $\theta < 0.3$ where the two patches are filling simultaneously), the overall kinetics in this region could be described by the Elovich equation

$$\frac{d\theta}{dt} = a \exp(-\alpha\theta) \qquad \text{(IV)–(30)}$$

Adsorption rate data are frequently found to conform to this equation which can be derived in a number of ways from various continuous models for surface heterogeneity. However, the present example shows that, in fact, the Elovich equation has no diagnostic value in the study of the fine structure of surface heterogeneity.

In section III of this Chapter and in the present section we have illustrated the ways in which a heterogeneous surface controls both the nature of the adsorption isotherm and the rate equation for adsorption. It was shown that a simple two-patch surface could, on the one hand, give a Freundlich isotherm, while on the other hand it could result in the rate of adsorption following the Elovich equation. This type of correspondence between isotherm and rate expressions has been discussed in detail by Temkin (1967) for the case of a continuously distributed surface. First it should be noted that in formulating either adsorption equilibrium or adsorption kinetics on a heterogeneous surface, the three variables in each case (e.g. $_iq_{st}$, $\Delta \bar{S}_i$, χ_i: E_i, A_2, χ_i) can be reduced to two if the heat (energy) and entropy (frequency factor) terms are combined in the obvious way into the corresponding free energies. Temkin assumes that the number of adsorption sites with a standard free energy of adsorption in the range ΔG_a° to $(\Delta G_a^\circ + d\Delta G_a^\circ)$ is proportional to $\exp(n\Delta G_a^\circ/RT)$. With the constant $n = 0$ a Temkin isotherm follows ($\theta \propto \log p$), and with $o < n < 1$, a Freundlich isotherm is obtained ($\theta \propto p^n$). The reasonable assumption is made that a linear relation exists between ΔG_a° and the free energy of activation, $\Delta G_a^\ddagger$ and the resulting rate equation is approximated by the Elovich form provided $\theta \to 1.0$.

In view of the discussion throughout this section, it will be clear that the procedure sometimes adopted of using the constants from Elovich plots to obtain the functional dependence of activation energy on coverage cannot be accepted. Furthermore, the criticism given in section III of the use of continuous distributions to describe surface heterogeneity on metal films apply also to the treatment of reaction kinetics.

C. GENERAL COMMENTS ON SURFACE HETEROGENEITY

Although the surface heterogeneity of most metal films has been long recognized, in the past this has not often been used in an explicit or quantitative manner in surface studies. However, we have discussed the consequences of heterogeneity in some detail in this Chapter with the conviction that this aspect of the surface chemistry of metal films will be of increasing importance in the immediate future, and an exploration of this field offers the best hope of relating the surface chemistry of metals to surface structure.

One final point must, however, be emphasized again, and at the risk of tedious repetition. Although, in principle, surface heterogeneity can be accurately conceived in terms of the crystallographic and quasi-crystallographic features of the surface, in practice heterogeneity only has significance in relation to the chemistry of the method used to measure it. An example which will suffice further to illustrate this point is the quite different specificity of the different crystal planes of tungsten, particularly of the (110) plane, for the adsorption of nitrogen (Delchar and Ehrlich, 1965) and of oxygen (Bell et al., 1968). Or again, it is likely that a patch analysis of rare gas adsorption does not *necessarily* provide a patch distribution the details of which will be automatically applicable to (say) the kinetics of chemisorption or catalysis. This is a general problem which will only be resolved as part of a better understanding of the whole area of the influence of structure on surface chemistry.

D *

REFERENCES

Allen, J. A. and Mitchell, J. W. (1950). *Discuss. Faraday Soc.* **8**, 309.

Allen, J. A., Evans, C. C. and Mitchell, J. W. (1959). *In* "Structure and Properties of Thin Films" (C. A. Neugebauer, J. B. Newkirk, D. A. Vermilyea, eds.). Wiley, New York, p. 46.

Allpress, J. G., Jaeger, H., Mercer, P. D. and Sanders, J. V. (1966). *In* "Proc. 6th Int. Cong. on Electron Microscopy" (R. Uyeda, ed.). Maruzen, Tokyo, p. 489.

Anderson, J. R. and Kemball, C. (1954). *Proc. R. Soc.* **A223**, 361.

Anderson, J. R. and Baker, B. G. (1962). *J. Phys. Chem.* **66**, 482.

Anderson, J. R., Baker, B. G. and Sanders, J. V. (1962). *J. Catal.* **1**, 443.

Anderson, J. R. and Tare, V. B. (1964). *J. Phys. Chem.* **68**, 1482.

Anderson, J. R. and Avery, N. R. (1966). *J. Catal.* **5**, 446.

Anderson, J. R. and McConkey, B. H. (1967). *Proc. I.R.E.E. Australia*, 132, April. (Proceedings Radio Research Board Symposium "The Physics of Thin Films", Adelaide, 1966.)

Anderson, J. R. and Macdonald, R. J. (1969). *J. Catal.* **13**, 345.

Baker, B. G. and Fox, P. G. (1965). *Trans. Faraday Soc.* **61**, 2001.

Baker, B. G. (1966). *J. Chem. Phys.* **45**, 2694.

Baker, B. G. and Bruce, L. A. (1968). *Trans. Faraday Soc.* **64**, 2533.

Baker, B. G. and Head, A. K. (1968). Appendix to Baker, B. G., Bruce, L. A. and Fox, P. G. *Trans. Faraday Soc.* **64**, 477.

Baker, B. G. and Hoveling, A. W. (1969). *In* "22nd International Congress of Pure and Applied Chemistry", Sydney.

Beeck, O., Smith, A. E. and Wheeler, A. (1940). *Proc. R. Soc.* **A177**, 62.

Bell, A. E., Swanson, L. W. and Crouser, L. C. (1968). *Surface Sci.* **10**, 254.

de Boer, J. H. (1953a). "The Dynamical Character of Adsorption." Oxford University Press, London, p. 119.

de Boer, J. H. (1953b). "The Dynamical Character of Adsorption." Oxford University Press, London, pp. 123–132.

Bolder, H., Dallinga, G. and Kloosterziel, H. J. (1964). *Catal.* **3**, 312.

Bond, G. C. (1962). "Catalysis by Metals." Academic Press, London.

Boreskov, G. K. (1961). *In* "Actes du Deuxième Congrès International de Catalyse." Editions Tecknip, Paris, p. 1095.

Brennan, D., Hayward, D. O. and Trapnell, B. M. W. (1960). *Proc. R. Soc.* **A256**, 81.

Campbell, K. C. and Duthie, D. T. (1965). *Trans. Faraday Soc.* **61**, 558.

Clasing, M. and Sauerwald, F. (1952). *Z. Anorg. Allg. Chem.* **271**, 88.

Constable, F. H. (1925). *Proc. R. Soc.* **A108**, 355.

Crawford, E. and Kemball, C. (1962). *Trans. Faraday Soc.* **58**, 2452.

Crawford, E., Roberts, M. W. and Kemball, C. (1962). *Trans. Faraday Soc.* **58**, 1761.

van Craen, J., Orban, J. and Bellemans, A. (1968). *J. Chem. Phys.* **49**, 1988.

Delchar, T. A. and Ehrlich, G. (1965). *J. Chem. Phys.* **42**, 2686.

Dell, R. M., Klemperer, D. F. and Stone, F. S. (1956). *J. Phys. Chem.* **60**, 1586.

Drechsler, M. and Nicholas, J. R. (1967). *J. Phys. Chem. Solids* **28**, 2609.

Duell, M. J., Davis, B. J. and Moss, R. L. (1966). *Discuss. Faraday Soc.* **41**, 43.

Dwyer, F. G., Eagleston, J., Wei, J. and Zahner, J. C. (1968). *Proc. R. Soc.* **A302**, 253.

Ehrlich, G. (1963). "Metal Surfaces." American Society for Metals, Metals Park, Ohio, U.S.A., p. 221.

Ehrlich, G. (1966). *Discuss. Faraday Soc.* **41**, 7.

Fedak, D. G. and Gjostein, N. A. (1967). *Acta Met.* **15**, 827.
Fowler, R. and Guggenheim, E. A. (1949). "Statistical Thermodynamics." Cambridge University Press, London.
Gault, F. G. and Kemball, C. (1961). *Trans. Faraday Soc.* **57**, 1781.
Gaunt, D. S. and Fisher, M. E. (1965). *J. Chem. Phys.* **43**, 2840.
Grover, S. S. (1968). *In* "Catalysis Reviews" (H. Heinemann, ed.), Vol. 1. Dekker, New York, p. 153.
Gundry, P. M. and Tompkins, F. C. (1957). *Trans. Faraday Soc.* **53**, 218.
Gundry, P. M. (1961). *In* "Actes du Deuxième Congrès International de Catalyse." Editions Tecknip, Paris, p. 1081.
Halsey, G. D. and Taylor, H. S. (1947). *J. Chem. Phys.* **15**, 624.
Harris, G. M. (1951). *Trans. Faraday Soc.* **47**, 716.
Harris, L. B. (1968). *Surface Sci.* **10**, 129.
Henning, C. A. O. (1968). *Surface Sci.* **9**, 277 (and references cited therein).
Hill, T. L. (1946). *J. Chem. Phys.* **14**, 441.
Jaeger, H. (1967). *J. Catal.* **9**, 237.
Kemball, C. (1951). *Proc. R. Soc.* **A207**, 539.
Kemball, C. (1952). *Proc. R. Soc.* **A214**, 413.
Kemball, C. (1953). *Proc. R. Soc.* **A217**, 376.
Kemball, C. (1954). *Trans. Faraday Soc.* **50**, 1344.
Kemball, C. (1966). *Discuss. Faraday Soc.* **41**, 190.
Klemperer, D. F. and Stone, F. S. (1957). *Proc. R. Soc.* **A243**, 375.
Laidler, K. J. (1954). *In* "Catalysis" (P. H. Emmett, ed.), Vol. 1. Reinhold, New York, Chapters 3, 4, 5.
Lander, J. J. and Morrison, J. (1967). *Surface Sci.* **6**, 1.
Langmuir, I. (1940). *J. Chem. Soc.* **1940**, 511.
Lennard-Jones, J. E. (1937). *Proc. R. Soc.* **A163**, 127.
Logan, S. R. and Kemball, C. (1959). *In* "Structure and Properties of Thin Films" (C. A. Neugebauer, J. B. Newkirk, D. A. Vermilyea, eds.). Wiley, New York, p. 495.
Morgan, A. E. and Somorjai, G. A. (1969). *J. Chem. Phys.* **51**, 3309.
Melmed, A. A. (1965). *J. Appl. Phys.* **36**, 3585.
McConkey, B. H. (1965). Ph.D. Thesis, University of Melbourne.
Neugebauer, C. A. and Webb, M. B. (1962). *J. Appl. Phys.* **33**, 74.
Nicholas, J. F. (1959). *J. Chem. Phys.* **31**, 922.
Nicholas, J. F. (1968). *Aust. J. Phys.* **21**, 21.
Palmberg, P. W. and Rhodin, T. N. (1967). *Phys. Rev.* **161**, 586.
Palmberg, P. W. and Rhodin, T. N. (1968). *J. Appl. Phys.* **39**, 2425.
Palmberg, P. W. (1969). *In* "Proc. Fourth Int. Materials Symposium." Wiley, New York.
Phillips, M. J., Crawford, E. and Kemball, C. (1963). *Nature, Lond.* **197**, 487.
Porter, A. S. and Tompkins, F. C. (1953). *Proc. R. Soc.* **A217**, 544.
Preece, J. B., Wilman, H. and Stoddart, C. T. H. (1967). *Phil. Mag.* **16**, 447.
Primak, W. (1955). *Phys. Rev.* **100**, 1677.
Rhodin, T. N. (1953). *Advances in Catalysis* **5**, 40.
Rideal, E. K. (1968). "Concepts in Catalysis." Academic Press, London.
Roberts, M. W. (1960). *Trans. Faraday Soc.* **56**, 28.
Roginsky, S. A. (1944). *Dokl. Akad. Nauk. SSSR* **45**, 61.
Ross, S. and Olivier, J. P. (1964). "On Physical Adsorption." Interscience, New York.

Sachtler, W. M. H., Dorgelo, G. and van der Knapp, W. (1954). *J. Chim. Phys.* **51**, 491.

Sachtler, W. M. H., Kiliszek, C. R. and Nieuwenhuys, B. E. (1968). *Thin Solid Films* **2**, 43.

Satterfield, C. N. and Sherwood, T. K. (1963). "The Role of Diffusion in Catalysis." Addison-Wesley, Reading, U.S.A.

Schay, G. (1956). *J. Chim. Phys.* **53**, 691.

Schay, G., Fejes, P. and Szathmary, J. (1957). *Acta Chim. Acad. Sci. Hung.* **12**, 299.

Sosnovsky, H. M. C. (1959). *J. Phys. Chem. Solids* **10**, 304.

Stebbins, J. P. and Halsey, G. D. (1964). *J. Phys. Chem.* **68**, 3863.

Stranski, I. N. (1949). *Discuss. Faraday Soc.* **5**, 13.

Suhrmann, R., Heyne, H. and Wedler, G. (1962). *J. Catal.* **1**, 208.

Suhrmann, R., Gerdes, R. and Wedler, G. (1963). *Z. Naturf.* **18a**, 1208, 1211.

Takaishi, T. (1958). *Z. Phys. Chem. Frankf. Ausg.* **14**, 164.

Temkin, M. I. (1967). *Kinet. and Catal.* **8**, 865.

Thiele, E. W. (1939). *Ind. Eng. Chem.* **31**, 916.

Thun, R. E. (1963). *In* "Physics of Thin Films" (G. Hass, ed.). Academic Press, New York, p. 187.

Tonks, L. (1940). *J. Chem. Phys.* **8**, 477.

Toya, T. (1960). *J. Res. Inst. Catalysis Hokkaido Univ.* **8**, 209.

Toya, T. (1961). *J. Res. Inst. Catalysis Hokkaido Univ.* **9**, 134.

Trapnell, B. M. W. (1953). *Proc. R. Soc.* **A218**, 566.

Vand, V. (1953). *Proc. Phys. Soc.* **A55**, 222.

Wei, J. and Prater, C. D. (1962). *Advances in Catalysis* **13**, 204.

Wheeler, A. (1951). *Advances in Catalysis* **3**, 249.

Wranglen, G. (1955). *Acta Chem. Scand.* **9**, 661.

Young, D. M. and Crowell, A. D. (1962). "Physical Adsorption of Gases." Butterworths, London.

Zeldowitch, J. (1934). *Acta Phys.-chim. URSS* **1**, 961.

Chapter 8

Catalytic Reactions on Metal Films

J. R. ANDERSON* and B. G. BAKER

School of Physical Sciences,
Flinders University, Adelaide, Australia

*Present address: CSIRO Division of Tribophysics, University of Melbourne.

I. Introduction

This chapter discusses the main classes of catalytic reactions which have been studied over evaporated metal film catalysts, and since information about the adsorbed state is highly germane to a discussion of the reaction mechanisms, a separate section has been devoted to this. Since most work has been performed with films of nominally random polycrystalline structure, the convention has been adopted that unless otherwise specified, film catalysts are of this type, and evaporated on to a substrate at about 0 °C. Films with an oriented structure or prepared under special conditions are specifically designated as such. Details of reaction on supported catalysts have also been given where this information is relevant, but generally this has been limited to providing ancillary data, or providing a contrast to the behaviour of film catalysts.

Throughout this Chapter, literature references to material contained in tables are given in the body of the text: in the few cases where this was not possible, or where the origin of the material might otherwise be ambiguous, the references are given in the tables themselves.

II. The Adsorbed State

A. simple saturated hydrides

The chemisorptions of methane, ethane, ammonia and hydrogen sulphide exhibit a number of common features and have been most extensively studied.

Two adsorbed states exist for saturated hydrocarbons, physisorption involving the unfragmented molecule, and chemisorption which requires at least a C–H bond to be ruptured, but which may proceed to more extensive C–H and C–C bond rupture. With molecules such as ammonia and hydrogen sulphide which possess at least one lone pair of electrons, there is the additional possibility of coordinative bonding, such as $H_3N \rightarrow$ metal. This type of bonding has been emphasized by Maxted (1951) in connection with the action of catalyst poisons, but there is not much evidence with clean surfaces. On general grounds we would expect coordinative bonding to occur widely, and to act as a precursor to dissociative chemisorption. However, when the latter is a very rapid process with a very low activation energy (as is observed on a number of clean transition metals), the life-time of the coordinated state will be very short and the surface concentration negligibly small.

1. C_1–C_3 Hydrocarbons

Evidence for chemisorption of saturated hydrocarbons has been observed over a wide range of transition metals, and it is reasonable to assume that all transition metals possess this behavior. On the other hand, nontransition metals such as cadmium (Wright, 1968) are inactive.

Most work has been carried out with C_1–C_3 hydrocarbons. The presence of an activation energy for adsorption depends both on surface coverage and on the temperature range. If the surface coverage is high enough, activated adsorption is generally observed. However, the temperature at which adsorption is carried out influences the nature of the adsorbed residues, that is the extent of dehydrogenation, so that the magnitude of the activation energy may be temperature dependent. The assessment of the nature of the adsorbed residues is an important problem, and may be approached in a number of ways. Measurement of adsorption stoichiometry with respect to gas phase composition will give the overall C/H ratio on the surface; however, this measurement alone provides no indication of how much of the surface hydrogen is bound to carbon, or how much is directly chemisorbed as $H_{(s)}$. Two techniques have been used for this; Galwey and Kemball (1959) made use of the much faster exchange of $H_{(s)}$ with deuterium than the exchange rate of hydrogen bound in a residue, while Anderson and Baker (1963) saturated the surface with $H_{(s)}$ prior to hydrocarbon adsorption so that hydrogen removed from the hydrocarbon would result in the return of hydrogen to the gas phase.* Clearly the two techniques give information about the nature of the adsorbed hydrocarbon residues under quite different regimes of surface coverage. However, the techniques only provide information about the average composition of the residues. Under nonequilibrium adsorption conditions one may expect a spectrum of surface residues of varying degrees of dehydrogenation, and this will be even more apparent on crystallographically heterogeneous surfaces such as random polycrystalline films. Furthermore if, for instance, adsorption occurs for a C_3 hydrocarbon, the determination of the average residue composition gives, of itself, no conclusive evidence about possible C–C bond rupture.

Adsorption data for methane and ethane on a range of surfaces are collected into Table 1. Two classifications of adsorption rates are given. Those cases where rapid adsorption occurred at about room temperature or below, that is adsorption complete in less than about 60 seconds or so are designated F: cases where adsorption occurred at a slower rate are designated S. The numerical value for the activation energy of adsorption is given where available. However, F processes are probably generally associated with low values of the activation energy, perhaps of the order of 1 kcal mole^{-1} or less.

*In this Chapter a subscript s designates an adsorbed species.

TABLE 1

Adsorption of Saturated Hydrocarbons on Metals

System	Type of Surface*	Temp. (°C)	Percentage surface occupation (σ), and character; percentages give number of adsorbed molecules per 100 surface metal atoms†	Reference
CH_4/tungsten	polycrystal film; deposited 0 °C (HV)	0 to 43	0–16%, F** 16–20%, S**	1
CH_4/tungsten	polycrystal film; deposited 0 °C (HV)	−78 70	0–15%, F 15–18%, S	2
CH_4/tungsten	annealed wire (UHV)	30	adsorption occurred	3
C_2H_6/tungsten	polycrystal film; deposited 0 °C (HV)	0 90	0–24%, F	1
C_2H_6/tungsten	polycrystal film; deposited 0 °C (HV)	−78 70	0–13%, F	2
C_2H_6/tungsten	annealed wire (UHV)	−63	adsorption occurred	3
CH_4/molybdenum	polycrystal film; deposited 0 °C (HV)	−78 70	0–3%, F 3–9%, S	2
C_2H_6/molybdenum	polycrystal film; deposited 0 °C (HV)	−78 70	0–14%, F	2
CH_4/chromium	polycrystal film; deposited 0 °C (HV)	−78 70	0–3%, F 3–8%, S	2
CH_4/chromium	polycrystal film; deposited 0 °C (UHV)	27 100	0–12%, F 12–14%, S	5
C_2H_6/chromium	polycrystal film; deposited 0 °C (HV)	−78 70	0–8%, F	2
C_2H_6/chromium	polycrystal film; deposited 0 °C (UHV)	27 100	0–12%, F 12–14%, S	5
CH_4/tantalum	polycrystal film; deposited 0 °C (HV)	−78 70	0–3%, F 3–5%, S	2

Reaction/metal	Film	Temperature	Observations	Ref.
C_2H_6/tantalum	polycrystal film; deposited 0 °C (HV)	−78 70	0–6%, F / 6–7%, S	2
CH_4/nickel	polycrystal film; deposited 0 °C (HV)	−78 70	no adsorption	2
CH_4/nickel	polycrystal film; deposited 0 °C (HV)	140 260	0–10%, average activation energy ~ 11 kcal mole^{-1}, S	6
CH_4/nickel	polycrystal film; deposited 0 °C (HV)	140 200	0–9%, average activation energy ~ 10 kcal mole^{-1}, S	1
CH_4/nickel	polycrystal film; deposited 20 °C (UHV)	>0	<1%, S	4
CH_4/nickel	polycrystal film; deposited 250 °C (UHV)	50 250	0–0.1% average activation energy ~ 1 kcal mole^{-1}, S	8
CH_4/nickel	(111) nickel, single crystal (UHV)	20 75	adsorption occurred, S, but probably electron bombardment assisted	7
C_2H_6/nickel	polycrystal film; deposited 0 °C (HV)	−78 70	0–2%, F / 2–6%, S	2
C_2H_6/nickel	polycrystal film; deposited 0 °C (HV)	0 22	0–3%, F / 3–10%, S	1
C_2H_6/nickel	polycrystal film; deposited 250 °C (UHV)	100 250	0–0.1%, average activation energy ~ 3 kcal mole^{-1}, S	8
C_2H_6/nickel	(111) nickel, single crystal (UHV)	20	adsorption occurred, S, but probably electron bombardment assisted	7
CH_4/cobalt	polycrystal film; deposited 0 °C (HV)	−78 70	no adsorption	2
C_2H_6/cobalt	polycrystal film; deposit 0 °C (HV)	−78 70	no adsorption	2
CH_4/iron	polycrystal film; deposited 0 °C (HV)	−78 70	no adsorption	2
CH_4/iron	polycrystal film; deposited 0 °C (HV)	170	0–18%, S	1

Table 1—*cont.*

System	Type of Surface*	Temp. (°C)	Percentage surface occupation (σ), and character; percentages give number of adsorbed molecules per 100 surface metal atoms†	Reference
C_2H_6/iron	polycrystal film; deposited 0 °C (HV)	−78 70	0–0.4%, F	2
C_2H_6/iron	polycrystal film; deposited 0 °C (HV)	77 99	0–8%, S	1
C_2H_6/iron	polycrystal film; deposited 0 °C (HV)	77	0–6%, S	9
CH_4/iridium	polycrystal film; deposited 0 °C (UHV)	27	0–14%, average activation energy ~1–2 kcal mole^{-1} 14–25%, S	10
C_2H_6/iridium	polycrystal film; deposited 0 °C (UHV)	27 100	0–10%, F 10–30%, S	10
CH_4/rhodium	polycrystal film; deposited 0 °C (HV)	−78 70	0–1%, F 1–6%, S	2
C_2H_6/rhodium	polycrystal film; deposited 0 °C (HV)	−78 70	0–11%, F 11–15%, S	2
CH_4/palladium	polycrystal film; deposited 0 °C (HV)	−78 70	0–3%, S	2
C_2H_6/palladium	polycrystal film; deposited 0 °C (HV)	−78 70	0–12%, F 12–16%, S	2
CH_4/titanium	polycrystal film; deposited 0 °C (HV)	−78 70	0–1%, F 1–3%, S	2
C_2H_6/titanium	polycrystal film; deposited 0 °C (HV)	−78 70	0–5%, F 5–9%, S	2

*UHV Refers to a surface prepared under ultrahigh vacuum conditions (typically $<10^{-9}$); HV Refers to a surface prepared under ordinary high vacuum conditions.

† Computing percentage surface occupation required film areas and number of surface metal atoms per unit area; data provided by individual authors were used.

** F, fast; S, slow adsorption: See text for details.

(1) Wright et al. (1958); (2) Trapnell (1956); (3) Rye (1967); (4) McCarroll (1969); (5) McElligott et al. (1967); (6) Kemball (1951); (7) Maire

We shall on occasions use two criteria to express surface coverages. We define the percentage surface occupation σ_x, as the number of adsorbed x per 100 adsorption sites, the latter being arbitrarily set equal to the number of surface metal atoms; the surface coverage, θ, is defined in the usual way in relation to some defined monolayer capacity.

The data provided by the various sources are in general in quite reasonable agreement although we incline to the belief that, because of the likely presence of impurity in the evaporation source and the difficulty of thorough outgassing, the cleanliness of the H.V. chromium and probably the iron and cobalt films quoted in Table 1 is open to question.

The table shows that on all metals ethane is adsorbed more readily than methane in the same coverage range. There are no strictly comparable data for the adsorption of saturated hydrocarbons higher than C_2, however deuterium exchange behaviour (see later) strongly suggests that the main discontinuity in reactivity occurs between methane and ethane. This may well be due to the greater bond energy for CH_3-H (101 kcal mole^{-1}) than for C_2H_5-H (96 kcal mole^{-1}) (Cottrell, 1958).

As the surface coverage of adsorbed hydrocarbon becomes appreciable, desorbed hydrogen may become readily measurable. The composition of the adsorbed phase in terms of the gross $(H/C)_{(s)}$ ratio gives an upper bound to the hydrogen-to-carbon ratio of the average adsorbed residue. Table 2 lists these data (prefixed by $\not>$) as $(H/C)_{(s)}$ values obtained when the rate of hydrogen desorption had become negligible. The table also includes more direct estimates for x in the average residue written as CH_x. All of these data are of only semi-quantitative accuracy, and the most important single conclusion which emerges is that chemisorption of saturated hydrocarbons on transition metals at room temperature and above, is likely to yield an average adsorbed residue stripped of more than one hydrogen atom, and the extent of dehydrogenation generally increases as the temperature increases, at least for a modest temperature increment. It should be noted that, both here and in subsequent usage, x is merely the hydrogen/carbon ratio for the average adsorbed residue; of itself it carries no implications for the degree of carbon catenation.

Both C–C bond rupture and dehydrogenation are activated processes, and in a kinetically controlled regime lower temperatures favour the retention of more fully hydrogenated and less fragmented surface species. The comparative aspects of this have not been quantitatively delineated, but we may summarize the processes by writing for ethane

TABLE 2

Average composition of adsorbed hydrocarbon residues

Values of x for CH_x at temperatures (T °C), obtained when desorption had become negligibly slow

	Tungsten	Iron	Nickel	Rhodium	Iridium	Platinum		
methane	3(0)[3]	↗2(170)[3]	↗2(150)[3] ↗1(200)				*	metal films
							*	
ethane	0.4(150)[2] ↗5(40)[3] ↗3(70)	↗5(100)[3]	2.1(200)[2] ↗5(40)[3] ↗4(100)	1.7(120)[2] ↗1.5(27–100)[5] 1.0(100)[5]	↗2.8(27)[4] ↗2.4(100)	1.3(290)[2]	† * * * ‡‡	
propane	1.5(130)[2]		1.2(210)[2]	1.8(90)[2] ↗2.5(27)[6] ↗1.8(100)		1.8(270)[2]	† * *	
n-butane				↗1.7(27)[6] ↗2.1(100)			* *	
n-pentane			1.8(0)[1] 1.4(60–200)				§§	supported nickel
cyclopentane			0.6(0)[1]				§	
2-methylbutane			2.2(0)[1] 1.4(60–200)				§§	
2-3 diemthylbutane			1.7(0–200)[1]				§	
cyclohexane			1.8(0)[1] 1.3(60–200)[1][7]			1.8[8]	§§	

Field emitter tip

(1) Galwey and Kemball (1959); (2) Anderson and Baker (1963); (3) Wright et al. (1958); (4) Roberts (1962a); (6) Roberts (1962b); (7) Selwood (1957); (8) Block (1963).   (5) Roberts (1963).

* from adsorption stoichiometry   ‡ from adsorption of C_2H_6 and C_2D_6

† from adsorption onto hydrogen-covered surface   § from treatment of adsorbed layer with D_2

for instance,

$$C_2H_{6(g)} \rightleftarrows ----- \rightleftarrows C_2H_{6-x(s)} \rightleftarrows ----$$

$$\left. \begin{array}{c} \downarrow \quad\quad (H) \\ CH_{a(s)} + CH_{b(s)} \rightleftarrows CH_{4(g)} \end{array} \right\} \quad\quad (II)-(1)$$

and the surface carries, in principle, a range of C_2 residues with x ranging from 1–6, as well as CH_y residues with y ranging from 0–3. The reverse arrows in reaction (II)–(1) indicate steps (dehydrogenation hydrogenation) that are possibly reversible, but not necessarily in established equilibrium. There is evidence from the istopic labelling experiments of Anderson and Avery (1966) that C–C bond rupture at a metal surface is totally irreversible, and it is written in reaction (II)-(1) as such. However, while reaction (II)-(1) proceeds, adsorbed hydrogen is certainly in equilibrium with gas phase hydrogen (Taylor, 1957; Anderson, 1957a).

Anderson and Baker (1963) concluded from a knowledge of the rates of the hydrocracking reactions that from the adsorption of ethane and propane on tungsten, nickel, rhodium and platinum at the temperatures indicated in Table 2 (well above room temperature), only C_1 residues were likely to be present after the time taken for adsorption to occur— a period of 1–2 minutes. Roberts (1962a) concluded from treating the adsorbed residues with deuterium that those formed from adsorbed ethane on rhodium at 0–27 °C were mainly C_1 with a minor proportion of C_2 after the adsorbed hydrocarbon had resided on the surface for about 100 hours; at 100 °C they were all C_1 after about 30 minutes. Since carbon-carbon bond rupture is slow compared with the rate of hydrocarbon adsorption, the proportion of C_1, C_2, etc. residues will depend on the time scale of the measurement in relation to the temperature. There are no data for the behaviour of other metals, but one might expect the clean surfaces of metals which form stable carbides (e.g. tungsten) to be efficient for carbon-carbon bond rupture during adsorption.

It will be seen from Table 1 that on many metals there is a methane adsorption process proceeding with very low activation energy up to a few percent surface occupation, above which an activated adsorption occurred. We associate this low activation energy process with adsorption by the first simple dissociative step, for instance

$$CH_{4(g)} + 2S \rightarrow CH_{3(s)} + H_{(s)} \quad\quad (II)-(2)$$

which is then followed by the formation of more extensively dehydrogenated species as an activated process.* Chemisorbed $CH_{3(s)}$ groups are

*In this Chapter an adsorption site in a reaction is designated by the symbol S.

clearly indicated from the results of Kemball (1953) for exchange reactions over random polycrystalline metal films in which CH_3D is
an important initial exchange product. Furthermore, the flash filament
results of Rye (1967) with tungsten also point to $CH_{3(s)}$ as the first
mode of chemisorption.

The low apparent activation energies found by Baker and Hoveling
(1969) (Table 1) for the adsorption of methane and ethane (50–250 °C)
on UHV nickel films at very low pressures ($<10^{-6}$ torr) and very
low surface occupations (0–0.1%) require comment. These adsorption
processes were slow by the time scale criterion we have adopted in this
discussion, and a compensation effect also operated so that the frequency factors were low. This behaviour is probably because the rate
controlling step is connected to the gas phase via an established adsorption equilibrium involving a relatively weakly bound species so
that, as described in section IV.A.1 of Chapter 7, the heat and entropy
of adsorption contribute, respectively, to the apparent activation
energy and frequency factor. This can only occur if the coverage of the
weakly bound species is low (pressure low enough and temperature high
enough). We note, by comparison, that the measurements of Kemball
(1951) and Wright et al. (1958) for the rate of methane adsorption on
nickel films which gave an activation energy of 10–11 kcal mole^{-1}, were
taken at pressures of the order of 1 torr. The precise rate controlling
step in Baker and Hoveling's work is not known, but it is very likely to
involve the formation of an extensively dehydrogenated species.

Maire (1969) has studied the adsorption of methane, ethane and neopentane on clean (111), (110) and (100) nickel surfaces by LEED.
With all hydrocarbons there was obvious evidence for substantial
adsorption at 20 °C. On films of random polycrystalline nickel (which
undoubtedly exposed a substantial proportion of (111) surface) adsorption of methane is activated and does not become appreciable until
about 140 °C (Table 1). The behaviour of methane in the LEED work
is probably because methane adsorption is facilitated by bombardment
of the surface with the electron beam; an analogous sort of behaviour
has been observed by Anderson and Estrup (1968) in the adsorption of
ammonia on (100) tungsten. Initial adsorption at 20 °C was structurally
disordered, but further exposure resulted in a sequence of ordered structures. Since methane, ethane and neopentane give the same ordered
structures, it is likely that C–C bond rupture has occurred in adsorption. The ordered residues were probably $CH_{2(s)}$, although the LEED
data are not able to distinguish between $CH_{3(s)}$, $CH_{2(s)}$, $CH_{(s)}$ or $C_{(s)}$.

Very little is known of the thermodynamics of adsorption of saturated
hydrocarbons since direct measurements of heats of adsorption are

complicated by the difficulty of establishing gas-surface equilibrium, and also the difficulty of establishing equilibrium with respect to the desired dehydrogenation states of the hydrocarbon residue. Let us consider the adsorption of methane on tungsten and nickel by way of example, as occurring by reaction (II)-(2). The only direct piece of experimental thermochemical data appears to be a measurement of the binding heat for a CH_3 group chemisorbed on tungsten and obtained by Ritchie and Wheeler (1966) by a high temperature method (1400–1800 °K) on a thoroughly annealed tungsten wire. A value of 47 kcal mole^{-1} was obtained. This may be compared with a figure of 60 kcal mole $^{-1}$ calculated by Eley's (1950) semi-empirical method, and with 42 kcal mole^{-1} for the mean tungsten-carbonyl bond energy (Skinner, 1964). Ritchie and Wheeler obtained self-consistent results starting with a range of methyl-containing compounds, and also obtained a value for the heat of binding for hydrogen on tungsten that agrees well with existing data as tabulated for instance by Ehrlich (1961). Nevertheless, it seems possible that Ritchie and Wheeler's figure is somewhat low, and we propose to use a value of 55 kcal mole^{-1}. Although this figure is clearly subject to considerable uncertainty, we believe it to be sufficiently accurate for the conclusions we shall draw.

If we write the thermochemical scheme

$$CH_{4(g)} \rightarrow CH_{3(g)} + H_{(g)} \ (\Delta H_1) \qquad\qquad \text{(II)–(3)}$$

$$S + H_{(g)} \rightarrow H_{(s)} \ (\Delta H_2) \qquad\qquad \text{(II)–(4)}$$

$$S + CH_{3(g)} \rightarrow CH_{3(s)} \ (\Delta H_3) \qquad\qquad \text{(II)–(5)}$$

the enthalpy change for adsorption, ΔH_a, for reaction (II)–(2) will be

$$\Delta H_a = \Delta H_1 + \Delta H_2 + \Delta H_3 \qquad\qquad \text{(II)–(6)}$$

ΔH_2 and ΔH_3 contain terms for any energy that may be required to "uncouple" the surface orbitals of the metals, but it is not necessary to know this contribution explicitly. Schuit et al. (1961) argue that these are substantial terms. Using $\Delta H_1 = 101$ kcal mole^{-1} (Cottrell, 1958) and for the tungsten system we take $\Delta H_2 = -75$ kcal mole^{-1} (Ehrlich, 1961), and $\Delta H_3 = -55$ kcal mole $^{-1}$, we obtain ΔH_a for methane on bare tungsten equal to -29 kcal mole^{-1}. For localized adsorption one can expect the entropy change to make a substantial contribution to the free energy change. To estimate the entropy change we first adopt the gas phase standard of 1 atmosphere pressure, and the surface standard state of $\theta = 0.01$ at 300 °K. A straightforward analysis leads to the following contributions to the entropy change: (a) the loss of three degrees of translational freedom (34.2 cal deg^{-1} mole^{-1}), and three

degrees of rotational freedom (10.2 cal deg^{-1} mole^{-1}); (b) an overall gain
of six degrees of vibrational freedom. In fact, there are 15 vibrational
modes associated with the chemisorbed CH_3 and H, of which six are
deformation modes in the plane of the surface. Assigning reasonable
values to the vibrational contributions in gas phase methane (0.1 cal
deg^{-1} mole^{-1}) and in the adsorbed state (4.6 cal deg^{-1} mole^{-1}), and
evaluating the configurational entropy (Kemball, 1950) at $\theta = 0.01$
(16 cal deg^{-1} mole^{-1}), we find ΔS^0 for adsorption about -24 cal deg^{-1}
mole^{-1}. This yields ΔG^0 as -21.8 kcal mole^{-1}, and this is equivalent to
saying that the surface layer would be in equilibrium with a methane
pressure of about 1.9×10^{-13} torr, which is obviously thermodynamically favourable for adsorption, and is at least consistent with Rye's
observation that methane adsorption commenced on clean tungsten
at about 300 °K at an equivalent pressure of about $10^{-7} - 10^{-6}$ torr by a
process that was kinetically controlled (Rye, 1967).

The estimation of ΔH_a for reaction (II)–(2) on a nickel surface is very
uncertain because of lack of data. However, we may note that the mean
nickel-carbonyl bond energy is about 7 kcal mole^{-1} less than the tungsten-carbonyl figure (Skinner, 1964), while the heat of binding for
methyl chemisorbed on nickel calculated by Eley's method is about 8
kcal mole^{-1} less than the calculated tungsten figure; thus as a rough
estimate we propose to use 48 kcal mole^{-1} for the heat of binding of
chemisorbed methyl on nickel. By the same process as before, this gives
about -22 kcal mole^{-1} for ΔH_a, and assuming the same entropy change
as on tungsten, we obtain -14.8 kcal mole^{-1} for ΔG° at 300 °K and
$\theta = 0.01$.

Now consider the feasibility of adsorption by reaction (II)–(2) at
higher coverages on tungsten. It is known that in the adsorption of
hydrogen alone on tungsten, the magnitude of the heat of adsorption
falls with increasing θ_H. Although there may be some uncertainty about
defining the scale for θ_H, it is clear that in the region of $\theta_H \simeq 0.9$, ΔH_a
for hydrogen will be in the vicinity of -20 kcal mole^{-1} (Beeck, 1950;
Trapnell, 1951), and thus the corresponding heat of binding for $H_{(s)}$
($-\Delta H_2$) is about 62 kcal mole^{-1}. One would expect a change in the same
direction for the heat of binding of $CH_{3(s)}$ ($-\Delta H_3$), and if we make the
conservative assumption of 45 kcal mole^{-1} for $-\Delta H_3$ at high coverage,
the enthalpy change for methane adsorption (ΔH_a) by reaction (II)–(2)
would be only -6 kcal mole^{-1}. At this coverage the configurational
entropy of the surface layer will be about 4 cal deg^{-1} mole^{-1}. The
entropy contributions from the vibrational modes of the surface species
will be somewhat greater when the latter are more loosely bound, however, the increment cannot be large, and we add an extra 4 cal deg^{-1}

mole^{-1} to take this into account. On this basis $\Delta S°$ becomes -31.8 cal deg^{-1} mole^{-1} and $\Delta G°$ is $+ 3.5$ kcal mole^{-1}. A similar calculation for adsorption on nickel at high coverage leads to an even more positive value for $\Delta G°$. It should be noted that a site exchange adsorption process such as

$$CH_{4(g)} + H_{(s)} \rightarrow CH_{3(s)} + H_{2(g)} \qquad \text{(II)–(7)}$$

results in a $\Delta G°$ for adsorption at least as unfavourable as that calculated for process (II)–(2). Despite the inaccuracy of the data, we believe that the following conclusions follow from these calculations: (1) adsorption of methane on tungsten or nickel to give a high coverage of $CH_{3(s)}$ at best is only marginally feasible, (2) when a high coverage of adsorbed methane is formed, most of this is likely to be present as more strongly bound residues of the type CH_y with $y < 3$; (3) under exchange conditions only a small part of the total adsorbed methane is present as $CH_{3(s)}$, and this may be assisted by the operation of one or more of the following factors: (a) the configurational entropy then becomes large enough to make $\Delta G°$ favourable, (b) the surface is energetically heterogeneous with a small proportion of high energy binding sites for $CH_{3(s)}$.

2. Ammonia

It has been known for many years (Frankenburger and Hodler, 1932) that adsorption of ammonia on tungsten (powder) at 90–200 °C yields surface residues $NH_{2(s)}$, $NH_{(s)}$, $N_{(s)}$. Wahba and Kemball (1953) studied the adsorption of ammonia on films of tungsten, iron and nickel, and showed that at 20 °C adsorption was rapid with an activation energy which we estimate at < 1 kcal mole^{-1} up to the point $\sigma_{NH3} \simeq 50\%$. At about this point too, hydrogen was returned to the gas phase on tungsten and iron with further ammonia adsorption. Slow hydrogen desorption proceeded to an extent indicating the formation of more extensively dehydrogenated surface species than $NH_{2(s)}$. Thus the adsorption processes may be summarized as

$$NH_{3(g)} + 2S \rightarrow NH_{2(s)} + H_{(s)} \qquad \text{(II)–(8)}$$

which has a very low activation energy over most of the coverage range, and

$$NH_{2(s)} + S \rightarrow NH_{(s)} + H_{(s)} \qquad \text{(II)–(9)}$$

$$NH_{(s)} + S \rightarrow N_{(s)} + H_{(s)} \qquad \text{(II)–(10)}$$

which are activated, and which probably correspond to the processes observed by Frankenburger and Hodler on tungsten powder to have an activation energy of 12 kcal mole^{-1}. The field emission measurements of Dawson and Hansen (1966) suggest that decomposition of adsorbed ammonia, perhaps as far as $N_{(s)}$ occurs on (100) tungsten at 500 °K. At high surface coverage the following adsorption process becomes important (Wahba and Kemball, 1953),

$$NH_{3(g)} + H_{(s)} \rightarrow NH_{2(s)} + H_{2(g)} \qquad (II)-(11)$$

and this is probably also activated.

The surface potential results by Gundry *et al.* (1962) confirm that $NH_{2(s)}$, $NH_{(s)}$ and $N_{(s)}$ residues are produced. At the temperatures considered (-78 to 140 °C) $N_{(s)}$ may not have been present on nickel, but of course at temperatures where ammonia decomposition occurs (> 300 °C), $N_{(s)}$ is a likely surface precursor to nitrogen desorption.

Anderson and Estrup (1968) have reported that the adsorption of ammonia on a (100) tungsten surface at 500 °K yields initially a diffraction pattern equivalent to residues adsorbed in a $C(2 \times 2)$ structure with $\sigma_{NH_3} \simeq 50\%$, and on treatment with further ammonia and mild heating, more ammonia is adsorbed together with hydrogen desorption and a (1×1) structure is formed with $\sigma_{NH_3} \simeq 100\%$. At the temperatures used it is possible that some $NH_{(s)}$ and $N_{(s)}$ and $N_{(s)}$ may have been formed, but these would presumably not have been distinguishable from $NH_{2(s)}$ in the LEED data. The conversion of the $C(2 \times 2)$ structure to the (1×1) structure was accelerated by the electron beam, and illustrates the danger of accepting kinetic data from LEED unless this possibility is explicitly considered.

Calorimetric heats of adsorption of ammonia on a number of metals have been measured in the rapid adsorption regime by Wahba and Kemball (1953). These are set out in Table 3 at two coverages, and both refer to reaction (II)–(8). By an analogous thermochemical calculation to the one described for methane, we may calculate the heats of binding of $NH_{2(s)}$ residues at the metal surfaces, defined to be $-\Delta H$ for

$$NH_{2(g)} + S \rightarrow NH_{2(s)} \qquad (II)-(12)$$

These values are also included in Table 3. The high coverage values were obtained using ΔH_a for hydrogen adsorption at $\theta_H = 0.9$ of -20, -18, -19 kcal mole^{-1} for tungsten, iron and nickel respectively (Ehrlich, 1961) so the corresponding heats of binding for $H_{(s)}$ are 62, 60, 61 kcal mole^{-1}. Gundry *et al.* (1962) concluded that for reaction (II)–(9) on nickel at high coverage the enthalpy change is numerically positive (endothermal reaction). For this to be so, and assuming the bond energy in $NH-H_{(g)}$ to be 94 kcal mole^{-1}, the mean of the values for H_2N-H and $N-H$, and using the value for the heat of binding of $NH_{2(s)}$ from Table 3, we may estimate the heat of binding of $NH_{(s)}$ to the surface (at $\sigma_{NH_3} = 45\%$) to be not greater than 92 kcal mole^{-1}.

TABLE 3

Heats of adsorption of Ammonia on metal films

	Percentage surface occupation (σ_{NH_3}): number of adsorbed molecules per 100 surface metal atoms	Heat of adsorption ($-\Delta H_a$) (kcal mole^{-1})	Heat of binding of $NH_{2(s)}$ (kcal mole^{-1})
tungsten	0	72	99
	45	38	78
iron	0	45	79
	45	25	67
nickel	0	37	72
	45	18	59

3. *Ammonia decomposition*

The catalytic decomposition of ammonia has been studied over films of iron, nickel, cobalt, rhodium, platinum, ruthenium, rhenium, vanadium and tungsten (Logan *et al.*, 1958, 1960), while Tamaru (1961, 1964) studied the reaction over powdered tungsten and nickel. The decomposition on various iron synthetic ammonia catalysts has been widely studied in relation to ammonia synthesis (Bokhoven *et al.*, 1955). All of these workers consider that decomposition begins by dissociative ammonia adsorption [e.g. reactions (II)–(8) to (10)] and that nitrogen desorption is rate controlling. On all the metals studied with the exception of tungsten, the decomposition rate varies with both ammonia and hydrogen pressure. If the rate is written as proportional to the product

$$p_{NH_3}^a \, p_{H_2}^b$$

the values of a and b lie in the ranges 1.5 to 1.35 and -0.7 to -2.45 respectively, so that $-b/a$ is in the region of 1.5 to 2.0. The value of $-b/a$ required by the simple form of the theory of Temkin and Pyzhev (1940) is 1.5. Essentially, this considers the formation of $N_{(s)}$ using a virtual pressure argument (cf. Chapter 7, Section III). If NH_3, H_2, N_2, $N_{(s)}$ and $H_{(s)}$ are all in established equilibrium, the concentration of $N_{(s)}$ obtained from reaction with ammonia will be the same as from the adsorption of nitrogen. However, if $N_2 \rightleftarrows 2N_{(s)}$ is not established, but if $N_{(s)}$ is in equilibrium with NH_3 and H_2, the concentration of $N_{(s)}$ will take a value corresponding to a virtual N_2 pressure equal to $K_p\, p_{NH_3}^2\, p_{H_2}^{-3}$, where K_p is the equilibrium constant for

$$2NH_3 \rightleftharpoons N_2 + 3H_2$$

and the virtual nitrogen pressure may then be made very large. Thus the concentration of $N_{(s)}$ will be proportional to $p_{NH_3}\, p_{H_2}^{-3/2}$; the desorption rate is apparently first order in the concentration of $N_{(s)}$, since the expression $p_{NH_3} p_{H_2}^{-3/2}$ is a general approximation to the pressure dependence of the rate of nitrogen production.

The existence of a pre-desorption equilibrium means that the surface concentration of $N_{(s)}$ will be temperature dependent. Thus, if desorption is the slow step, the measured activation energy will contain the terms ΔH_1 and ΔH_2, where ΔH_1 is the enthalpy change for

$$NH_3 \rightleftharpoons \tfrac{1}{2}\nu N_2 + \tfrac{3}{2}H_2$$

and ΔH_2 the enthalpy change for

$$\tfrac{1}{2}\nu N_2 \rightleftharpoons N_{(s)}$$

the subscript ν designating a virtual rather than an established nitrogen pressure.

Elaborations of the theory occur in two directions. It may be necessary to consider the rates of both synthesis and decomposition reactions simultaneously, and this was done in the original theory, although for the present purpose the synthesis rate can be ignored. More importantly, the adsorption isotherms for the reactants should be considered, and this can modify a and b. Thus if we write the decomposition rate proportional to

$$(p_{NH_3}^1\, p_{H_2}^{-3/2})^\delta$$

then $\delta = 1$ when both gases are independently adsorbed and for both the heat of adsorption and the pressure are low enough so that $\theta \propto p$. As the heat of adsorption or the pressure rises so that θ becomes less than directly proportional to p, δ decreases towards zero. This appears to be the case in the reaction on tungsten, where the rate is independent of ammonia pressure (Tamaru, 1961). If chemisorbed hydrogen is able to compete with adsorbed nitrogen atoms, the value of $-b/a$ will rise above 1.5, as has been found on most metal films (Logan et al., 1958, 1960).

It is generally found that systems with a high activation energy are associated with high values for the ammonia pressure exponent a. This detail is related to the pre-desorption equilibria and to the way the terms vary with coverage; a detailed account is given by Logan and Kemball (1960).

This model of nitrogen desorption as the rate controlling step has been questioned by Dawson and Hansen (1966) who point out that on tungsten, the ammonia decomposition rate is much faster than the rate of nitrogen desorption from tungsten measured by Ehrlich (1965). This particular criticism does not appear to be well founded, since Tamaru (1961) showed, in fact, that on a freshly prepared tungsten surface, nitrogen desorption does not occur, but that the steady-state catalytic reaction proceeds on a tungsten nitride surface. The same is true for reactions over tantalum, vanadium and, depending on the conditions, iron. Tamaru showed that on his catalyst the rate of nitrogen production corresponded to the rate of nitrogen desorption separately measured in the absence of the ambient reactant.

However, there is a feature of Dawson and Hansen's field emission results which suggests that the reactions of ammonia at a tungsten surface may be more complex, since they observed that reaction with gaseous ammonia can convert chemisorbed β-nitrogen (the strongly bound state) on the (100) face of tungsten at 400–500 °K to another chemisorbed form. A somewhat analogous conclusion was reached by Frankenberger and Hodler (1932) that a reaction could occur between ammonia and the nitrogen of a nitrided tungsten surface. These processes may be a hydrogen transfer reaction such as

$$N_{(s)} + NH_{3(g)} \rightarrow NH_{(s)} + NH_{2(s)} \text{ etc.} \qquad \text{(II)–(13)}$$

and this would agree with the observed work function changes. Moreover, it is well to recall that on tungsten NH_3 decomposes more rapidly than ND_3 (Taylor and Jungers, 1935; Barrer, 1936) and there is a hydrogen isotope effect in the ammonia synthesis reaction (Ozaki et al., 1960). These observations have been interpreted by a scheme in which $NH_{(s)}$ and $(HN-NH)_{(s)}$ residues are important, and the latter may also be a product of the reaction between NH_3 and $N_{(s)}$.

The decomposition of the analogous molecules stibine and arsine, and also of stannane and germane proceeds catalytically over films of their own elemental decomposition product; the reactions are generally first order in hydride pressure indicating weak adsorption, and hydrogen desorption is probably rate controlling. Bond (1962) provides a summary.

4. Hydrogen Sulphide and Water Vapour

On nickel, tungsten and iron films, the adsorption of hydrogen sulphide is rapid and dissociative at temperatures as low as −80 °C, so that at surface occupations up to $\sigma_{H_2S} \simeq 23\%$ the activation energy for adsorption is low, probably <1 kcal mole^{-1} (Gundry et al., 1962; Roberts and Ross, 1966). At $\sigma_{H_2S} \simeq 23\%$ hydrogen was returned to the gas phase and the maximum reached in σ_{H_2S} before bulk incorporation became significant was about 50%. On non-transition metals such as silver and lead (Saleh et al., 1961, 1964), adsorption also occurred at -80 °C, but this was mainly reversible and apparently non-dissociative and reached $\sigma_{H_2S} \simeq 100\%$; in this case the hydrogen sulphide molecule is probably bound coordinatively to the metal via a lone electron pair. On lead at temperatures above 100 °C dissociative adsorption occurred with the return to the gas phase of about half of the total amount of hydrogen originally associated with the adsorbed hydrogen sulphide. The fast uptake limit of $\sigma_{H_2S} \simeq 23\%$ corresponds very well with the value expected if the adsorbed entities are $H_{(s)}$ and $S_{(s)}$, with the latter occupying two surface sites as required by the magnetic data of Den Besten and Selwood (1962) (that is, a total of four surface sites per molecule), and in agreement with the stoichiometry for hydrogen sulphide adsorption on germanium (Boonstra and van Ruler, 1966).

At $\sigma_{H_2S} > 60\%$, treatment of the surface with deuterium shows that one surface hydrogen per adsorbed sulphur is resistant to exchange (Saleh et al, 1961). This resistant hydrogen is also probably chemisorbed $H_{(s)}$ but in sites sterically hindered to exchange, but some $SH_{(s)}$ may also be present.

In general terms then, the adsorption processes are

$$H_2S_{(g)} \rightarrow H_{(s)} + SH_{(s)} \rightarrow 2H_{(s)} + S_{(s)} \qquad (II)-(14)$$

In the region $\sigma_{H_2S} > 23\%$, heating caused desorption of hydrogen that was not re-adsorbed on cooling to the original temperature, presumably because of sulphide incorporation.

Saleh et al. (1963) examined the crystallographic specificity of H_2S adsorption on a tungsten field emitter tip and showed that adsorption occurred preferentially around the (112) and (110) planes, and then extended on heating along the rest of the (111) zone. This result is readily understood since the initial adsorption regions are atomically rough planes where the sulphur may achieve maximum coordination, and a generally similar specificity has been observed with oxygen, no doubt for similar reasons (Good and Muller, 1956). This is in agreement with the results of Benard et al. (1965) who measured the heat of adsorption of hydrogen sulphide on single crystal planes of silver in the range 300–600 °C. In this work the sulphur activity was controlled by controlling the gas phase pressure ratio p_{H_2S}/p_{H_2}, and at a nominal coverage of $\theta = 0.5$ the heats on the (111), (100) and (110) planes were 27, 29 and 33 kcal per half mole of S_2, with some evidence for selective adsorption at surface steps which are present when the surface deviates slightly from the low index ideal. Benard assumes that this heat of adsorption refers to an equilibrium between gaseous hydrogen sulphide, hydrogen and adsorbed sulphur ($S_{(s)}$). Using the data from Cottrell (1958) for the bond energy of $S_{2(g)}$ we may compute the binding energy of $S_{(s)}$ on silver as 69, 71 and 75 kcal mole^{-1} on (111), (100) and (110) planes respectively.

Saleh *et al.* also estimated the activation energy for sulphur desorption from tungsten at 80–90 kcal mole^{-1}. In the regime of negligible activation energy of adsorption, we take this figure to be equal to the binding energy of $S_{(s)}$ on tungsten.

Water vapour adsorption shows features reminiscent of hydrogen sulphide. On tungsten films at 0 °C (Imai and Kemball, 1968) $H_{(s)}$ and $OH_{(s)}$ form initially, while further treatment gives hydrogen desorption and a high coverage of $OH_{(s)}$. Heating to 200 °C causes further hydrogen desorption and conversion of some $OH_{(s)}$ to $O_{(s)}$. The behaviour of nickel, iron and copper is qualitatively similar (Suhrmann *et al.*, 1964, 1968).

B. UNSATURATED HYDROCARBONS

The olefins, alkynes and aromatics are all strongly adsorbed on the transition metals.

The adsorption of ethylene has been measured volumetrically on films of nickel (Beeck, 1950; Jenkins and Rideal, 1955) tungsten (Trapnell, 1952) and palladium (Stephens, 1959). In each case the initial adsorption was found to be very fast. At higher coverage the residual gas phase was found to contain ethane, showing that some of the ethylene adsorbs dissociatively. Jenkins and Rideal (1955) showed that the ethane observed in the gas phase results mainly from reaction of ethylene with the adsorbed hydrogen resulting from this dissociative adsorption. The nett process at room temperature and above is a self-hydrogenation,

$$2C_2H_{4(g)} \rightarrow C_2H_{2(s)} + C_2H_{6(g)} \qquad (II)-(15)$$

At lower temperatures ethylene is mainly non-dissociatively adsorbed.

The heat of adsorption of ethylene has been measured on evaporated films of Ta, W, Cr, Fe, Ni and Rh (Beeck, 1950). In each case the initial heat was found to fall as the coverage increased. While this is almost certainly due in part to surface heterogeneity it is also likely that both dissociative and non-dissociative adsorptions occur. Non-dissociative adsorption would be expected to increase with coverage and have the lower heat and it is therefore difficult to interpret the measured heats of adsorption. However, the values at low coverage can be taken as an indication of the general order of reactivity towards olefin adsorption: (Ta, W, Cr) > 100 kcal mole^{-1}, (Fe, Ni, Rh) 50–70 kcal mole^{-1} and (Cu, Au) ~20 kcal mole^{-1}.

The nature of the adsorbed species is also dependent on the presence of hydrogen. Eischens and Pliskin (1958) have shown by infrared spectroscopic study of the intensity of C–H bonds on a nickel-silica surface that the non-dissociative form of adsorbed ethylene results when hydrogen is preadsorbed on the surface. This form also results when hydrogen is added to a surface covered by dissociatively adsorbed ethylene. Even at the lowest hydrogen to carbon ratio of the adsorbed species, the C-H bonds were observed to be paraffinic rather than olefinic. This has been taken as evidence for

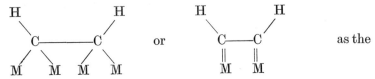

(schematic) structure of dissociatively adsorbed ethylene rather than the possible alternative

The infrared work has also shown evidence for polymeric structures consistent with the detection of gas phase butane and higher hydrocarbons along with ethane when the surface complex is hydrogenated.

Roberts (1963) has shown that the adsorption of ethylene at low pressures on iridium films at 27° C results in some self-hydrogenation to ethane. On an iridium field emitter tip ethylene, and also ethane and acetylene, were adsorbed at <400 °C on all of the high index faces as an immobile layer, whereas hydrogen was readily desorbed at 100 °C (Arthur and Hansen, 1962). Adsorbed ethylene and acetylene were found to be similar above 180 °C and it is concluded that dehydrogenation of ethylene to acetylene can occur and that this process is followed at higher temperatures by the formation of a crystalline carbon residue.

Taylor et al. (1968) have studied the behaviour of ^{14}C-ethylene and tritium on alumina supported palladium, rhodium and platinum. In the range 20–200 °C it was found that ^{14}C-ethylene is adsorbed to form a strongly bound species which is inactive in molecular exchange and hydrogenation and is not displaced by acetylene. Using tritium, the relative abilities of these metals to promote ethylene self-hydrogenation were determined to be Rh > Pt > Pd. They also found that the same catalyst surface showed differing degrees of surface heterogeneity for different hydrocarbon molecules but there was evidence that the hydroxyl groups of the alumina support were participating in the reaction.

Rapid adsorption of acetylene occurs on the transition metals. While the experimental studies are less numerous and detailed than those for ethylene it is apparent that the adsorption of acetylene is much stronger than ethylene. Beeck (1950) showed that nickel films were effectively poisoned by adsorbed acetylene so that species resistant to hydrogenation must form on the surface: it is presumed that these result from dissociative adsorption and have a lower hydrogen-carbon ratio than the species from ethylene. At higher coverages and in the presence of hydrogen, that is conditions existing in a catalytic hydrogenation reaction, it is likely that acetylene adsorbs non-dissociatively yielding diadsorbed structures similar to ethylene.

Ethylene and acetylene adsorption has also been measured on copper, gold and aluminium. It has been suggested that copper and gold may have d-band vacancies created by the chemisorption since both metals have small energies for d-s promotion of electrons (Hayward and Trapnell, 1964).

Benzene adsorption has been found to be rapid on the transition metals. Although there is some evidence for dissociative adsorption on nickel, iron and platinum (Suhrmann, 1957) the loss of hydrogen is much less than for ethylene and acetylene. The important adsorbed species is considered to be that arising from non-dissociative adsorption. The heat of adsorption of benzene is generally about half of that for an olefin or acetylene which has been taken to indicate that the aromatic resonance energy is lost on adsorption. This has led to much speculation on the nature of the surface species for benzene and also for the non-dissociative forms of adsorbed ethylene and acetylene. Since these are the likely species in a catalytic hydrogenation where the coverage is high and excess hydrogen is present, a knowledge of the structure is of obvious importance.

Most of the earlier considerations were based on the formation of σ bonds to the surface. Non-dissociatively adsorbed ethylene was represented as

$$
\begin{array}{c}
\diagdown \qquad \diagup \\
\mathrm{C}\!-\!\mathrm{C} \\
\diagup \qquad \diagdown \\
\mathrm{M} \qquad\quad \mathrm{M}
\end{array}
$$

E

while benzene was represented by either 6-point or 2-point attachment by σ bonds, the former implying that the molecule lies flat on the surface. The directional nature of these covalent bonds invited much speculation on the importance of the arrangement and spacing of the surface atoms. This geometric approach to adsorption and catalysis has been extensively reviewed (see Balandin, 1958; Bond, 1962) but generally the evidence presented is ambiguous since the electronic properties of the metal are related to the geometric parameters being considered. Much of the experimental evidence relating to oriented surfaces must be discounted now that we have more detailed knowledge of film structures.

The more recent discussions of the mechanism of hydrogenation reactions have considered π-bonded structures to represent the non-dissociatively adsorbed state. The idea arose from the numerous π-bonded organometallic compounds described since the discovery of ferrocene. The application of the principle to the hydrogenation of unsaturated hydrocarbons is discussed by Rooney and Webb (1964) and by Bond and Wells (1964). It is considered that while σ-bonded olefins and alkynes are possible, the more likely intermediate in the reaction is π-bonded to a single metal site as represented in Figure 1. Thus in an

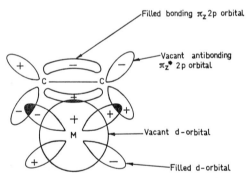

FIG. 1. Diagrammatic representation of the π-bonding of an olefin to a metal atom having a vacant d-orbital.

olefin one π-bond is formed while in acetylene there are two π-bonds. Because only one site is involved the geometrical arguments regarding site spacing are irrelevant and many of the inconsistencies of the geometric theory vanish.

For benzene the molecule is considered to lie flat with the π-orbitals interacting predominantly with a single site but possibly dispersing to neighbouring sites. Most of the evidence for π-bonds is based on mechanistic arguments from the reactions of substituted hydrocarbons particularly isomerization reactions. It has, however, been noted that the surface potentials of adsorbed ethylene and acetylene are positive (Mignolet, 1950) indicating a strong donation of charge to the acceptor orbitals of the metal.

This concept of a π-bonded complex as the reactive state of an adsorbed unsaturated hydrocarbon has been further developed by Bond (1966). A molecular orbital model was used to describe the direction of the emerging orbitals at the metal surface and the degree of their occupation. Bond discusses the possible π-adsorbed states of olefins, diolefins and alkynes and concludes that the (111) plane of a face centred cubic metal is least suited for their adsorption. This raises again the question of surface geometry and despite the uncertainties of the present molecular orbital treatment it is apparent that future observations of surface specificity in adsorption will be explained in terms of this type of theory rather than any simple consideration of lattice parameters.

C. OXYGEN-CONTAINING MOLECULES

1. Carboxylic Acids

The decomposition of formic acid has been studied most frequently as a reaction to test the influence of catalyst structure. There have been numerous direct studies of the adsorption in order to elucidate the structure of the adsorbed species. Infrared measurements on the adsorption of formic acid on supported nickel and gold (Sachtler and Fahrenfort, 1961) and on supported silver and copper (Hirota et al., 1961) have indicated the presence of formate ions $HCOO^-$ on the surface. At low temperatures there is also evidence for covalently bound formic acid on nickel (Eischens and Pliskin, 1961). Even where formate ions were detected there was evidence that some of the formic acid was undissociated.

The coverage of formic acid achieved at room temperature was found to be high on nickel but low on silver (Tamaru, 1959). However, Lawson (1968a) has shown that the adsorption on silver is dependent on the surface structure, the coverage being high on atomically rough surface but low on the close packed (111) plane.

2. Alcohols

No direct adsorption data for clean metal surfaces are available. On supported nickel, adsorption has been studied by Blyholder and Neff (1963) using infrared methods. Dehydrogenation of methyl and ethyl alcohols occurs during adsorption, and at 20 °C adsorbed carbon monoxide was revealed as the only identifiable surface dehydrogenation product from methyl alcohol, and this was identified by the intense absorption bands at 2040 and 1900 cm^{-1}. A similar dehydrogenation path for methyl alcohol on platinum is suggested by the course of the methyl alcohol decomposition reaction studied by McKee (1968). Adsorption of ethyl alcohol on supported nickel at 20 °C results in C–C bond rupture as well as C–O bond rupture, to give adsorbed carbon monoxide together with adsorbed residues $CH_{3(s)}$ and $CH_{2(s)}$ which were identified by adsorption bands at 2960 and 2920 cm^{-1}, respectively (Blyholder and Neff, 1963). By comparison, the decomposition reaction of methyl alcohol on ruthenium shows that C–O bond rupture is important, and McKee suggests the formation of $CH_{2(s)}$ residues.

Blyholder and Neff could find no infrared evidence for C–O bond rupture in the adsorption of diethyl ether on nickel.

3. *Aldehydes and Ketones*

Again, there is no direct adsorption data available for clean metal surfaces. However, there is evidence from the mechanisms deduced from catalysis studies that the transition metals generally adsorb the carbonyl group probably at both oxygen and carbon atoms. The question is discussed further in section IV.

D. ALKYL HALIDES

Adsorption of methyl chloride was studied by Anderson and McConkey (1968) on transition metal films of nickel, platinum, palladium, tungsten, cobalt, manganese and titanium in the temperature range 0–250 °C. Rapid adsorption occurred at 0 °C with a low activation energy, and this was followed by a slower uptake at higher surface occupations; at 0 °C the surface occupation (σ_{CH_3Cl}) typically reached about 20%. Adsorption of CCl_2F_2 is known to occur on iron (Roberts, 1962c); from bond strength considerations rupture of the C–Cl bond should occur before the C–F bond. The behaviour on nontransition metals depended on the metal and the conditions. On sodium and copper, methyl chloride adsorption occurred at 0 °C, but on aluminium and silver no adsorption occurred at 0 or 100 °C, but was substantial at >200 °C (Anderson and McConkey, 1967, 1968). Clearly, at the moment the behaviour of individual nontransition metals cannot be generalized.

Adsorption results first in the rupture of the C–Cl bond, followed by rupture of some C–H bonds; the latter occurs particularly on transition metal surfaces, but is also found on copper and on aluminium at 200 °C. This order of bond reactivity agrees with Galwey and Kemball's conclusions for alkyl halide adsorption on supported nickel, with Robert's (1962c) observation that adsorption of CCl_2F_2 occurred on iron films in

preference to ethane and the observations on the adsorption of chloro-
form on a tungsten field emitter (Duell *et al.*, 1966). For methyl chloride
adsorption on metal films the $(H/C)_{(s)}$ values were such that the composition of the average hydrocarbon residue expressed as CH_x must
have $x < 3$, and $(H/C)_{(s)}$ values decreased with increasing temperature.
For instance, adsorption at 0 °C gave the following values for $(H/C)_{(s)}$;
nickel, 2.2; tungsten, 2.7; platinum, 1.2; cobalt, 2.4; manganese, 1.0.
Both methyl chloride and methylene chloride adsorption were examined
on palladium and titanium, and at 0 °C the $(H/C)_{(s)}$ values were respectively: palladium, 1.4, 1.8; titanium, 2.9, 1.9. It is clear therefore
that adsorption of methyl chloride gives both $CH_{3(s)}$ and $CH_{2(s)}$ residues, and methylene chloride both $CH_{2(s)}$ and $CH_{(s)}$ residues. No
hydrogen chloride is returned to the gas phase under these conditions,
and hydrogen is removed from the surface by desorption of H_2 and
methane. By comparison, Galwey and Kemball (1961) used treatment
with deuterium to show that adsorption of alkyl chlorides on supported
nickel at 0–200 °C gave the following residue compositions, expressed
as x in CH_x: ethyl chloride, 1.5; 1– and 2–propyl chloride, 1.7; n-butyl
chloride, 2.0; t-butyl chloride, 1.7. It is likely that the presence of a high
surface concentration of adsorbed chlorine will sterically impede the
exchange of some $H_{(s)}$, in much the same way as happens with the
hydrogen sulphide system, so Galwey and Kemball's values for x may
be somewhat high. Nevertheless, these results are generally consistent
with the behaviour of the hydrocarbons given previously.

In a somewhat analogous fashion to methane adsorption, adsorption
of carbon tetrachloride on tungsten probably leads to the formation of
$CCl_{x(s)}$ residues with $x \simeq 3$ (Duell *et al.*, 1966); however, no chlorine
desorption occurs at these temperatures, and dissociated chlorine must
all be held in the surface chloride layer.

The behaviour of methyl chloride over non-transition metals requires
comment. On copper at 0 °C or silver at 230 °C, methyl chloride adsorption results in the rupture of C–H bonds as well as C–Cl bonds,
yet at these temperatures dissociative adsorption of methane or ethane
does not occur. This is not a unique situation; for instance, deuterium
exchange within the methyl group occurs in the exchange of methyl
alcohol on a silver film at 262 °C despite the complete inactivity of
silver as a catalyst for methane exchange. We believe that when a silver
or copper surface is heavily covered with a strongly electronegative
adsorbent such as chlorine or oxygen, the binding energy for chemisorbed hydrogen is so modified that C-H bond rupture becomes possible.

The rupture of the C–Cl bond in methyl chloride was shown by the
use of C^{13} and Cl^{35} labelling to be completely irreversible on titanium

(Anderson and McConkey, 1968). However, Cockelbergs *et al.* (1959) have shown that it is reversible on tungsten and molybdenum. Furthermore, Morrison and Krieger (1968) have shown that exchange occurs between ethyl iodide and other alkyl iodides over supported copper, iron, nickel, palladium, rhodium and iridium catalysts at <200 °C. The exchange mechanism is probably dissociative, indicating reversible C–I bond rupture. The reason for the exceptional behaviour of titanium may be the very high Ti–Hal bond energy.

E. INCORPORATION PROCESSES

Complete dehydrogenative adsorption of hydrocarbons, ammonia or hydrogen sulphide can yield $C_{(s)}$, $N_{(s)}$ or $S_{(s)}$ and oxygen or halogen-containing molecules can yield in the same way $O_{(s)}$ or $Hal_{(s)}$. If thermodynamically allowed, the reaction may proceed to form a bulk product phase. Knowledge of this should clearly be important in catalysis because if it occurs the surface exposed to the gas will no longer be that of the metal. For instance, there is a fourfold reduction in the activity of vanadium for ammonia decomposition when the surface is converted to vanadium nitride, and with iron the corresponding reduction is 17-fold (Logan and Kemball, 1960). We may consider as examples the reactions

$$M + xCH_{4(g)} \rightleftarrows MC_x + 2xH_{2(g)} \qquad \text{(II)–(16)}$$

$$M + xNH_{3(g)} \rightleftarrows MN_x + 3/2xH_{2(g)} \qquad \text{(II)–(17)}$$

$$M + xH_2S_{(g)} \rightleftarrows MS_x + xH_{2(g)} \qquad \text{(II)–(18)}$$

$$M + xH_2O_{(g)} \rightleftarrows MO_x + xH_{2(g)} \qquad \text{(II)–(19)}$$

$$M + xHCl_{(g)} \rightleftarrows MCl_x + \frac{x}{2}H_{2(g)} \qquad \text{(II)–(20)}$$

and adopt as criteria the values for $p_{CH_4}/p_{H_2}^2$, etc. corresponding to equilibrium at 400 °C, which are given in Table 4. Gas phase compositions giving ratios higher than the listed values correspond to conditions of product stability. Although no quantitative data are available, it is clear that the carbides and nitrides of the noble metals are highly unstable. The data in Table 4 may be used as a guide for possible incorporation reactions. Evidence for bulk carbide or nitride formation on the surface of evaporated films as a result of catalytic experiments is relatively meagre particularly with carbides; indeed the detection of a layer only a few unit cells in thickness may be quite a difficult task, and yet such a layer will completely modify the catalytic properties of the metal. When the formation of such a layer is kinetically rather than thermodynamically controlled, it may happen that

	$(p_{CH_4} p_{H_2}^{-2})$ equil at 400°C (atmos^{-1})	$(p_{NH_3} p_{H_2}^{-\frac{3}{2}})$ equil at 400°C (atmos^{-1})	$(p_{H_2S} p_{H_2}^{-1})$ equil at 400°C (atmos^0)	$(p_{H_2O} p_{H_2}^{-1})$ equil at 400°C (atmos^0)	$(p^2_{HCl} p_{H_2}^{-1})$ equil at 400°C (atmos^1)
nickel	$\sim7 \times 10^3$(Ni₃C)*	$(\sim10^2)$(Ni₃N)*	4×10^{-3}(NiS)	5×10^2(NiO)	6×10^{-2}(NiCl₂)
iron	3×10^1(Fe₃C)*	2×10^{-1}(Fe₄N)*	1×10^{-4}(FeS) 6×10^{-3}(FeS₂)	1×10^{-1}(FeO) 7×10^{-1}(Fe₂O₃)	3×10^{-5}(FeCl₂) 6×10^2(FeCl₃)
cobalt	$\sim10^2$(Co₂C)*			5×10^1(CoO)	3×10^{-3}(CoCl₂)
tungsten	$(\sim10^{-1})$(W₂C)* $(\sim10^{-1})$(WC)*		$\sim10^{-4}$(WS₂)	1×10^{-1}(WO₃)	
molybdenum	1×10^{-8}(Mo₂C)*	$\sim10^{-8}$(Mo₂N)*	2×10^{-5}(MoS₂)	2×10^{-2}(MoO₂) 4×10^1(MoO₃)	
chromium	3×10^{-7}(Cr₄C)*	$\sim10^{-13}$(Cr₂N)*		3×10^{-9}(Cr₂O₃)	2×10^{-9}(CrCl₂) 6×10^{-6}(CrCl₃)
vanadium	$(\sim10^{-8})$(VC)*	$\sim10^{-23}$(VN)*	$\sim10^{-11}$(VS)	2×10^{-11}(VO) 6×10^{-4}(V₂O₅)	2×10^{-12}(VCl₂) 4×10^{-7}(VCl₃)
tantalum		$\sim10^{-27}$(TaN)*		4×10^{-12}(Ta₂O₅)	
titanium	$\sim10^{-13}$(TiC)*	$\sim10^{-45}$(TiN)*		2×10^{-19}(TiO) 4×10^{-16}(TiO₂)	$\sim10^{-16}$(TiCl₂) $\sim10^{-15}$(TiCl₃)
manganese	8(Mn₃C)*		6×10^{-14}(MnS)	2×10^{-10}(MnO) 1×10^1(MnO₂)	1×10^{-14}(MnCl₂)
rhenium			$\sim10^{-2}$(ReS₂)	$\sim10^4$(ReO₂) $\sim10^5$(ReO₃)	
ruthenium			$\sim10^{-3}$(RuS₂)	$\sim10^{12}$(RuO₂)	$\sim10^{12}$(RuCl₃)
osmium			~1(OsS₂)		
rhodium				$\sim10^{13}$(RhO)	$\sim10^9$(RhCl₂) $\sim10^{10}$(RhCl₃)
iridium				$\sim10^{13}$(IrO₂)	$\sim10^{11}$(IrCl₂)
palladium				$\sim10^{14}$(PdO)	
platinum			5×10^{-1}(PtS) 7(PtS₂)		$\sim10^{13}$(PtCl₂)
copper			6×10^{-1}(CuS)	2×10^6(Cu₂O) 2×10^8(CuO)	4×10^6(CuCl₂)
silver				3×10^{17}(Ag₂O)	4×10^1(AgCl)

Values in brackets are estimates or interpolations subject to considerable uncertainty. Richardson (1953); Pearson and Ende (1953); Juza (1966); Mah (1960); Alekseev and Shvartsman (1962, 1963); Gleiser and Chipman (1962); Solbakken and Emmett (1969); Landolt-Bornstein (II. Band, 4.Teil; 1961). Compounds marked * are more stable at 400° C than at 25° C; for the remaining compounds the reverse is true. Compounds are indicated in terms of the nominal stoichiometry only.

the time scale of a laboratory experiment using a film catalyst is sufficiently short to avoid extensive incorporation. The implications for catalyst stability in technical catalysts are obvious. Saleh *et al.* (1961) and Roberts and Ross (1966) have shown that sulphide incorporation occurs readily from the reaction of hydrogen sulphide on iron and nickel, and was significant at temperatures as low as 0 °C. Logan and Kemball (1960) observed nitride formation over films of tantalum, tungsten and vanadium during an ammonia decomposition reaction, but on a wide range of other metals, including rhodium, cobalt, palladium, ruthenium, rhenium, platinum and nickel, no bulk nitride could be detected. Out of this list, the behaviour of vanadium, tantalum and nickel is in agreement with the known thermodynamic properties. This is also true in the case of iron, and the free energy of formation of Fe_4N is such that its formation can be easily controlled by adjustment to the composition of the ammonia/hydrogen reaction mixture (Logan *et al.*, 1958).

It is generally found that the extent of incorporation is very dependent on the nature of the reacting molecule. Thus, judging by the data in Table 4, the formation of W_2C and Mo_2C should be thermodynamically favourable under the conditions used for methane/deuterium exchange and ethane hydrocracking; however, the examination of film catalysts after these reactions has never given evidence for the presence of carbide. On the other hand, the reaction of methylcyclopentane with hydrogen over a tungsten film rapidly gave a completely poisoned catalyst which certainly had strongly held carbon residues and possibly had carbide on the surface (Anderson and Baker, 1963). By comparison with the behaviour of saturated hydrocarbons, the reaction of methylamine and hydrogen on film catalysts readily results in the formation of Ni_3C, Co_2C, W_2C, and VN (Anderson and Clark, 1966). As another example, it is often observed that nitride formation proceeds much more readily by reaction of the metal with ammonia than with nitrogen itself at the same pressure. It will be obvious from Table 4 that there are many metals of which sulphides, oxides or chlorides are of high stability and on which incorporation to give these product phases will be possible under appropriate catalytic conditions.

The extent of an incorporation reaction may thus be controlled by both thermodynamic and by kinetic factors. In a regime of kinetic control, the incorporation reactions can be understood from the theory of metal tarnishing reactions in which the general reactant is $X_{(s)}$; the main difference from ordinary tarnishing reactions is thus the way in which the $X_{(s)}$ concentration is controlled by the ambient gas pressures. The rate of the incorporation reaction will increase with increasing $X_{(s)}$ concentration. In the simplest case where the rate of layer growth is controlled by diffusion down a linear concentration gradient, the growth rate will be directly proportional to the concentration, and the layer thickness will grow according to a parabolic law (thickness $\propto t^{1/2}$), and this will be true irrespective of whether the diffusing entity is the metallic or non-metallic component. In the same way as described previously in the section dealing with ammonia decomposition, the existence of a pre-incorporation equilibrium means that even if diffusion is the slow step, the measured activation energy will not equal that for diffusion because the former includes a ΔH term for the pre-incorporation equilibrium.

For very thin layers the form of the growth law will be determined by the form of the field gradient in which the ions move or by the nature of the reaction at the gas/solid interface, and various growth laws can result. In fact, various rate controlling steps for incorporation are possible. Dobychin et al. (1964) who studied the formation of molybdenum carbide from molybdenum and methylamine measured an activation energy of ~ 40 kcal mole^{-1} and concluded that carbon diffusion was rate controlling. On the other hand, if dissociation of the reactant molecule at the gas/solid interface is slow, this can control the rate, and the conclusion that this is the case for the reaction between hydrogen sulphide and lead is well founded (Saleh et al., 1964).

By way of illustration, consider the incorporation of nitride. We saw from a virual pressure argument in a previous section that on the simplest model the concentration of $N_{(s)}$ will be proportional to $p_{NH_3} p_{H_2}^{-3/2}$, and the incorporation rate should thus vary with pressure in this way also. In a generally similar way, if $C_{(s)}$ is in equilibrium with CH_4 and H_2, the rate of carbide layer formation would be proportional to $p_{CH_4} p_{H_2}^{-2}$. Under most catalytic conditions it is unlikely that this equilibrium can be properly established, and this expression in methane and hydrogen pressure is then only of use as a guide. The assumption that H_2 and $H_{(s)}$ are in established equilibrium is usually a good one, particularly with hydrocarbons. However, with strongly adsorbed molecules such as amines or hydrogen sulphide, this may not always be so, and the facility with which methylamine forms metal carbides compared with hydrocarbon reactions may well be due to this, as well as to the greater ease of C–N bond rupture compared to C–C bond rupture; in this case Anderson and Clark (1966) have suggested that rapid ammonia desorption reduces the $H_{(s)}$ concentration to below the equilibrium value. In a quite analogous way, the reaction of methylamine on platinum and palladium results in the formation of a substantial amount of dissolved carbon, bulk carbides of these metals being unstable. Differences in the reactivity of various hydrocarbons for

E*

carbide formation are thus mainly associated with their ability to form $C_{(s)}$ residues, and in detail this will be controlled by quite complex kinetic factors including the effect of molecular geometry on the rate of dehydrogenative adsorption and C–C bond rupture.

On the model adopted above, the difference between, for instance, $C_{(s)}$ and incorporated carbon is one of coordination number. However, $C_{(s)}$ is not considered to be merely carbon existing in a normal crystallographic position in the metal carbide surface, since a carbon which reaches such a position is incorporated. We assume that $C_{(s)}$ is equivalent to "adsorbed" carbon which is crystallographically distinct with a still lower coordination, and presumably with a higher reactivity. It should be remembered that just as $N_{(s)}$ has an alternative path to incorporation, that is the formation of elemental nitrogen (N_2), there is an alternative path to carbon incorporation which is the formation of graphite, and graphite formation is often observed in the reaction of hydrocarbons with metals at high temperatures.

The reverse of incorporation, that is the reaction between nickel carbide (Ni_3C) and hydrogen, has been investigated by Galwey (1962). The rate was first order in hydrogen pressure, indicating that the slow step involved the reaction between H_2 and a surface species, but the mechanism is uncertain. In addition to methane, some 5% of the reaction produce was ethane, showing that during the hydrogenation, some propensity for carbon–carbon bond formation exists, presumably by reaction between two partly hydrogenated residues.

III. Exchange Reactions of Saturated Molecules

A. H_2/D_2 Exchange

The exchange of hydrogen and deuterium (H_2/D_2) over metals, including metal films has been extensively reviewed, and the reader is referred to Bond (1962), Taylor (1957), Eley (1948, 1966) and to Hayward and Trapnell (1964). For the present purpose we need only remark that two exchange mechanisms are possible. In both, "hydrogen" must be chemisorbed as atoms, but the reacting molecule may also produce chemisorbed atoms which link up randomly in pairs and evaporate as molecules, thus

$$\tfrac{1}{2} H_2 + \tfrac{1}{2} D_2 + 2S \rightarrow H_{(s)} + D_{(s)} \rightarrow HD + 2S \qquad \text{(III)-(1)}$$

The reaction either requires surface diffusion of the adatoms if adsorption is random, or the use of four adjacent sites of suitable geometry.

Alternatively, the molecule may be physically adsorbed and, as a molecule, react with a chemisorbed atom.

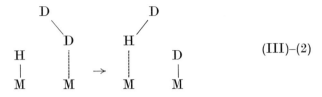

$$(III)-(2)$$

and as a variant of this, the reaction may require three sites rather than two. Finally, there is the suggestion due to Boreskov and Vassilevitch (1961) that in reaction (III)–(2), or in its three-site alternative, the "hand-on" process results in atom migration across the top of the chemisorbed layer, until a molecule is desorbed from a low energy site. This mechanism is not likely to be important except on surfaces of high energetic heterogeneity. For transition metals the activation energy for chemisorption of hydrogen as atoms is very low, and H_2/D_2 exchange occurs easily. At low temperatures (say $<100\ °C$) non-transition metals do not catalyse the exchange, and this is undoubtedly due to their inability to chemisorb hydrogen in this temperature range. However, these metals are active for exchange at 300–450 °C probably because thermal promotion of electrons from the normally full d–band to the s–band makes hydrogen chemisorption possible (Mikovsky et al., 1954). Exchange could also probably be initiated on these metals at low temperatures by directly providing atoms (rather than molecules) for initial adsorption either by thermal dissociation of molecular hydrogen at a hot filament, or by atomic diffusion through palladium in the manner described by Wood and Wise (1966).

The H_2/D_2 exchange occurs over most of the surface exposed by a polycrystalline specimen (Eley, 1941). We know of no determination of the dependence of the kinetic parameters on crystal face, but since this exchange does not make severe stereochemical demands on the catalyst it seems likely that for a given metal the exchange will not be particularly sensitive to crystal face.

An illuminating study of this exchange has been made by Eley and Norton (1966) over clean nickel wire (UHV conditions). They concluded that reaction (III)–(1) occurred at 330–400 °K with an activation energy of 5–7.5 kcal mole^{-1} and with a kinetic order of about 0.6, while at 200–300 °K the mechanism followed (III)–(2) or its three site alternative, with an activation energy of 1.5–2.7 kcal mole^{-1} and a variable kinetic order in the range 0.04–0.4. These results are in accord with what may be expected for the dependence of the mechanism on surface coverage, the latter being controlled by the reaction temperature (Anderson, 1957a).

B. SATURATED C_1–C_4 HYDROCARBONS

The subject has previously been reviewed by Anderson (1957a), Kemball (1959) and Bond (1962). Exchange involves dissociative chemisorption of the hydrocarbon on the metal surface, requiring the rupture of one or more C–H bonds. For instance, the formation of CH_3D from methane and deuterium follow from the formation of $CH_{3(s)}$ according to a reaction such as (II)–(2). Exchange between hydrogen and deuterium is known to be rapid compared with the rate of hydrocarbon exchange, and thus, in the presence of a large excess of deuterium the reverse (desorption) reaction occurs with near unit chance of forming a C–D bond. The principle of microscopic reversibility requires that in a single path process, the free energy profile through the transition state should be symmetrical about its mid-point: in other words the desorption step should be exactly the reverse of the adsorption step save for the replacement of $H_{(s)}$ by $D_{(s)}$. However, there is no need to require that exchange shall necessarily occur by a single reaction path only. Examples are the proposals by Anderson (1957a) and Rooney et al. (1960) for the deuterium to enter the desorbing species from either the "top" or the "bottom", these terms being defined relative to the surface. In a general way we may illustrate this idea thus

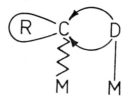

where the mode of binding of the carbon atom to the surface is left unspecified at this stage (wiggley line), and the carbon atom shown explicitly is supposed to be bonded in some way to the rest of the residue R. The important thing is that the direction from which the deuterium enters may, in principle, have specific stereochemical consequences. We shall encounter some specific examples subsequently. When such a two-path desorption process is proposed, the principle of microscopic reversibility only requires that for each process adsorption and desorption shall be exactly reciprocal. The application of the principle of microscopic reversibility to reaction kinetics has been recently discussed more fully by Burwell and Pearson (1966). Although this principle is discussed in this section, these conditions are, of course, quite general and apply to all of the exchange mechanisms we shall be concerned with.

In most catalytic deuterium exchange work, kinetic isotope effects are usually ignored. Although the assumption does not lead to serious error when one is concerned with the broad pattern of catalytic behaviour, it is in fact known that exchange of C–D with H_2 is somewhat slower than the corresponding exchange of C–H with D_2. Thus Meyer and Kemball (1965) found that the exchange of C_6D_{14} with H_2 over a palladium film had an activation energy some 2 kcal mole^{-1} in excess of that for the exchange of C_6H_{14} with D_2. This may be compared with the results of Flanagan and Rabinovitch (1956) who showed that in hydrocarbon (ethylene) adsorption on nickel, a C–H bond was broken more easily than a C–D bond. This agreement provides good evidence in support of an assumption that it is the rate of the hydrocarbon adsorption which governs the rate of exchange. Qualitatively, the influence of this kinetic isotope effect on exchange product distributions is clear: if, for instance the initial exchange products contain highly deuterated compounds in excess of the amounts corresponding to final exchange equilibrium, as equilibrium is approached these products will be retained at concentrations above those that would be expected in the absence of the isotope effect. On the other hand, the equilibrium isotope effect in these systems is quite small, and the composition of an equilibrium mixture of deuterated hydrocarbons can be obtained with reasonable accuracy by a classical statistical model.

The rate of the exchange reaction of "light" hydrocarbon equals the rate of hydrocarbon adsorption which also equals the rate of desorption, since under catalytic exchange conditions a steady state in surface coverage is reached. This rate is a convenient measure of that for the overall exchange process, and will be called the rate of adsorption/desorption. Table 5 compares the overall behaviour of C_1–C_4 hydrocarbons over three typical metals. In the case of methane, the reaction goes by two simultaneous processes, one giving CH_3D, the other giving CH_2D_2, CHD_3 and CD_4. Kinetic parameters for both processes are listed, the first for each metal being for CH_3D production.

Using as a criterion of hydrocarbon reactivity T_r, the temperature at which the rate of adsorption/desorption equals the arbitrary but convenient figure of 2×10^{12} molec sec^{-1}cm^{-2} (near the middle of the experimentally accessible range), we see that methane is much less reactive than the other hydrocarbons. There are some apparent differences in the reactivity of the other hydrocarbons, but the extent to which the exchange rate is influenced by self-poisoning reactions due to strongly adsorbed hydrocarbon residues makes it unreasonable to place much weight upon such differences. For instance, the data in Table 5 for nickel show ethane to be much less reactive than propane. However, in fact ethane does exchange at 0 °C, that is with a reactivity not much

TABLE 5

Relative reactivities of some hydrocarbons on three typical random polycrystalline metal films

	tungsten			rhodium			nickel		
	T_r	$E^{\ddagger}$	$\log_{10} A^{\ddagger\ddagger}$	T_r	$E^{\ddagger}$	$\log_{10} A^{\ddagger\ddagger}$	T_r	$E^{\ddagger}$	$\log_{10} A^{\ddagger\ddagger}$
methane*	110	9.1 11.7	17.5 18.3	130	23.6 28.0	25.2 27.9	220	28.4 33.8	24.9 28.0
ethane	−48	8.2	20.2	16	11.7	21.1	187	18.0	20.8
propane	−67	9.0	21.8	−45	13.3	25.0	−33	10.4	21.7
isobutane	−54	7.9	20.1	—	—	—	−35	9.0	20.5

T_r (°C), temperatures at which rate of adsorption/desorption equals 2×10^{12} molecules cm^{-2} sec^{-1}, corrected to uniform reactant pressures
‡E, activation energy, kcal $mole^{-1}$.
‡‡A, frequency factor, molecules cm^{-2} sec^{-1}.
*The reaction of methane goes by two simultaneous processes, and kinetic parameters for each are given (see text).

different from that of propane, but this exchange reaction is self-poisoned and recommences at higher temperatures. In other words, the apparent difference in reactivity between ethane and propane on nickel shown in Table 5 is really an indication of the extent to which a facile, low temperature exchange with ethane has been inhibited by the formation of strongly bonded residues.

The formation of strongly bonded residues is certainly a general phenomenon, or at least potentially so, as judged by the data given in a previous section from hydrocarbon adsorption. Thus, exchange may well occur on only a relatively small section of the surface, a fact previously emphasized by Kemball (1959). In general, self-poisoning during exchange becomes more important as the size of the hydrocarbon molecule increases. The activation energy for extensive dissociative adsorption appears to decrease with increasing molecular size: however, in addition to this, with larger molecules there are probably steric factors which limit the accessibility of hydrogen to the adsorbed carbon, thus resulting in a situation favouring more extensive dehydrogenation. Since dissociative adsorption is an activated process, pre-treatment of the catalyst with hydrocarbon at a temperature in excess of that used in the subsequent exchange can lead to very extensive poisoning, and Gault and Kemball (1961) have in this way observed reduction factors approaching 10^4 for the exchange rate of n-hexane on rhodium. Their results also show that these strongly bound residues are only slowly and partly removed by deuterium under exchange conditions, and confirm the great difficulty in the establishment of surface dehydrogenation equilibria. Self-poisoning is also a function of the crystal face of the metal, and in the exchange of propane and ethane appears to be more extensive on a (100) than on a (111) or a polycrystalline surface (Anderson and Macdonald, 1969a). Thus, self-poisoning is undoubtedly an important effect, and it should be clear as a consequence that correlations of catalytic activity, either between varying metals or varying reactants, are subject to a considerable degree of uncertainty. In this respect we believe that, as currently measured, only big differences in reactivity or quite general trends are likely to retain significance in terms of the exchange mechanisms themselves. The low reactivity of methane in exchange correlates with the relative difficulty for the adsorption of methane compared to ethane noted in section II.

The rate of hydrocarbon adsorption/desorption has been measured over a range of metals with methane and ethane, from which two main conclusions emerge: (a) There is a general trend that high activation energies are associated with high frequency factors, the reasons for this include the presence of deuterium as an inhibiting gas as outlined in section IV of Chapter 7. The inhibition by deuterium occurs in virtually all systems, as indicated by the way in which the rate of adsorption/ desorption varies with gas pressure, and some results showing this are collected into Table 6. The trend between activation energy and frequency factor is illustrated in Figure 2. (b) There is a trend for T_r to

TABLE 6

Pressure dependence of exchange reactions on metal film catalysts

	Methane				Ethane adsorption/ desorption		Propane adsorption/ desorption	
	CH_3D		CD_4, CD_3H, CD_2D_2					
	a	b	a	b	a	b	a	b
nickel	1.0	−0.9	1.0	−1.4	—	—	1.0	−0.6
rhodium	1.0	−0.8	1.0	−1.2	0.8	−0.7	0.5	−0.8
tungsten	0.6	−0.4	0.9	−0.9	0.1	−0.1	−0.4	−0.4
platinum	0.4	−0.3	0.5	−0.9	—	—	—	—
palladium	0.3	−0.1	0.3	−0.6	0.9	−0.8	—	—

a = exponent of hydrocarbon pressure.
b = exponent of deuterium pressure.

increase with decreasing metal–metal bond strength. This trend is shown in Figure 3 in which the heat of vaporization of the metal is used as a measure of the metal–metal bond strength. This is the correlation used by Schuit et al. (1961) in the sense that they correlate high activation energy for exchange with low metal–metal bond energy. This latter correlation probably originates in the following way. A high metal–metal bond energy is to be associated with a high metal–carbon and metal–hydrogen binding energy, this being an example of Pauling's semi-empirical bond energy relations (Pauling, 1960). Now, if we picture the transition state for the process in which dissociative adsorption at a C–H bond occurs, it is reasonable to suppose that the surface orbitals from adjacent metal atoms must interact with the $1s$ orbital of hydrogen and with the sp^3 carbon orbital,

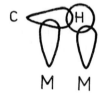

The greater the extent to which $1s$–metal and sp^3–metal overlap occurs

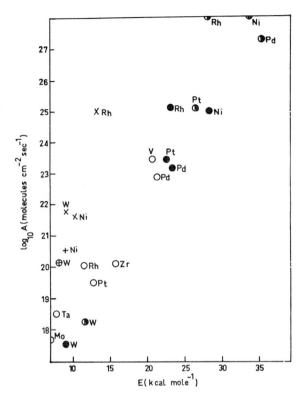

FIG. 2. Relation between activation energy (E) and frequency factor (A) for methane and ethane exchange with deuterium: ●, methane (CH_3D formation); ◑, methane (CH_2D_2, CHD_3, CD_4 formation); O, ethane; X, propane; +, isobutane.

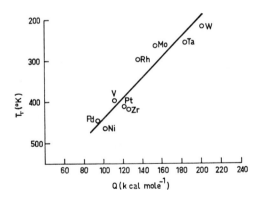

FIG. 3. Relation between T_r(°K, temperature for rate of adsorption/desorption to equal 2×10^{12} molecules cm^{-2} sec^{-1}) and Q (kcal $mole^{-1}$, energy of vapourization of the metal), for exchange of ethane with deuterium.

the lower can be the extent of $1s$–sp^3 overlap, provided the metal orbitals are of high electronegativity. Thus, if the binding energy for metal–carbon and metal–hydrogen is used as an index of the degree of overlap possible in the transition state, we may at least rationalize the observed behaviour.

In few reaction systems with hydrocarbons is the monodeutero-molecule the sole exchange product: that is, M, the mean number of hydrogen atoms exchanging initially per molecule is seldom exactly unity. For instance, with ethane M ranges from 1.15 on tantalum to 5.0 on rhodium. The extent to which M exceeds unity is an index of the occurrence of multiple exchange, which is the process by which more than one hydrogen atom may be exchanged in a molecule during one residence on the surface. A principal aim of deuterium exchange experiments is to interpret multiple exchange evidence in terms of the nature of the adsorbed species.

Multiple exchange can be strongly influenced by the nature of the metal film, particularly its thermal history, although the importance of this depends both on the metal and the hydrocarbon. For this reason it is convenient to discuss multiple exchange results under two headings: on "low temperature" films, by which we mean films deposited onto a substrate at about 0 °C and not subject to further treatment before the reactions, such films being essentially randomly polycrystalline; and on "high temperature" films, which have been deposited on a substrate at an elevated temperature, or else have been sintered at an elevated temperature before reaction.

1. Low Temperature Films

Consider the behaviour of methane and ethane as the main examples. Kemball (1951, 1953) has shown that exchange of methane over rhodium, nickel, platinum, tungsten and palladium gave CH_3D and CD_4 as the two principal initial products, with lesser amounts of CH_2D_2 and CHD_3: a similar result has been reported over a cobalt/thoria catalyst (Thompson et al., 1951). Examples of initial product distributions are shown in Table 7. Clearly CH_3D arises from $CH_{3(s)}$, while the more highly deuterated products must require the presence of more extensively dehydrogenated surface residues such as $CH_{2(s)}$, $CH_{(s)}$ and possibly $C_{(s)}$. It was found that CH_2D_2, CHD_3 and CD_4 are all produced with the same activation energy (cf. Table 5) and this is substantially higher than the activation energy for CH_3D: this has two consequences. First, it means that the formation of CH_2D_2, CHD_3 and CD_4 by independent desorption processes from, respectively, $CH_{2(s)}$, $CH_{(s)}$ and $C_{(s)}$ is highly improbable; second, the formation of the more highly

TABLE 7

Initial distributions of products from the exchange of
methane on metal films

catalyst	Ratio of D_2/CH_4 in initial reaction mixture	percentages of isotopic species				
		d_1	d_2	d_3	d_4	M
Rh, 162 °C	1.0	21	5	29	45	2.98
Rh, 172 °C	3.4	21	0	16	63	3.21
Ni, 237 °C	0.75	12	3	24	61	3.34
Pt, 259 °C	1.0	36	12	25	27	2.43
W, 151 °C	1.0	76	1	9	14	1.61
Pd, 254 °C	1.0	95	1	2	2	1.11

deuterated species by a rapid interconversion between $CH_{3(s)}$ and
$CH_{2(s)}$ is unlikely since if this were so, the amount of CD_4 should increase relative to CH_2D_2 or CHD_3 as the temperature is increased, and
this was not found. Kemball (1953) suggested that the highly deuterated
methanes are formed by a rapid interconversion between $CH_{2(s)}$ and
$CH_{(s)}$. Now the fact is that CH_2D_2, CHD_3 and CD_4 are produced
initially in proportions roughly corresponding to statistical equilibrium.
In other words, any molecule undergoing this part of the reaction leaves
the surface with a random collection of deuterium and hydrogen atoms:
however, the apparent pool from which these atoms are drawn cannot
have a composition simply equal to that of the ordinary chemisorbed
"hydrogen" which under initial conditions must be virtually entirely
deuterium, and CD_4 would then be the only product in addition to
CH_3D. Furthermore, the proportion of CD_4 increased as the D_2/CH_4
reactant ratio increased.

Kemball (1959) has suggested that CH_2D_2 and CHD_3 arise from inter-
residue hydrogen transfer reactions, these being effectively direct, that
is, without proceeding via chemisorbed hydrogen which could be
rapidly equilibrated with gas phase "hydrogen". Rather more direct
evidence for inter-residue hydrogen transfer reactions involving $CH_{3(s)}$
and $CH_{2(s)}$ residues has been obtained by Anderson and McConkey
(1968) who generated these residues by adsorption of the corresponding
alkyl halides. The mechanism by which such a hydrogen transfer
occurs may involve two residues with a third adsorption site between

them. For instance, a hydrogen transfer which overall may be written as

$$2 \ CH_{2(s)} \rightarrow CH_{(s)} + CH_{3(s)} \qquad\qquad (III)-(3)$$

may pass through the configuration

The pattern of initial exchange products from ethane falls into one of three classes depending on the metal: (a) C_2H_5D as the major product with monotonically decreasing proportions of the more highly deuterated molecules, (b) C_2D_6 as the major product with monotonically decreasing proportions of the less deuterated molecules, (c) a combination of (a) and (b). Table 8 gives a typical distribution of each class,

TABLE 8

Observed and calculated initial product distributions
from the exchange of ethane on metal films

catalyst		percentages of isotopic species						
		d_1	d_2	d_3	d_4	d_5	d_6	M
molybdenum	observed, $-12°$ C	82	14	3	0.7	0	0	1.23
	calc., P $= 0.25$	80	17	2.5	0.3	0	0	1.23
palladium	observed, 177° C	5	6	8	11	19	51	4.80
	calc., P $= 18$	5.3	7.3	8.7	11.2	16.9	50.6	4.79
platinum	observed, 142° C	20	17	12	10	15	26	3.61
	calc. $\begin{cases} P_1 = 2.0\ (54\%) \\ P_2 = 20.0\ (46\%) \end{cases}$	20	18	14	11	11	26	3.53

Mean deuterium numbers for some other metals:
  W(-54 °C), M $= 1.30$; Ta(-14 °C), M $= 1.15$; Zr(145 °C), M $= 2.3$;
  V(125 °C), M $= 2.6$; Rh(26 °C), M $= 5.0$

together with a summary of the behaviour of other metals in terms of their M values. Unlike the behaviour of methane, with ethane the initial value of M decreases with increasing deuterium pressure. Since deuterium is a kinetically inhibiting gas (cf. Table 6) the chance of obtaining a free site is diminished at higher deuterium pressures, and thus one concludes that multiple exchange requires more sites over and above the number required for single point adsorption as $C_2H_{5(s)}$. Anderson and Kemball (1954a) explained multiple exchange of ethane by repeated interconversions between $C_2H_{5(s)}$ and $C_2H_{4(s)}$, that is

$$C_2H_{6(g)} \rightleftarrows \underset{\underset{M}{|}}{C_2H_5} \rightleftarrows \underset{\underset{M}{|}\ \underset{M}{|}}{H_2C\!-\!CH_2}$$

where the chemisorbed species are here written without regard to iso-topic composition. The interconversion between $C_2H_{5(s)}$ and $C_2H_{4(s)}$ may occur more than once before the molecule is finally desorbed, and the more often it occurs, the greater is the chance that on desorption the ethane will be highly deuterated. We shall refer to this mechanism as 1–2 multiple exchange. The experimental M values for ethane are inde-pendent of temperature, and this, or only a small temperature depend-ence of M, is a characteristic of multiple exchange in other systems that proceed by an analogue of reaction (III)–(4).

Theoretical product distributions were calculated for this model. This was done in terms of a parameter P, this being defined as

$$P = \frac{\text{chance of } ''C_2H_{5(s)}'' \to ''C_2H_{4(s)}'' \to ''C_2H_{5(s)}''}{\text{chance of } ''C_2H_{5(s)}'' \to ''C_2H_{6(g)}''} \qquad \text{(III)–(5)}$$

and P depends on the catalyst. Table 8 gives theoretical distributions for comparison with experiment. Distributions that have a minimum between C_2H_5D and C_2D_6 (type c distributions above) cannot be generated by a single value of P, and it was suggested that these arose from the simultaneous occurrence of two simple distributions, possible on different crystal faces of the film. Further discussion of this point will be given after a description of exchange on "high temperature" films.

It has been noted previously by Anderson (1957a) and by Addy and Bond (1957) that the extent of multiple exchange of the ethane type is generally higher on f.c.c. metals than on b.c.c. metals, and it also in-creases with increasing % d-character of the intermetallic bond. It is possible, as Kemball (1959) has suggested, that these correlations arise from a common origin, namely the dependence of multiple exchange on the available lattice spacings at the surface, but in fact it is difficult convincingly to demonstrate such a correlation. One of the problems may well be the extent to which this multiple exchange is influenced by the presence of strongly bound self-poisoning residues. If multiple exchange occurs by a mechanism which requires the conversion of (say) a singly to a multiply bound residue (of which 1–2 multiple exchange would be a special case), one would expect poisoning to bring about a reduction in the value of M, since it will reduce the chance of finding the necessary extra adsorption sites. Such a reduction in M has been

observed by Gault and Kemball (1961) for the exchange of n–hexane on
rhodium films which had been subjected to pre-treatment with reactant
(Table 9). This effect is clearly substantial and it may well be that a

TABLE 9

Effect of poisoning on M-value for n-hexane exchange on
rhodium films at 0 °C

pre-treatment mixture	temperature of pre-treatment (°C)	poisoning factor*	M
$C_6H_{14} + D_2$	0	1	7.4
$C_6H_{14} + D_2$	50	6	2.4
$C_6H_{14} + D_2$	100	23	3.7

* Ratio of unpoisoned to poisoned exchange rate.

reliable and realistic correlation between multiple exchange and a
characteristic property of the metal will await exchange measurements
made at very low coverage, probably in a regime where a hydrocarbon
adsorption steady state has not been reached.

The exchange of propane and isobutane has been studied over a range
of metal films by Kemball (1954a) and Gault and Kemball (1961).
Over tungsten the product distributions were similar to those obtained
with ethane in that the monodeuteromolecule was the main product:
exchange followed equation (IV)–(8) of Chapter 7, indicating that all
the hydrogen atoms in the molecule were equally susceptible to exchange. Over nickel, both exchange reactions occurred at about room
temperature and below, and the monodeuteromolecule was again the
main product: self-poisoning was evident but was not so severe as with
ethane. The exchange data did not fit equation (IV)–(8) and by the use
of the methods outlined in section IV of Chapter 7 for exchange reactions of molecules with hydrogens of unequal reactivity, it was shown
that on nickel the relative hydrogen reactivities are in the ratio
90:30:1 for 3°:2°:1° hydrogen atoms. Over rhodium, both molecules
exchange to give the perdeuterocompound as the major product, and on
this catalyst all the hydrogen atoms in the molecules were equally
susceptible to exchange. The initial product distributions over rhodium
films are given in Table 10.

It is interesting that an enhanced reactivity of the 2° relative to the
1° hydrogen in propane was found by Kauder and Taylor (1951) on
platinized platinum. Exchange of propane on palladium films gave a

TABLE 10

Initial product distributions from propane and isobutane
on rhodium films

| | percentages of isotopic species | | | | | | | | | | |
	d_1	d_2	d_3	d_4	d_5	d_6	d_7	d_8	d_9	d_{10}	M
propane ($-24°$ C)	6	8	10	8	5	14	22	27	0	0	5.6
isobutane ($-27°$ C)	9	12	10	12	6	4	5	7	11	24	6.0

distribution of initial products rather similar in character to that obtained from ethane (Gault and Kemball, 1961).

The important question which remains is whether with larger molecules than ethane there are other mechanisms which may also operate. The theoretical distributions calculated for propane by Kemball and Woodward (1959) for a 1–2 multiple exchange model are generally in satisfactory agreement with experiment, although in some cases (supported palladium, rhodium and platinum) a combination of two simple distributions is required. However, the distributions obtained on rhodium films have subsidiary maxima in the region of the d_3– and d_4– species, and these cannot be satisfactorily accounted for in this way. From Kemball and Woodward's calculations it appears that the subsidiary maxima could result from a modest enhancement in the reactivity of 2° and 3° hydrogens relative to 1° hydrogens, an enhancement that may be insufficient to be detectable when tested for by the fit of the exchange data to equation (IV)–(8) of Chapter 7. Furthermore, with a molecule containing more than two carbon atoms, the possibility arises that multiple exchange may involve the formation of a residue bonded to the surface at more than two carbon atoms. If this is σ–bonded, as for the ethane multiple exchange intermediate of reaction (III)–(4), then the residue from propane will be (III)–(6a): however, evidence considered in a later section from the exchange behaviour of alicyclic hydrocarbons leads to the conclusion that, at least on some catalysts but principally on palladium, an adsorbed π–allylic species is the most likely multiple exchange intermediate (III)–(6b),

$$\begin{array}{ccc} CH_2 & CH & CH_2 \\ | & | & | \\ M & M & M \end{array} \qquad \text{(III)–(6a)}$$

$$\begin{array}{c} CH_2 \text{---} CH \text{---} CH_2 \\ \downarrow \\ M \end{array} \qquad \text{(III)–(6b)}$$

while a subsequent section also shows that under certain conditions, very extensively dehydrogenated surface species may take a direct part in multiple exchange. For the present, we merely note that if multiple exchange proceeds via (III)—(6a), the process may be written

$$C_3H_{8(g)} \rightleftarrows C_3H_{7(s)} \rightleftarrows C_3H_{6(s)} \rightleftarrows C_3H_{5(s)} \qquad \text{(III)–(7)}$$

This mechanism requires two parameters for its definition, P defined quite analogously to expression (III)–(5), and X defined by

$$X = \frac{\text{chance of } ''C_3H_6''_{(s)} \rightarrow ''C_3H_5''_{(s)} \rightarrow ''C_3H_6''_{(s)}}{\text{chance of } ''C_3H_6''_{(s)} \rightarrow ''C_3H_7''_{(s)}} \qquad \text{(III)–(8)}$$

Kemball and Woodward's results show that such a mechanism can produce a distribution with maxima at C_3H_7D and C_3D_8, such as is required for supported rhodium and platinum.

The possibility of multiple exchange via adsorption in the 1–3 mode, that is

$$
\begin{array}{ccc}
 & CH_2 & \\
 & \diagup \quad \diagdown & \\
CH_2 & & CH_2 \\
| & & | \\
M & & M
\end{array}
$$

appears to be slight at the lower end of the exchange temperature range, since work with neopentane (Kemball, 1954b) shows that multiple exchange by this mechanism is considerably more difficult than 1–2 multiple exchange, in agreement with the conclusions of Burwell and Briggs (1952) who worked with more complex hydrocarbons such as 3,3–dimethylhexane and 2,2,3–trimethylbutane over a nickel/kieselguhr catalyst. They concluded that multiple exchange could not be "propagated" past a quaternary carbon atom. That is, in 3,3–dimethylhexane for instance

$$
\begin{array}{c}
CH_3 \\
| \\
CH_3 - CH_2 - C - CH_2 - CH_2 - CH_3 \\
| \\
CH_3
\end{array}
$$

multiple exchange could occur either with the ethyl or the propyl group, but it is improbable for the molecule to transfer its point of attachment to the surface from (say) the ethyl to the propyl group within a single residence time: at the same time there was negligible exchange in the quaternary methyl groups. Analogous results were obtained by Rowlinson et al. (1955) with 3,3–dimethylpentane and

1,1–dimethylcyclohexane over various nickel catalysts. However, at higher temperatures, adsorption in the 1–3 mode becomes more favourable particularly on platinum, and this mode of adsorption becomes eventually an important intermediate for the skeletal isomerization of saturated hydrocarbons, a topic that is dealt with in detail in a later section of this Chapter.

For the present it suffices to say that for molecules more complex than ethane, the number of physically realistic variables which must be taken into account in the construction of theoretical exchange product distributions from an assumed model is such that, apart from the complexity of the calculations, the number of disposable parameters becomes so large that the value of the calculation as a means of validating the proposed model becomes quite dubious.

Although the relative strengths of adsorption of the reactants may be inferred from the pressure dependence data such as is given in Table 6, this may be seen more directly from a calculation of the dependence of the proportion of free surface sites, θ_F, on the gas pressures. Let us assume that the two slow processes in methane exchange are

$$\begin{array}{ccc} CH_4 & & CH_3 \quad H \\ | & \rightleftarrows & | \qquad | \\ M & M & M \qquad M \end{array} \qquad \text{(III)–(9a)}$$

$$\begin{array}{ccc} CH_3 & & CH_2 \quad H \\ | & \rightleftarrows & \| \qquad | \\ M & M & M \qquad M \end{array} \qquad \text{(III)–(9b)}$$

remembering that once $CH_{2(s)}$ has been formed, interconversion with $CH_{(s)}$ will rapidly replace H by D in the residue. For reaction (III)–(9a) proceeding from left to right, and assuming the surface concentration of physically adsorbed methane, $\overset{\displaystyle CH_4}{\underset{\displaystyle M}{|}}$, is directly proportional to pressure

$$p_{CH_4} \, \theta_{F,Ni}^2 \propto (p_{CH_4}^{1.0} \, p_{D_2}^{-0.9}) \qquad \text{(III)–(10)}$$

where the expression $(p_{CH_4}^{1.0} \, p_{D_2}^{-0.9})$ is the experimental pressure dependence of the rate of reaction (III)–(9a) on nickel. Thus

$$\theta_{F,Ni} \propto (p_{CH_4}^{0} \, p_{D_2}^{-0.45})$$

Similarly, for reaction (III)–(9b)

$$(p_{CH_4} \, p_{D_2}^{-0.5}) \, \theta_{F,Ni} \propto (p_{CH_4} \, p_{D_2}^{-1.4}) \qquad \text{(III)–(11)}$$

where the term $(p_{CH_4} \, p_{D_2}^{-0.5})$ is the pressure dependence of θ_{CH_3}, obtained assuming established gas/surface equilibria as described in section II of Chapter 7 and the term ($p_{CH_4}\, p_{D_2}^{-1\cdot4}$) is the experimental pressure dependence of the rate of reaction (III)–(9b). The θ_F obtained in this way for a given system should be the same for (III)–(9a) and (III)–(9b), and this is a useful criterion for the selection of mechanisms. In this way the following values of θ_F have been obtained by Kemball (1953, 1959).

$$\theta_{F,\,Ni} \propto (p_{CH_4}^0\, p_{D_2}^{-0\cdot45})$$

$$\theta_{F,\,Rh} \propto (p_{CH_4}^0\, p_{D_2}^{-0\cdot38})$$

$$\theta_{F,\,W} \propto (p_{CH_4}^{-0\cdot13}\, p_{D_2}^{-0\cdot2})$$

$$\theta_{F,\,Pt} \propto (p_{CH_4}^{-0\cdot28}\, p_{D_2}^{-0\cdot18})$$

$$\theta_{F,\,Pd} \propto (p_{CH_4}^{-0\cdot55}\, p_{D_2}^{-0\cdot05})$$

In an entirely analogous way for ethane exchange, if one assumes the mechanisms

$$\begin{array}{ccccc} C_2H_6 & & C_2H_5 & H & \\ | & \rightleftharpoons & | & | & \qquad\qquad \text{(III)–(12a)}\\ M & M & M & M & \end{array}$$

$$\begin{array}{cccccc} C_2H_5 & & & H_2C\!\!-\!\!-\!\!CH_2 & H & \\ | & & \rightleftharpoons & |\qquad\quad| & | & \qquad \text{(III)–(12b)}\\ M & M & M & M\qquad M & M & \end{array}$$

we obtain, using the experimental pressure dependence data of Anderson and Kemball (1954a)

$$\theta_{F,\,Rh} \propto (p_{C_2H_6}^{-0\cdot05}\, p_{D_2}^{-0\cdot36})$$

$$\theta_{F,\,Pd} \propto (p_{C_2H_6}^{-0\cdot04}\, p_{D_2}^{-0\cdot40})$$

It is to be emphasized that these expressions for θ_F refer only to the surface entities active in the exchange reactions. Nothing is said about strongly absorbed residues derived from the hydrocarbons but which take no direct part in the exchange reactions.

2. High Temperature Films

The properties of evaporated films prepared, or subjected to treatment at high temperatures, have been discussed in Chapter 7. Anderson and Macdonald (1969a) have compared the exchange of propane, ethane and methane over high temperature films with that obtained on random polycrystalline low temperature films. Some comparative data are given in Table 11. These films were prepared under ultrahigh

TABLE 11

Initial product distributions for exchange of propane, ethane and
methane on high temperature nickel films

		percentages of isotopic species								
		d_1	d_2	d_3	d_4	d_5	d_6	d_7	d_8	M
propane (120 °C)	(111)	0	19	0	0	0	0	0	81	6.86
propane (150 °C)	(111)	0	4	0	7	0	4	5	80	7.35
propane (160 °C)	(100)	0	37	0	0	0	10	0	53	5.58
propane (127 °C)	sintered random poly-crystal, deposited 0 °C, sintered 350 °C)	40	6	2	9	0	5	5	33	4.23
propane (162 °C)		10	4	3	4	0	6	17	56	6.46
propane (0 °C)	low temperature random polycrystal, deposited at 0 °C	100	0	0	0	0	0	0	0	1.00
ethane (170 °C)	(111)	0	48	0	11	5	36			3.81
ethane (189 °C)	(100)	0	53	0	6	0	41			3.76
ethane (168 °C)	sintered random polycrystal, deposited 0 °C, sintered 340 °C	45	10	0	0	0	45			3.34
ethane (0 °C)	low temperature random polycrystal, deposited 0 °C	98	2	0	0	0	0			1.04
methane (247 °C)	(111)	41	0	0	59					2.77
methane (268 °C)	(100)	31	0	0	69					3.07
methane (242 °C)	low temperature random polycrystal, deposited 0 °C)	37	0	1	62					2.88
methane (251 °C)		27	3	8	62					3.05

vacuum conditions so that some assessment of the effect of possible surface contamination of films used in earlier work is also possible.

Low temperature nickel and platinum films prepared under UHV conditions behaved the same as earlier low temperature films prepared under ordinary vacuum conditions, and thus there is no evidence that the earlier data are subject to serious surface contamination problems. On the other hand, low temperature UHV chromium films, prepared from a metal specimen of exceptional purity, proved to be inert for ethane exchange, and the only reaction to be observed was hydrocracking with the formation of deuteromethanes above about 250 °C. This is to be compared with the previous result of Anderson and Kemball (1954a) showing ethane exchange in the region of 140 °C. Thus, the behaviour of clean chromium towards ethane resembles that of iron, and probably for similar reasons. The earlier chromium films were undoubtedly contaminated, probably by oxide, so that the exchange reaction occurred on a chromium oxide surface. Chromium is a notoriously difficult metal to obtain in a very high state of purity, so this result may not be typical, but it does point to the need for stringent control in vacuum conditions during film preparation, control that has only too frequently been lacking in the past.

Reactions of ethane and propane over nickel films lead to the following conclusions. The low temperature (about 0 °C) exchange is absent on high temperature films, and the reactions first become established above about 120 °C. On (111) or (100) nickel surfaces (prepared at about 350 °C), the retention of a mono-adsorbed residue under exchange conditions is unlikely because of the ready formation of a double bond between the carbon and the nickel. Surface sites for the retention of a mono-adsorbed residue are present on low temperature random poly-crystalline films, but they are thermally unstable in that they are much reduced by sintering, and are largely absent on (111) or (100) surfaces. It seems impossible to explain the product distributions obtained on high temperature nickel films by reactions (III)–(4) or (III)–(7). This is most evident at the lower end of the reaction temperature range where the d_2- and perdeuteromolecules are the dominant products. We may also note from Table 11 the quite strong dependence of M on exchange temperature, and this is a feature that is not a characteristic of mechanism (III)–(4). It is interesting to compare these data with those obtained by Thompson *et al.* (1951) for ethane exchange at 183 °C over a cobalt/thoria catalyst. Here the product distribution (essentially initial) consisted mainly of $C_2H_4D_2$ and CD_6, with C_2H_5D apparently absent. Anderson and Macdonald (1969a) suggested that an important contribution to this exchange is the direct desorption from a variety of dehydrogenated surface residues, while the very highly deuterated products may be formed by a "methane type" mechanism at the individual carbon atoms. One may speculate that the thermally unstable sites are features such as steps, terrace edges, or crystal corners, and possibly self-adsorbed metal atoms.

With methane, high temperature and low temperature nickel films gave the same initial exchange product distributions. This marked difference in behaviour compared to ethane and propane may result from differences in C–H bond strengths for successive dissociations. If, as the existing methane exchange mechanism proposes, there is a substantial energy barrier for further dehydrogenation of $CH_{3(s)}$ to $CH_{2(s)}$, the result is that $CH_{3(s)}$ is retained as an important surface species. It is also interesting to see from Table 11 that the methane product distributions are very near to ideal for the model given previously in that, particularly at the lower end of the reaction temperature range, the products are almost exclusively CH_3D and CD_4: that is, the complication of the presence of CH_2D_2 and CHD_3 and ascribed to an inter-residue hydrogen transfer, appears to be absent.

It is clear that the behaviour of individual molecules can be governed by quite subtle bond energy differences, and that at this stage behavioural predictions are hazardous. The behaviour of nickel should not be extrapolated to other metals, since both high and low temperature platinum films have given the same product distribution for ethane (Anderson and Macdonald, 1969b). This was essentially the same distribution to that previously reported by Anderson and Kemball (1954a).

The participation of extensively dehydrogenated residues in ethane exchange is not a new suggestion. However, the manner of participation suggested by Anderson and Macdonald is considerably different from previous suggestions. For instance, Miyahara (1956) has proposed a model for ethane exchange and this uses the assumption that all of the following surface species exist, and that rapid interconversion exists between them

$$C_2H_5{}_{(s)} \rightleftarrows C_2H_4{}_{(s)} \rightleftarrows C_2H_3{}_{(s)} \rightleftarrows C_2H_2{}_{(s)} \qquad (III)–(13)$$

While the structures for $C_2H_5{}_{(s)}$ and $C_2H_4{}_{(s)}$ are the same as occur in mechanism (III)–(4), $C_2H_3{}_{(s)}$ and $C_2H_2{}_{(s)}$ were assumed to be

$$\begin{array}{cc} CH_2 & \\ \parallel & \\ CH & HC=CH \\ | & | \quad | \\ M & M \quad M \\ (A) & (B) \end{array}$$

which are quite different from the structures suggested as participating in ethane exchange over high temperature nickel films, and which contained carbon–metal double bonds and carbon–carbon single bonds. The possibility of the formation of $C_2H_4{}_{(s)}$ and $C_2H_2{}_{(s)}$ from ethane adsorption is readily examined thermodynamically with the general result that an appreciable surface concentration of $C_2H_2{}_{(s)}$ is possible at 155 °C provided the free energy of adsorption of C_2H_2 is at least 25 kcal mole^{-1} more favourable than for C_2H_4 adsorption. This is probably more reasonable if the adsorbed species contain carbon–metal double bonds rather than structures (A) and (B).

C. LARGER HYDROCARBONS

1. Aliphatics

The exchange of n-pentane and n-hexane has been studied over evaporated films of palladium and rhodium by Gault and Kemball

(1961). The pattern of exchange products on palladium generally resembles that obtained from propane. It also closely resembles the product distributions found by Burwell *et al.* (1957) for n-hexane exchange on a supported palladium catalyst. However, on rhodium films, in addition to the normal multiple exchange process there exists a specific process responsible for the introduction of two deuterium atoms, and which must involve a diadsorbed species that does not take part in the normal multiple exchange. This type of distribution is illustrated in Table 12. The exchange of n-heptane on supported nickel

TABLE 12

Initial product distributions from exchange of n-pentane
and n-hexane on rhodium films

	percentages of isotopic species														
	d_1	d_2	d_3	d_4	d_5	d_6	d_7	d_8	d_9	d_{10}	d_{11}	d_{12}	d_{13}	d_{14}	M
n-pentane (0 °C)	8.5	23.5	5.0	3.6	2.4	1.9	2.4	3.1	4.7	5.4	11.5	28.0			7.1
n-hexane (− 7 °C)	8.0	15.0	8.5	7.2	6.6	6.4	6.0	5.6	4.4	4.9	5.1	5.3	7.4	9.6	6.8
n-hexane (0 °C)*	24.5	47.5	12.0	5.0	4.2	3.0	2.3	1.5	0	0	0	0	0	0	2.4

* After poisoning by reaction of hexane and deuterium at 50 °C.

catalysts studied by Burwell and Briggs (1952) and Burwell and Tuxworth (1956) also revealed distributions with a peak at the d_2 compound. While it is possible, as Gault and Kemball have suggested, that this specific mechanism for d_2 product formation occurs by direct adsorption into the 1–2 mode, it is then not clear why this mechanism should be differentiated from the other multiple exchange process. It is obvious from Table 12 that on a poisoned rhodium surface the ordinary multiple exchange is strongly suppressed, but adsorption in the mode responsible for the d_2 product is much less affected. Clearly the latter requires fewer sites than the former, and it is possible that the specific diadsorption occurs in a 1–1 mode at a single carbon atom: rhodium is known to be a relatively good catalyst for the formation of 1–1 diadsorbed alkane, as shown for instance by its exchange behaviour with methane and neopentane. Moreover, it is significant that self-poisoning is severe with rhodium, and this probably involves the formation of extensively dehydrogenated residues with at least two hydrogens removed from more than one carbon atom. This exchange intermediate would then resemble those proposed by Anderson and Macdonald (1969a) to account for the pairwise introduction of deuterium into propane and ethane over high temperature nickel films. However, the proposal for a 1–1 diadsorbed exchange intermediate on rhodium is not without difficulty, since the ordinary multiple exchange component is of higher apparent activation energy by about 9 kcal mole^{-1}. Furthermore, it is not clear why the phenomenon should not then also be observed with propane. The problem may best be resolved by a direct determination of the positions of the deuterium atoms in the d_2 component.

2. *Alicyclics*

We shall first discuss unsubstituted alicyclic hydrocarbons. The exchange of cyclopentane and cyclohexane has been examined on a variety of metal films by Anderson and Kemball (1954b), while these hydrocarbons and cycloheptane and cyclooctane have been studied over supported palladium and nickel catalysts by Burwell *et al.* (1957) and Burwell and Tuxworth (1956). The main feature of the exchange of cyclopentane and cyclohexane on a metal such as palladium which is efficient for multiple exchange is the division of the N hydrogens in the molecule into two groups of N/2 hydrogens, as illustrated by the examples given in Table 13. The carbon ring in cyclopentane is nearly

TABLE 13

Initial product distributions for exchange of
cyclopentane and cyclohexane on palladium films

| | percentages of isotopic species | | | | | | | | | | | | |
	d_1	d_2	d_3	d_4	d_5	d_6	d_7	d_8	d_9	d_{10}	d_{11}	d_{12}	M
cyclopentane (25 °C)	18.2	9.6	3.8	5.3	17.7	0.3	0.8	5.8	6.2	32.3	0	0	5.91
cyclohexane (44°)	3.4	1.9	0.9	1.5	3.3	11.8	2.8	2.9	4.2	10.2	13.6	43.5	9.58

flat, and therefore the stereochemistry of this molecule is the most straightforward to discuss. One must obviously identify each group of five exchangeable hydrogens with one of the groups of five hydrogens on either side of the ring, and the product pattern shows that multiple exchange can occur within each group. It is easy to see how initial adsorption into a group on one side of the ring can give a multiple exchange distribution in the product range d_1–d_5. From the discussion given previously one would anticipate that in a molecule of this size it would be possible for the adsorbed multiple exchange intermediate to involve more extensive adsorption than that required for 1–2 multiple exchange. We shall for the moment proceed by assuming that multiple exchange occurs by a 1–2 σ-bonded intermediate, and the immediate problem is to understand on this model how the multiple exchange distribution in the range d_6–d_{10} arises.

It was originally suggested by Anderson and Kemball that a transition from group (d_1–d_5) to group (d_6–d_{10}) was effected via a higher energy 1–1 diadsorbed surface intermediate. Schrage and Burwell (1966) have criticized this proposal, arguing that since one would expect the 1–1 diadsorbed intermedate to be kinetically symmetrical with respect to transition either into the group (d_1–d_5) or group (d_6–d_{10}), the d_6 product should be a major rather than a minor one. However, this conclusion is only valid if excursion into the 1–1 diadsorbed structure were of comparable facility to multiple exchange itself, and this is certainly not a feature of the proposed model, which we believe remains valid. For example, from the temperature dependence of the distributions, an extra activation energy of about 5 kcal mole^{-1} is required for transition from the group (d_1–d_5) to group (d_6–d_{10}). The four-site adsorbed intermediate suggested by Schrage and Burwell to account for this sort of group-to-group transition does not appear to offer any special advantages, and indeed it cannot satisfactorily explain complete exchange within an isolated —CH_2—CH_2—CH_2— segment of a ring, as in bicyclo-3,3,0-octane (Roth et al., 1968).

The exchange behaviour of cyclohexane on a multiple exchange catalyst generally resembles that of cyclopentane. However, C_6, C_7 and C_8 alicyclic rings are buckled, and this introduces the possibility that the transition from one group of exchangeable hydrogens to the other can occur via an alternative 1–2 diadsorbed intermediate of lower energy than the 1–1 diadsorbed intermediate, and formed using two hydrogens lying trans to each other in the molecule. This possibility was pointed out by Anderson and Kemball (1954b) for cyclohexane. However, it is reasonable to suppose that maximum orbital overlap and thus maximum (σ-bond) binding energy would occur for a 1–2 diadsorbed multiple exchange intermedate in an eclipsed conformation, and Burwell et al. (1957) have pointed out that for trans hydrogens this first becomes possible without unreasonably large ring strain with cycloheptane, but it cannot occur with cyclohexane. These workers also obtained convincing evidence to support the contention that if multiple exchange involves a 1–2 diadsorbed intermediate, this must be an eclipsed rather than a staggered conformation, by a study of the exchange of bicyclo-2,2,1-heptane which contains within it both rigid eclipsed and staggered adjacent hydrogens, and more recent work with bycyclo-3,3,1-nonane by Schrage and Burwell (1966) is in agreement with this. Now the expected effect of being able to make a transition between the "two groups" of exchangeable hydrogens in this way (that is without an energy barrier) would be that multiple exchange could be propagated through the molecule and the division between the two groups would disappear. In fact, cycloheptane and cyclooctane show a negligible discontinuity in the exchange products between $N/2$ and $(N/2 + 1)$ in agreement with the model (Table 14). Of course even with

TABLE 14

Initial product distributions for exchange of cycloheptane and cyclooctane on supported palladium catalyst

| | percentages of isotopic species | | | | | | | | | | | | | | | | |
	d_1	d_2	d_3	d_4	d_5	d_6	d_7	d_8	d_9	d_{10}	d_{11}	d_{12}	d_{13}	d_{14}	d_{15}	d_{16}	M
cycloheptane (60 °C)	13	25	7.0	3.0	2.0	1.6	4.0	1.3	1.0	1.3	1.3	2.0	5.6	32			7.4
cyclooctane (65 °C)	7.4	11	2.4	1.6	1.0	1.0	0.7	1.2	1.2	1.2	1.2	1.5	1.5	2.2	8.0	57	12.0

cyclopentane or cyclohexane a reduction or elimination of this discontinuity can also be produced at a sufficiently high exchange temperature.

Recent work by Schrage and Burwell (1966) with supported palladium catalysts shows that even the exchange of cyclopentane is sensitive to catalyst history, although there is no separate knowledge about the details of surface topography which can be correlated with the observed reaction details. Some of the varied distributions found by Schrage and Burwell are given in Table 15, and attention is im-

TABLE 15

Some initial product distributions from the exchange of cyclopentane on supported palladium catalysts

| catalyst* | percentages of isotopic species | | | | | | | | | | |
	d_1	d_2	d_3	d_4	d_5	d_6	d_7	d_8	d_9	d_{10}	M
0.5% Pd on Al_2O_3, 50 °C	3	7	4	3	5	1	4	4	2	67	8.2
5% Pd on Al_2O_3, 40 °C	11	18	7	5	26	0	2	4	2	25	5.3
5% Pd on Al_2O_3, 40 °C	20	8	4	3	23	0	1	6	1	34	5.8

* The three catalysts are different batch preparations.

mediately drawn to the exceptional amounts of the d_2 and d_8 products which can be formed. With regard to the d_2 product we need add no further comment to those offered earlier. The origin of the d_8 material is far from clear, although it presumably requires some type of transition between the groups of hydrogen. The most important conclusion to come from this work and from a rather similar study by Burwell and

F

Tuxworth (1956) and Rowlinson *et al.* (1955) with supported nickel catalysts, is that there are a variety of sites with differing exchange properties, and the proportions of these sites also depend on the methods of catalyst preparation. In this respect, these supported catalysts are demonstratively more heterogeneous than even poly-crystalline evaporated films, and the factors include surface cleanliness and crystal size as well as the surface crystallographic features which are also potentially common to films. One would expect that crystal size itself, as distinct from surface crystallography and topography, would not become a significant catalytic variable unless the size be-comes sufficiently small directly to modify the electron energy or the proportion of corner sites. An extra complication with supported metal catalysts is the possibility that the exchange distributions (and kinetic parameters) are influenced by diffusion effects in the catalyst pores. The seriousness of this has recently been demonstrated by Dwyer *et al.* (1968), from whose work it also follows that it is improbable that any existing work with evaporated films is in serious error for this reason (cf. Chapter 7). We therefore feel that a detailed interpretation of much existing exchange data obtained with supported metal catalysts is insecurely based.

The exchange of a number of polymethylcyclopentanes and cyclo-hexanes has been studied over a range of metal films by Gault *et al.* (1962) and Rooney (1963). The nature of the problem introduced with this class of molecules may be well illustrated by comparing the ex-change products from 1,1,3,3-tetramethylcyclopentane and 1,1,3,3,4-pentamethylcyclopentane. Some examples are given in Table 16.

TABLE 16

Initial product distributions from exchange of polymethylcyclopentanes on metal films

system				percentages of isotopic species						
	d_1	d_2	d_3	d_4	d_5	d_6	d_7	d_8	d_9	M
C/Pd, (75 °C)	43	57	<1	0	0	0	0	0	0	1.6
C/Rh, (0 °C)	17.5	61	6	15.5	<0.1	0	0	0	0	2.3
C/Rh, (30 °C)	4.9	19.1	23.3	49.1	2.9	0.6	0.1	0	0	3.3*
D/Pd, (80 °C)	54.7	1.5	<1	<1	4.4	39.4	<1	0	0	3.1
D/Rh, (0 °C)	3.8	37	13.0	13.3	32	0.9	<0.2	0	0	3.4
D/Rh, (33 °C)	7.2	5.2	5.6	9.7	54.9	16.0	0.8	0.4	0.2	4.5*

* Extent of exchange greater than initial.
C = 1,1,3,3-tetramethylcyclopentane; D = 1,1,3,3,4-pentamethylcyclopentane.

First, consider the products to be expected from a 1–2 multiple exchange mechanism. In 1,1,3,3-tetramethylcyclopentane (C)

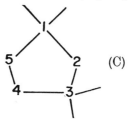

(C)

only 1–2 multiple exchange should be possible within each adjacent pair of hydrogens carried by carbon atoms 4 and 5: to propagate multiple exchange to all four hydrogens on carbons 4 and 5 would require a group-to-group transfer, via a 1–1 diadsorbed species, in a manner similar to that described for cyclopentane itself. The isolated methylene group at carbon 2 and the four methyl groups would be expected to be relatively inert. These expectations are in good agreement with the exchange results for (C) listed in Table 16. Thus it is seen that at 75 °C on palladium (multiple) exchange is almost entirely confined to an adjacent pair of hydrogens on one side of the ring at carbons 4 and 5. At this temperature there is negligible transfer into the other group of two hydrogens on the other side of the ring. On the other hand, on rhodium at 0 °C all four hydrogens on carbons 4 and 5 are exchanged, and this agrees with the known greater ease of formation of a 1–1 diadsorbed structure on rhodium than on palladium. On rhodium at 33 °C the small but appreciable amount of d_5 product indicates the extension of the exchange to other parts of the molecule, probably to the isolated methylene group. If we turn now to 1,1,3,3,4-pentamethylcyclopentane (D)

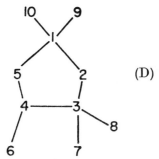

(D)

the pattern of expected multiple exchange products largely turns on the extent to which one would expect the hydrogens on carbon 6 to take part. So far as the ring hydrogens are concerned, the same 1–2 multiple exchange criteria as used for (C) would lead one to expect multiple exchange to be limited to the adjacent pair of hydrogens carried on carbons 4 and 5, with the possibility of exchange being propagated to the third hydrogen carried on carbons 4 and 5 if transition is made to the other side of the ring. The striking feature of the product distributions listed in Table 16 for (D) on palladium and rhodium is the extension of multiple exchange throughout the range d_1–d_6, and the conclusion is inescapable that in (D) multiple exchange can propagate throughout the six hydrogens which are made up from the two carried on carbon 5, the one on carbon 4, and the three methyl hydrogens on carbon 6. Now unlike the two gem-dimethyl structures at carbons 7, 8 and 9, 10 there is in fact no reason to suppose that methyl group 6 should not take part in a 1–2 multiple exchange process, except for one important proviso: because carbons 4 and 5 are part of a fairly rigid ring the accessible configurations available to carbons 4, 5 and 6 are more limited than would be the case for the three carbons in propane. Indeed, if multiple exchange among carbons 4, 5 and 6 involves more than a 1–2 mechanism, that is if a mechanism involving simultaneous adsorption at the three carbon atoms operates, then the three carbons must presumably acquire a co-planar (but not necessarily a linear) configuration, and Gault et al. (1962) suggest that this can be most readily achieved without undue ring strain by a rehybridization of the bonding orbitals of the carbon atoms. In somewhat more general terms, if part of a molecule

$$-C_aH_2-C_bH_2-C_cH_2-$$

is to be adsorbed as an exchange intermediate with C_a, C_b, and C_c each bonded to the surface, it is suggested that this can be achieved with greatest energetic economy not as a triadsorbed σ-bonded structure such as

$$\begin{array}{ccc} -CH & -CH & -CH- \\ | & | & | \\ M & M & M \end{array}$$

but as a π-bonded (π-allylic) structure

in which C_a, C_b, C_c, H_d, H_e, and H_f are all co-planar and each carbon atom is hybridized to sp^2. The π-molecular orbitals are constructed from the three p_z orbitals and bonding to the surface occurs by overlap of this π-system with a surface metal orbital (not shown in the figure) in rather the same way as in the formation of π-bonded complexes of the transition metals. If multiple exchange is then to occur, it will require rapid interconversion between the π-adsorbed species and one in which one of the carbon atoms has become "desorbed". Two suggestions have been made for this: Gault *et al.* suggested that the second species is 1–2 σ-diadsorbed, for instance

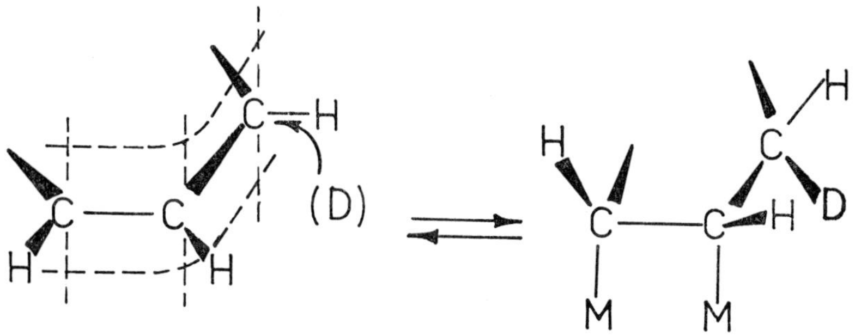

(III)–(14)

while Rooney suggests

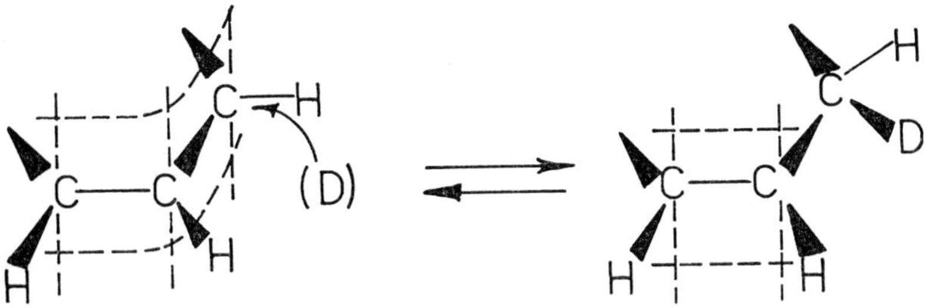

(III)–(15)

where both are π-bonded to a single metal atom.

A consequence of mechanism (III)–(15) would be an inability to exchange the final hydrogen carried on the middle carbon. With mechanism (III)–(14) all of the hydrogens can be exchanged, but to complete the exchange of both hydrogens on the centre carbon would require

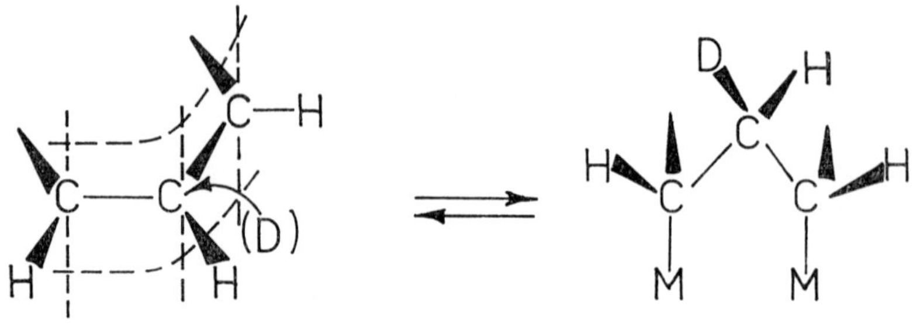

(III)–(16)

and since this involves a 1–3 σ-diadsorbed state would be expected to be relatively difficult at exchange temperatures. In (III)–(14), (III)–(15) and (III)–(16) attack by the deuterium has been written as occurring from beneath the plane of the carbon atoms, and this would plausibly be a chemisorbed deuterium atom in a surface crystallographic site. Alternatively, attack may occur from above the plane of the ring, possibly involving physically adsorbed molecular deuterium, the analogous process to reaction (III)–(14) then being

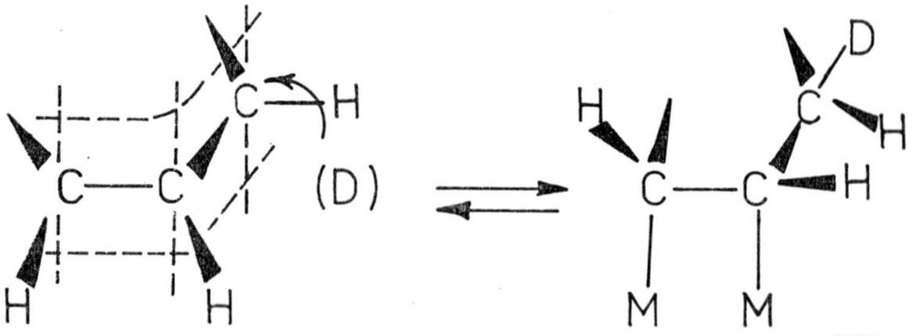

(III)–(17)

Process (III)–(14) and (III)–(17) have, in principle, specific stereo-chemical consequences. However, when multiple exchange occurs this becomes rapidly lost in the polydeutero product.

In the case of 1,1,3,3,4-pentamethylcyclopentane, the middle one of the three carbon atoms (carbon 4 in D) initially carries only a single hydrogen, so complete exchange of all the hydrogens carried on carbon

5, 4 and 6 would be possible by (say) process (III)–(14) alone. Rooney (1963) studied the exchange of 1,1,3,3-tetramethylcyclohexane

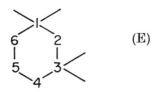

(E)

and found that on palladium at 42 °C multiple exchange was limited to the range d_1–d_5, in other words only five out of the six hydrogens carried on carbon atoms 4, 5 and 6 of (E) are exchanged. The problem that remains is whether this result is to be interpreted by the operation of reaction (III)–(15), or whether (III)–(14) and/or (III)–(17) operates and (III)–(16) is difficult. We believe that (III)–(15) is improbable because a consequence of adopting (III)–(15) is that the adsorbed species is never associated with more than a single adsorption site: it is then very hard to explain the inhibition of multiple exchange by increasing deuterium pressure. In the case of mechanism (III)–(14) an explanation for this is available.

The observation that the multiple exchange products from 1,1,3,3-tetramethylcyclohexane on palladium are limited to d_1–d_2 at low temperatures is however, an important result because it provides the strongest available piece of direct evidence in support of a π-allylic exchange intermediate. This result would be very difficult to explain on a pure σ-bonded mechanism. The remaining arguments are essentially ones of analogy, convenience and plausibility. The most useful general virtue of the proposal is that since the π-system interacts with only a single metal atom, its formation is attended with much less stringent stereochemical requirements than is the corresponding σ-bonded species. It should be noted that if several carbon atoms are simultaneously adsorbed in a molecule, it is not necessary to postulate that the π-system extends over them all. We believe that sterically favourable adsorption by σ-bonding is likely to be energetically more favourable and that the π-system will extend only as far as is needed to relax otherwise unacceptable steric limitations on pure σ-bond adsorption.

It is important to recognize that steric limitations can be important for π-bonded adsorption as well as for σ-bonded adsorption. This is well illustrated with molecules which are structurally locked in a configuration which will not allow carbon to rehybridize to sp^2. Thus, with adamantane

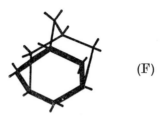

(F)

which has been studied by Schrage and Burwell (1966), rehybridization would be accompanied by a very large energetic penalty, (cf. Bredt's Rule) and a π-adsorbed intermediate is improbable. Moreover, with this molecule all adjacent pairs of hydrogens are rigidly held in staggered conformations, and thus an eclipsed conformation cannot be achieved for 1–2 σ-bonded adsorption either. As a consequence, multiple exchange cannot occur by either mechanism and this is in agreement with the product distribution observed for exchange on palladium. Schrage and Burwell also examined the exchange of bicyclo-3,3,1-nonane

(G)

which is structurally related to adamantane by the loss of a bridging methylene group. Again rehybridization at a bridgehead carbon (carbons 1 and 5) is energetically unfavourable, although not so unfavourable as in adamantane. Conformations with eclipsed hydrogens are also energetically unfavourable, but as Schrage and Burwell point out, the energetic penalty on rehybridization is the greater. As a consequence, this molecule probably undergoes multiple exchange via a 1–2 σ-bonded intermediate rather than by a π-bonded intermediate, and the conformation shown in (G) has been suggested to account for the observed main multiple exchange product at d_8, if all of the set of hydrogens marked with a dot can be approximately eclipsed and are thus exchangeable. Even if a π-system extended over only three adjacent carbon atoms in the set C_1–C_8 in (G), if this could be propagated through the bridgehead atoms one would expect the d_{14} product to be substantial; in fact it is absent.

In vic-dimethylcycloalkanes there exists the possibility of cis–trans isomerization. For instance, Gault et al. (1962) observed cis–trans isomerization during the exchange of cis- and trans-1,1,3,4-tetramethylcyclopentane (H) and (J), and of cis-1,2-dimethylcyclopentane (K)

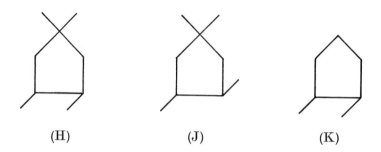

(H)   (J)   (K)

The exchange and isomerization reactions are kinetically related, and the greater the ability of a catalyst to promote multiple exchange the greater is the relative isomerization rate. In the isomerization process the molecule presumably passes through a conformation in which the two methyl-ring bonds are in the same plane, and this will be approximately the plane of the ring. The relation of the reaction to multiple exchange suggests that a π-bonded surface intermediate is involved. A generally similar conclusion for cis–trans isomerization with 1,2-dimethylcyclobutane follows from the work of Hillaire et al. (1967).

It is generally useful to draw a distinction between a π-allylic system which extends to a methyl ring substituent (exocyclic), and one which exists entirely within the ring (endocyclic). Because of the smaller steric restrictions to rehybridization, one expects the energy of an exocyclic π-allylic system to be the lower. The effect appears to be insignificant in the cyclopentane series, but it has important consequences in the cyclobutanes. Here the ring is already strained, and the lack of multiple exchange with 1,1-dimethylcyclobutane on palladium (M = 1.1) found by Rooney (1963) shows that an endocyclic π-allylic system is difficult to form. This result also shows that, again for reasons of strain, multiple exchange by a 1–2 σ-bonded intermediate cannot offer an alternative path. On the other hand, Hilaire et al. (1967) have shown that with methylcyclobutane multiple exchange occurs on palladium (M = 3.4) with the exchange distribution extending up to a peak at d_6,

F*

and this is the result to be expected for a π-allylic system extending over carbon atoms 5, 1 and 2

Cyclopropane possesses an inherent olefinic character, and it is reasonable to propose that this can be used in an important mode of adsorption

However, ring opening occurs so easily on film catalysts that deutero-cyclopropanes have only been observed on tungsten: on films of platinum, palladium, nickel, rhodium and iron the products are deutero-propanes, (Anderson and Avery, 1967b). On tungsten the d_1 compound is the dominant initial product, and this no doubt arises from a σ-mono-adsorbed species to which the π-adsorbed species is a precursor. Over palladium and rhodium films the distributions of deuteropropanes from ring opening closely resemble those from the exchange of propane itself. However, over tungsten and nickel the deuteropropanes from ring opening have a much higher mean deuterium number than the corresponding values for propane. This probably arises because the ring opening mechanism already produces a σ-diadsorbed species which can readily propagate multiple exchange.

Evidence for exchange by a π-bonded intermediate becomes more important relative to 1–2 σ-bonded multiple exchange as the temperature is increased, probably because of the increased ability to extend dissociative adsorption to further carbon atoms, and because of the energy barrier to the electronic reorganization required for rehybridization.

It should be emphasized that mere ability to promote multiple exchange is not of itself necessary evidence for the operation of a π-bonded intermediate. It is the nature of the product distributions that are (in suitable cases) of distinguishing importance. Thus the exchange distributions produced on rhodium films and given in Table 16 do not give clear support for a π-bonded intermediate, and probably result largely from σ-bonded multiple exchange. The following decreasing order of activity for multiple exchange by a π-bonded intermediate is found

$$Pd \gg (Pt, Ni) > Rh > W.$$

3. *Racemization*

It has been found by Burwell and his collaborators that, during exchange over nickel and palladium catalysts, (+)-3-methylhexane undergoes racemization. This was explained by Anderson (1957a) as resulting from a planar transition state when exchange at the asymmetric carbon atom occurred and which would require this carbon atom to be in sp^2 hybridization, with the entering deuterium being able to approach from the top or the bottom with approximately equal facility. While this mechanism is still possible there are two facts which also make it likely that on suitable catalysts, racemization proceeds via a π-bonded exchange intermediate. Exchange of (+)-3-methylhexane was studied over bulk nickel catalysts at fairly high temperatures (90–210 °C) where extensive multiple exchange occurred. Nickel is not a particularly good metal for the formation of a π-bonded exchange intermediate, which would account for the observed ratio of 1.6 for the rate of exchange to the rate of racemization. On supported palladium which is an efficient catalyst for exchange by a π-bonded intermediate, the ratio of the rate of exchange to racemization is unity. A useful mechanistic test would be to see if racemization occurred on tungsten or on a nickel film at low temperatures where exchange should be mainly limited to the hydrogen on the asymmetric carbon. The importance of this result is that it demonstrates that simultaneous reactions involving deuterium attack from "top" and "bottom" [e.g. reactions (III)–(14) and (III)–(17)] are likely to occur at least on these catalysts.

D. SIMPLE HYDRIDES OF N, O, AND S

The exchange of ammonia with deuterium over a wide range of metals gives NH_2D as the only initial product, so that $NH_{2(s)}$ is the only important surface entity for exchange (Kemball, 1952a). Table 17 lists the kinetic parameters and the reaction temperature range for each of the metals studied. Any mechanism such as operates with methane via a more extensively dehydrogenated adsorbed species to give a highly deuterated initial product is completely absent with ammonia. Consider the case of tungsten as an example. Ammonia exchange occurred at about 120 °C with a reaction mixture containing about 12 torr of each

TABLE 17

Kinetic parameters for ammonia exchange over metal films

	T_r (°C)†	E(kcal mole^{-1})	$\log_{10} A$ (A in molecules cm^{-2} sec^{-1})
platinum	-109	5.2	19.2
rhodium	-67	6.7	19.4
palladium	-30	8.5	19.9
nickel	25	9.3	19.1
tungsten	72	9.2	18.1
iron	105	12.5	19.5
copper	143	13.4	19.3
silver	172	14.1	19.2
zinc	$- - - - - - - - - - -$	inactive	$- - - - - - - - - -$

† T_r, temperature at which rate of adsorption/desorption equals 2×10^{12} molecules cm^{-2} sec^{-1}.

ammonia and deuterium. By comparison with the observed formation of tungsten nitride during the ammonia decomposition reaction, and the formation of $NH_{(s)}$ from ammonia adsorption on tungsten at about room temperature, we conclude that there is unlikely to be a thermo-dynamic restriction on the formation of $N_{(s)}$ or $NH_{(s)}$ during ammonia exchange: it is likely that both $N_{(s)}$ and $NH_{(s)}$ are present on the surface under exchange conditions, but they do not take significant part in the exchange process for kinetic reasons; that is, adsorption and desorption between ammonia and $NH_{2(s)}$ is much more rapid than any process involving $NH_{(s)}$ or $N_{(s)}$.

The activation energy for ammonia exchange is markedly increased by film sintering as shown in Table 18 and one must conclude that ener-

TABLE 18

Activation energies for ammonia exchange on
unsintered and sintered metal films

	Temperature Range (°C)	E(kcal mole^{-1})
unsintered iron films	111–232	12.5–15
sintered iron films	230–310	21
unsintered nickel and tungsten films	80–160	9
sintered nickel and tungsten films	230–310	15

Kemball (1952a); Singleton et al. (1951); Farkas (1936a).

getically favourable reaction sites are removed by thermal treatment of the film.

Water vapour exchanges over platinum wire in the temperature range 186–410 °C, with an activation energy that tends to increase with increasing temperature, the value is 13.5 kcal mole^{-1} above 250 °C (Farkas, 1936a). The exchange probably proceeds via the formation of $OH_{(s)}$ by dissociative adsorption, and this agrees with the analogous evidence for hydrogen sulphide exchange where $SH_{(s)}$ is the active exchange entity.

The pressure dependencies of the ammonia and water vapour exchange rates show some similarities. Over platinum wire, the rate of exchange of water vapor is proportional to $p_{D_2}^1$ and to $p_{H_2O}^0$, although the exponent of p_{H_2O} becomes negative at lower temperatures. Ammonia exchange on a low temperature random polycrystalline nickel film has the kinetic dependence $p_{NH_3}^{0.6} \, p_{D_2}^{0.5}$ (Kemball, 1952 a, b) while over a sintered nickel film it is $p_{NH_3}^0 \, p_{D_2}^{0.5}$ (Singleton et al., 1951). Over a low temperature iron film the dependence is $p_{NH_3}^0 \, p_{D_2}^{0.5}$ (Farkas, 1936a) and, over a sintered iron film $p_{NH_3}^{-ve} \, p_{D_2}^{0.5}$ (Singleton et al., 1951). These data may be compared with the result of Weber and Laidler (1951) for ammonia exchange over a synthetic ammonia catalyst, where the rate was proportional to

$$\frac{p_{NH_3} \, p_{D_2}^{\frac{1}{2}}}{(1 + a \, p_{NH_3})^2}$$

in which the denominator results from ammonia being an inhibiting gas and suggests that two sites are required in the rate controlling step. This sort of expression would often be written empirically as of the form $p_{NH_3}^x \, p_{D_2}^{\frac{1}{2}}$ with $x < 1$. Some of the possible kinetic schemes which are consistent with these data have been discussed by Kemball (1952a). Bearing in mind the mode of ammonia adsorption at high coverage described by reaction (II)–(6) the following exchange mechanism is probably the most reasonable:

$$NH_3 + D_{(s)} \rightarrow NH_{2(s)} + HD \qquad \text{(III)–(18a)}$$

$$NH_{2(s)} + D_2 \rightarrow NH_2D + D_{(s)} \qquad \text{(III)–(18b)}$$

and an exactly analogous process for water vapour

$$H_2O + D_{(s)} \rightarrow OH_{(s)} + HD \qquad \text{(III)–(19a)}$$

$$OH_{(s)} + D_2 \rightarrow HOD + D_{(s)} \qquad \text{(III)–(19b)}$$

If two sites are involved, (III)–(18) should be written

$$
\begin{array}{cc}
\mathrm{NH_3} & \mathrm{D} \\
| & | \\
\mathrm{M} & \mathrm{M}
\end{array}
\;\rightarrow\;
\begin{array}{cc}
\mathrm{NH_2} & \mathrm{HD} \\
| & | \\
\mathrm{M} & \mathrm{M}
\end{array}
\qquad \text{(III)–(20a)}
$$

$$
\begin{array}{cc}
\mathrm{NH_2} & \mathrm{D_2} \\
| & | \\
\mathrm{M} & \mathrm{M}
\end{array}
\;\rightarrow\;
\begin{array}{cc}
\mathrm{NH_2D} & \mathrm{D} \\
| & | \\
\mathrm{M} & \mathrm{M}
\end{array}
\qquad \text{(III)–(20b)}
$$

As is well known, it is extremely difficult to make an unambiguous decision about a catalytic mechanism from pressure dependence data alone. However, a comparison of the experimental pressure dependence exponents with the discussion in section IV of Chapter 7 makes reaction (III)–(18) more likely than the alternative

$$ \mathrm{NH_3} + 2\mathrm{S} \rightarrow \mathrm{NH_{2(s)}} + \mathrm{H_{(s)}} \qquad \text{(III)–(21a)} $$

$$ \mathrm{NH_{2(s)}} + \mathrm{D_{(s)}} \rightarrow \mathrm{NH_2D} + 2\mathrm{S} \qquad \text{(III)–(21b)} $$

The experimental exponents of the ammonia and water vapour pressures clearly show that these molecules are generally strongly adsorbed so that if (III)–(18a) is, for instance, rate controlling, the ammonia molecule is unlikely to be merely physically adsorbed, but adsorption as $\overset{\mathrm{NH_3}}{\underset{\mathrm{M}}{|}}$ is probable. Farkas (1936b) showed that, although over iron and platinum the rate of the parahydrogen conversion is reduced by ammonia and water vapour respectively, it is in each case more rapid than the exchange reaction. Consequently, despite the strong adsorption of the ammonia and water vapour, there is still opportunity for the surface to remain in equilibrium with the isotopic composition of the gas phase "hydrogen".

The incorporation reaction with the formation of metal sulphide would make it difficult in most cases to follow the exchange of hydrogen sulphide with deuterium on a metal surface *per se*. However, this exchange has been studied over powdered catalysts of molybdenum and tungsten sulphides (MoS_2 and WS_2) (Wilson *et al.*, 1962). On WS_2 the H_2/D_2 exchange rate was faster than the rate of hydrogen sulphide exchange while on MoS_2 the rates were comparable. It is reasonable to assume that on neither catalyst is chemisorbed hydrogen in serious disequilibrium with the gas phase H_2/D_2 composition, except in the initial adsorption stage of the reaction when strong dissociative sulphide adsorption gives a surface rich in chemisorbed $H_{(s)}$. It is not known if the initial reaction product is HDS or D_2S, but recalling that in exchange deuterium is present to excess, and adsorption data suggest the possibility of some $SH_{(S)}$, the former seems more likely with exchange proceeding mainly via $SH_{(S)}$. The rate of exchange is proportional to $p_{H_2S}^0 p_{D_2}^1$ on WS_2 and to $p_{H_2S}^{-0.3} p_{D_2}^1$ on MoS_2, and other kinetic parameters are contained in Table 19. The pressure dependences are consistent with

TABLE 19

Kinetic parameters for the exchange of sulfides on
MoS_2 and WS_2

	Adsorption/Desorption on MoS_2			Adsorption/Desorption on WS_2		
	Temperature Range (°C)	E‡	$\log_{10} A$‡‡	Temperature Range (°C)	E‡	$\log_{10} A$‡‡
H_2S	150–200	19.1	20.9	40–100	9.0	15.8
CH_3SH	150–190	15.0	19.1	40–100	9.1	15.7
C_2H_5SH	150–280	13	17.3	110–215	8.0	14.7

‡ E, activation energy, kcal mole $^{-1}$; ‡‡ A, frequency factor, molecules cm^{-2}sec^{-1}.

a reaction scheme analogous to (III)–(18),

$$
\begin{array}{cc}
H_2S & D \\
| & | \\
M & M
\end{array}
\rightarrow
\begin{array}{cc}
HS & HD \\
| & | \\
M & M
\end{array}
\qquad \text{(III)–(22a)}
$$

$$
\begin{array}{cc}
HS & D_2 \\
| & | \\
M & M
\end{array}
\rightarrow
\begin{array}{cc}
HDS & D \\
| & | \\
M & M
\end{array}
\qquad \text{(III)–(22b)}
$$

and again the requirement of strong hydrogen sulphide adsorption means that if (III)–(22a) is rate controlling, the reactant is likely to be adsorbed as $\overset{H_2S}{\underset{M}{|}}$. However, it has been pointed out that a scheme analogous to (III)–(21) can be made to fit the observed pressure dependence data if there exists an established equilibrium between $SH_{(S)}$ and $S_{(S)}$ (Wilson et al., 1962).

In all of these exchange experiments with ammonia, water vapour and hydrogen sulphide, the pressure dependence data are not sufficiently accurate to allow a firm decision as to whether the reactants are competing for the same or different surface sites, bearing in mind that even with competition for the same sites, the exponent on the inhibiting gas can be in the region of zero if the strength of adsorption is not sufficiently high to push the exponent to a more negative value.

E. SATURATED DERIVATIVES CONTAINING N, O, S AND Cl FUNCTIONS

1. *Molecules with an X–H function*

We include in this section primary and secondary amines, alcohols, mercaptans. These molecules show a number of points in common, and it is convenient to consider their behaviour in terms of the C-, N-, O- and S-functional groups, and by reference to the comparative behaviour of the saturated hydrocarbons, ammonia, water vapour, and hydrogen sulphide.

Platinum and palladium which are inefficient catalysts for the exchange of saturated hydrocarbons catalyse the exchange of N-hydrogens exclusively in methylamine and dimethylamine (Kemball and Wolf, 1955). On the other hand, tungsten which is efficient for the exchange of saturated hydrocarbons but inefficient for ammonia, catalyses the exchange of the C-hydrogens for preference. Films of nickel and iron catalyse simultaneous exchange of both C- and N-hydrogens. By comparison with the amine behaviour, methyl, ethyl, isopropyl and t-butyl alcohols exchange the O-hydrogen more rapidly than the C-hydrogens over films of nickel, iron, tungsten, rhodium, palladium, platinum and silver (Anderson and Kemball, 1955). Over all these metals some exchange of C-hydrogens occurred, particularly at the upper end of the reaction temperature range, but even so, the activation energy for C-hydrogen exchange is substantially in excess of that for the O-hydrogen, as shown in Table 20.

TABLE 20

Activation energies for exchange of C-hydrogens in alcohols, in excess of values for O-hydrogen exchange

System	Excess activation energy (kcal mole^{-1})	Temperature range (0 °C)
CH_3OH, W film	10	52 – 88
CH_3OH, Rh film	8	0 – 39
C_2H_5OH, Rh film	7	40 – 100

With both amines and alcohols, the exchange pattern is apparently dependent on the degree of surface contamination of the catalyst, because C-hydrogen exchange is much more sensitive to contaminant than is O- or N-hydrogen exchange. Thus, although some C-hydrogen exchange occurs with methylamine over a clean nickel film and with ethyl alcohol over a clean platinum film, the use of corresponding bulk catalysis with their inevitable surface contamination results in exchange entirely confined to the O- and N-functions respectively (Horiuti and Polanyi, 1934; Farkas and Farkas, 1937a; Roberts, et al., 1939).

Multiple exchange occurs to a relatively small extent with those catalysts which can exchange C- as well as N- or O-hydrogens. This is greatest on rhodium which is consistent with the high ability of this metal to promote multiple exchange in hydrocarbons, and suggests that a 1–2 diadsorbed multiple exchange intermediate is involved, that is for instance

$$\begin{array}{cc} CH_2\text{—}O & CH_2\text{—}NH \\ | \quad | & | \quad | \\ M \quad M & M \quad M \end{array}$$

from methyl alcohol and methylamine. For alkyl groups other than methyl, extensive exchange within that group may occur; for instance, multiple exchange in the ethyl group of ethyl alcohol occurs on rhodium, no doubt by a 1–2 multiple exchange mechanism. Multiple exchange probably also proceeds via adsorbed residues formed by the loss of more than one hydrogen at a carbon atom, and since this would be expected to have a higher activation energy than 1–2 multiple exchange, it would account for the marked increase in the extent of multiple exchange at higher temperatures.

The rate of adsorption/desorption of methyl alcohol over a nickel film, a process mainly involving adsorption at the OH group, is proportional to $p_{CH_3OH}^{0.6} \, p_{D_2}^{0.4}$. By comparison with the discussion in the previous section, this clearly invites the formulation of the mechanism as

$$\begin{array}{cccc} CH_3OH & D & CH_3O & HD \\ | & | \;\rightarrow & | & | \\ M & M & M & M \end{array}$$

$$\begin{array}{cccc} CH_3O & D_2 & CH_3OD & D \\ | & | \;\rightarrow & | & | \\ M & M & M & M \end{array}$$

(III)—(23)

with an analogous formulation for the amines.

Particularly at higher reaction temperatures and with reduced deuterium pressures the exchange reactions of methylamine and of methyl and ethyl alcohols are poisoned by strongly held residues. These involve extensive dehydrogenative adsorption, and in the case of the alcohols Blyholder's observations (section II) clearly demonstrate the reality of this. Just as adsorbed carbon monoxide is the identified end product of methyl alcohol dehydrogenative adsorption, adsorbed hydrogen cyanide is a likely product from methylamine. As would then be expected, poisoning is only slight with dimethylamine. The initiation of poisoning from the alcohols appears to require the loss of two hydrogen atoms from the O-carbon atom, since poisoning is slight with isopropyl alcohol and non-existent with t-butyl alcohol. The reason probably is that without the loss of more than a single hydrogen, the residue is not bound sufficiently strongly to function as a poison at the observed temperatures.

For the exchange of these alcohols and amines, the order of activity of the metals for adsorption/desorption falls in the sequence

$$Pt > Rh > Pd > Ni, W > Fe > Ag$$

which is the same activity sequence as for ammonia, but quite different from that for the saturated hydrocarbons. No thoroughly convincing explanation for this activity sequence is available, although the problem has been discussed by Schuit et al. (1961) in terms of (for instance) the metal–oxygen bond strength.

As with hydrogen sulphide, the ready incorporation reaction of sulphur would make it difficult to study exchange of alkyl mercaptans or sulphides on most metals. For instance, sulphidation of a nickel film by methyl mercaptan occurs at 0 °C. However, exchange of methyl and ethyl mercaptans has been followed over MoS_2 and WS_2 powdered catalysts (Wilson and Kemball, 1964; Kieran and Kemball, 1965). In all systems the S-hydrogen was exchanged more readily than the C-hydrogens, although the latter occurred to some extent. H_2/D_2 equilibrium was not established during the reactions, and for this reason initial product distributions are not of quantitative significance. However, the difference in rates of S-hydrogen and C-hydrogen exchange was substantial, being, for instance, a factor of 400 for methyl mercaptan at 155 °C and 7 for ethyl mercaptan at 215 °C over WS_2: in the latter case the activation energy for C-hydrogen exchange is about 2 kcal mole^{-1} higher than for S-hydrogen exchange. The activation energies and frequency factors are summarized in Table 19, together with the values for hydrogen sulphide exchange. The rate of methyl mercaptan exchange over WS_2 had the pressure dependence $p_{CH_3SH}^{-0.1} p_{D_2}^1$. Thus the kinetic parameters for S–H exchange are fairly similar for the three substances, and the mechanisms are all likely to be similar to reaction (III)–(22).

Exchange within the methyl or ethyl group may proceed in a manner analogous to that discussed in a previous section for alcohols and amines. However, exchange within the ethyl group of ethyl mercaptan gave an initial product distribution closely resembling that obtained from ethylene under similar conditions, and indeed ethylene was a product from ethyl mercaptan decomposition at somewhat higher temperatures than the exchange range. Mainly from these considerations, Kieran and Kemball suggest that exchange within the ethyl group proceeds by

$$C_2X_5S_{(s)} \underset{1'}{\overset{1}{\rightleftarrows}} C_2X_5{(s)} \underset{2'}{\overset{2}{\rightleftarrows}} C_2X_4{(s)} \qquad \text{(III)–(24)}$$

where X is D or H, and steps 2 and 2' give exchange within the ethyl group in a manner similar to ethylene. However, the critically important feature of this mechanism is the reversibility of C–S bond rupture and formation by reactions 1 and 1'. A reversible process of this sort might be thought unlikely on a metal surface because of the high heat of binding of the sulphur residue, but it may be more likely on MoS_2 and WS_2 where the heat of binding would presumably be less. Since the reversibility of steps 1 and 1' is based on the indirect evidence of initial product distributions, the conclusion can be only regarded as tentative. It is important that this conclusion should be checked by more direct means, and this could probably be done by the introduction into the reaction mixture of ethylene containing isotopically labelled carbon.

2. Tertiary Amines and Ethers

In these molecules, adsorption by $\overset{\backslash/}{\underset{M}{N}}$ and $\overset{\backslash/}{\underset{M}{O}}$ is still possible but dissociative adsorption and a facile exchange at the N- and O-functions cannot of course occur. As a consequence, Farkas and Farkas (1937a) found that diethyl ether exchanges at less than one-twentieth of the rate of ethyl alcohol over platinized platinum at 17 °C, while Clarke and Kemball (1959) found that the exchange of diethyl ether on nickel and palladium had a much higher activation energy than for the alcohols, and in a similar way, exchange of trimethylamine occurs on palladium, tungsten and iron with much higher activation energies than the corresponding figures for methylamine and dimethylamine (Kemball and Wolf, 1955).

Both of these molecules give initial product distributions which have important features in common with those from structurally related hydrocarbons. An important conclusion is that in neither case can attachment to the surface at the N or the O atom function as part of a

1–2 multiple exchange mechanism, probably because the activation energy for desorption at the N or O atom is relatively low. In accordance with this, we note that the exchange rate of trimethylamine is proportional to the first power of amine pressure, and the rate of ammonia exchange on iron is uninhibited by trimethylamine (Gutmann, 1955); and with trimethylamine exchange over palladium, tungsten and iron proceeds mainly by a simple exchange mechanism in which a single deuterium is introduced during each residence period on the surface. The exchange behaviour of each methyl group thus relates to that of methane, while the quite small amount of multiple exchange that was ancillary to the main process on tungsten and iron probably results from a 1–3 multiple exchange mechanism similar to that which occurs to a limited extent with neopentane.

On the other hand, the initial product distributions from diethyl ether resemble in their main features those to be expected by analogy with ethane. Thus the extremes of behaviour are given by tungsten and palladium. Exchange on tungsten gave mainly the d_1 product (M = 1.17) while palladium gave the d_4 and d_5 compounds as the most abundant products (M = 4.0). A maximum at d_4 probably results from steric hindrance during the 1–2 multiple exchange.

The activity of iron with both these molecules for C-hydrogen exchange is again a qualitative distinction from the behaviour of the alkanes.

Exchange of diethyl ether suffers from self-poisoning on films above the following temperatures; tungsten, 0 °C; iron 160 °C; nickel, 104 °C; palladium, 64 °C. This is no doubt due to the formation of strongly bonded residues formed by dehydrogenative adsorption, and this is probably the surface precursor of C–O bond rupture.

3. Alkyl Chlorides
The exchange of ethyl chloride has been studied by Campbell and Kemball (1961) over films of palladium, platinum, nickel, iron, tungsten and rhodium. Methyl chloride exchange over titanium films has been reported by Anderson and McConkey (1967), and t-butyl chloride exchange over films of palladium and platinum by Campbell and Kemball (1963). In no case with ethyl or methyl chlorides was any deuteroparent detected among the initial products, but deuterated products were formed following the rupture of C–Cl bonds. A generally similar result was obtained by Addy and Bond (1957) for the reaction between deuterium and n-propyl and isopropyl chlorides over supported palladium.

Exchange of methyl chloride on titanium at 275 °C gave exclusively

CH$_3$D from the formation of CH$_{3(s)}$ on the surface. The lack of any of the more highly deuterated methanes probably indicates that CH$_{2(s)}$ and CH$_{(s)}$ were not present on the surface, since with the reaction temperature at 275 °C it is unlikely they would be inert if present. The exchange of methane over titanium is unavailable for comparison.

The distributions of isotopic ethanes formed from ethyl chloride and deuterium are given in Table 21, from which two main points emerge.

TABLE 21

Initial distribution of isotopic ethanes formed by reaction of
ethyl chloride and deuterium on films

| | percentages of isoptic species | | | | | | |
	d_1	d_2	d_3	d_4	d_5	d_6	M
palladium (146 °C)	1	60	3	8	6	22	3.2
platinum (142 °C)	35	33	7	5	5	15	2.6
nickel (179 °C)	57	19	7	3	3	11	2.1
iron (181 °C)	60	10	1	3	4	22	2.5
tungsten (195 °C)	45	41	4	6	3	1	1.8
rhodium (226 °C)	0	61	1	9	7	22	3.3

In all cases the proportion of C$_2$H$_4$D$_2$ is much in excess of that which could result from a 1–2 multiple exchange mechanism. Apart from this, the distributions are not dissimilar to those formed from ethane, although on tungsten and rhodium the exchange temperatures are well above those required for ethane exchange itself. There can be little doubt that the large proportion of C$_2$H$_4$D$_2$ originates by desorption from

in confirmation of which it was shown mass spectrometrically that most of this compound was CH$_3$CHD$_2$. On the other hand, the most obvious route to the products from 1–2 multiple exchange is via the formation of C$_2$H$_{5(s)}$. Thus there are two reaction paths which proceed from C–Cl bond rupture.

Exchange of t-butyl chloride is distinguished from the above alkyl chloride reaction by the formation of some deutero-t-butyl chloride. On palladium this amounts to only some 2% of the total reaction product, but on platinum nearly 40% was formed. The reason for this different behaviour is readily understandable in terms of the influence of molecular geometry on the ease of C–Cl bond rupture. The initial distributions of isobutanes are similar to those obtained for the exchange of other simple hydrocarbons on these metals. The exchange product monodeutero-t-butyl chloride is no doubt produced by a simple dissociation of one of the methyl groups. The origin of the more highly deuterated products remains with a degree of uncertainty. If judged by the amounts of these to be expected from a 1–3 neopentane-type multiple exchange, there is rather too much of them present. Campbell and Kemball (1963) suggested that they arise from a degree of reversibility in C–Cl bond rupture, so that some of the highly exchanged isobutane could find its way back into deuteroparent. This is rather similar to the argument produced for the exchange of the mercaptans. However, in this case the proposal is substantiated by the recent work of Morrison and Krieger (1968) who showed that catalytic exchange of iodine occurred between various alkyl iodides over a range of metals, and by a dissociative mechanism.

IV. Hydrogenation of Unsaturated Molecules

A. hydrogenation of olefins

There have been many studies of the reactions of olefins on films of the transition metals. Most of these have been directed towards understanding the essential mechanism of the catalytic addition of hydrogen to the carbon–carbon double bond. Often the data have been limited to the determination of the activation energy and the order of the reaction. It is now realised that these data alone are not sufficient to establish a reaction mechanism.

Much of our present understanding of the reaction comes from studies of the exchange reactions with deuterium and the isomerization reactions of more complex olefins.

Reaction mechanisms have been sought to explain all of the processes that an olefin can undergo on reaction with hydrogen or deuterium. These are hydrogenation of the double bond, olefin exchange (that is the formation of deuto-olefin) and hydrogen exchange (defined as the formation of gaseous HD or H_2), double bond migration and cis–trans isomerization.

In general the proposed mechanisms concentrate on the nature of the adsorbed olefin and the possible intermediates in its reaction, but the role of hydrogen has been the matter of some speculation. The possibilities considered are that the reactive species is molecular hydrogen from the gas phase or weakly adsorbed at the surface, or adsorbed atomic hydrogen from the dissociative chemisorption of hydrogen molecules. Dissolved atomic hydrogen from the interior of the metal could also be important.

Adsorption and catalytic results indicate that the olefin is chemisorbed before hydrogenation. The nature of the adsorbed species formed when ethylene is adsorbed on a clean metal surface were discussed in section II. It seems unlikely that the dissociative chemisorption which occurs from an olefin itself on a clean metal surface leads to the important intermediate in the catalytic hydrogenation. Under usual reaction conditions the surface is highly covered, excess hydrogen is present in the gas phase and the more highly hydrogenated species are expected to be important.

Most recent discussions of the mechanisms of olefin reactions have been in terms of non-dissociatively adsorbed π-bonded intermediates (Bond and Wells, 1964; Garnett and Sollich-Baumgartner, 1966; Siegel, 1966). The σ-diadsorbed olefin is also possible, and olefins possessing one or more α-methylenic hydrogen atoms may lose one such hydrogen and form a π-allyl adsorbed species in which the π electrons extending over three adjacent carbon atoms interact with a metal site (Figure 4).

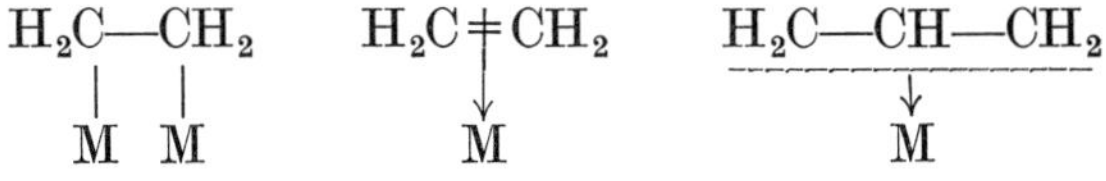

$$
\begin{array}{ccc}
\mathrm{H_2C-CH_2} & \mathrm{H_2C\!=\!CH_2} & \mathrm{H_2C-CH-CH_2} \\
\mid \quad \mid & \downarrow & \downarrow \\
\mathrm{M \quad M} & \mathrm{M} & \mathrm{M} \\
\sigma\text{-diadsorbed} & \pi\text{-adsorbed} & \pi\text{-allyl adsorbed}
\end{array}
$$

Fig. 4. Structural formula representations of olefin adsorption on a metal.

With such a range of possibilities many complexities can arise in a detailed reaction mechanism. Not all of these can be resolved; however, we now have some understanding of the important processes and a recognition of the differing mechanisms on different metals.

1. Reactions of Ethylene

(a) Nickel: The hydrogenation of ethylene on nickel films has been studied repeatedly. Beeck, et al. (1940, 1945, 1950) found a rate expression depending on the first power of the hydrogen pressure but independent of the ethylene pressure. This result was substantially confirmed by Jenkins and Rideal (1955), Kemball (1956) and Foss and Eyring (1958).

The catalytic activity of the evaporated films was found to be higher than wires or foils and the studies on films were generally carried out at much lower temperatures. Data from a number of studies are collected in Table 22. There is some disagreement in the values for the activation energy which may be partly explained by the different temperature ranges studied. It is likely that a compensation effect operating on surfaces of different heterogeneities is, at least in part, responsible for the

TABLE 22

Hydrogenation of ethylene on nickel catalysts

Catalyst Type	Activation Energy (kcal mole^{-1})	Temperature Range (° C)		Reference
Film	10.7	−80 to	150	Beeck (1950)
	10.2	20	150	Jenkins and Rideal (1955)
	7	−120	−100	Kemball (1956)
	8.0	0	96	Foss and Eyring (1958)
	9.7‡	−40	40	Crawford et al. (1962)
	10.5	0	72	Campbell and Emmett (1967)
Wire	8.2	60	110	Twigg (1939)
	5.3	−22	25	Masuda (1965)
Powder	7.5	∼ −70		Hall and Emmett (1959)
Ni/SiO$_2$	8.4	−78	0	Schuit and van Reijen (1958)
Ni/Al$_2$O$_3$	11.6	30	80	Wanninger and Smith (1960)

‡ Average value for films sintered at various temperatures.

various activation energies observed. However, Crawford et al. (1962) have studied the effect of sintering nickel films and find that although the rate of hydrogenation is lowered the activation energy is not dependent on sintering temperature.

There is general agreement that the reaction on nickel has no positive dependence on ethylene pressure. Beeck showed that ethylene is more strongly adsorbed than hydrogen and treated the system as a mixed adsorption–desorption equilibrium. This leads to the expectation that the coverage of hydrogen on the surface will be proportional to the pressure of hydrogen but inversely proportional to the pressure of ethylene. The observation that the overall kinetics are independent of ethylene pressure would then be consistent with the reaction rate being controlled by the arrival of ethylene at the surface. This has often been quoted in support of the Eley-Rideal mechanism which postulates that a gas phase or physically adsorbed molecule can react with a chemisorbed species.

Halsey (1963) has, however, pointed out that adsorption–desorption equilibrium might not be attained. Halsey suggests that it is easier to react the adsorbed species off the surface as ethane, than to evaporate them as ethylene. Thus a steady-state coverage of the surface is maintained by adsorption followed by reaction rather than desorption. Halsey discussed the consequences of a non-equilibrium mechanism operating on a heterogeneous surface and concludes that the observed kinetics could result if the relative adsorption energies of ethylene and hydrogen varied on different parts of the surface. In the limit of non-equilibrium adsorption where the strongly adsorbed ethylene could only be removed from the surface by reaction, it is apparent that an effectively constant coverage of ethylene would result and that the rate would then depend only on the pressure of hydrogen.

Twigg and Rideal (1939) found that above 150 °C desorption of ethylene from nickel becomes appreciable and the activation energy for hydrogenation falls. Twigg (1939, 1950) has discussed the variation of the activation energy with coverage and concludes that at ~ 90 °C the rates of hydrogenation and exchange are equal. Below this temperature hydrogenation is faster than exchange.

In a recent study of the coverages of the adsorbed species in the nickel catalysed hydrogenation of ethylene, Matsuzaki and Tada (1969) also conclude that the coverage of adsorbed ethylene varies with temperature and pressure. For a nickel wire catalyst they find that the optimum temperature for reaction is 130 °C and that at higher temperatures the coverage of ethylene becomes very low. For the reaction on a nickel film at much lower temperatures it seems likely that the ethylene coverage is high.

(b) Deuterium exchange: In order to formulate a reaction mechanism it is necessary to understand something of the surface reactions of the various adsorbed species. This information has been obtained from deuterium exchange experiments. The reaction of ethylene with deuterium has been studied on evaporated nickel films by Kemball (1956) and Crawford et al. (1962).

The initial products of the reaction of ethylene with deuterium at -100 °C on an evaporated nickel film are rich in deuterated ethylenes (Kemball, 1956) (Table 23). Hydrogen exchange, that is the appearance of HD and H_2, was found to be slight. This shows that there is no equilibrium between gas phase deuterium and the adsorbed hydrogen atoms and suggests that the hydrogen remaining from the olefin exchange must be used in the formation of the ethanes. Kemball's analysis of the

TABLE 23

Initial distribution of products
from the reactions of ethylene with deuterium at -100 °C on films

	Ethylenes (%)				Ethanes (%)							H_2 in total "D_2" (%)
	d_1	d_2	d_3	d_4	d_0	d_1	d_2	d_3	d_4	d_5	d_6	
Nickel	46.7	13.2	2.7	0.7	7.8	12.5	9.5	4.0	2.0	0.8	0.2	1.5
Iron	55.5	9.4	1.3	0.2	13.7	9.2	6.6	2.4	1.0	0.5	0.2	1.5
Rhodium	34.4	7.0	1.7	0.9	8.4	14.4	15.8	7.4	5.0	3.6	1.4	0.5
Tungsten	7.0	0	0	0	3.1	14.7	69.8	4.5	0.7	0.2	0	

distribution of deutero products is based on the solution of the simultaneous equations relating the probability parameters p, q, r and s in the following reaction scheme.

$$C_2X_{4(g)} \xleftarrow{1/(1+p)} C_2X_{4(s)} \underset{r/(1+r)}{\overset{p/(1+p)}{\rightleftharpoons}} C_2X_{5(s)} \xrightarrow{1/(1+r)} C_2X_6(g)$$

$$C_2X_{4(s)} + H \xrightarrow{1/(1+q)} C_2X_4H_{(s)}$$

$$C_2X_{4(s)} + D \xrightarrow{q/(1+q)} C_2X_4D_{(s)}$$

$$C_2X_{5(s)} + H \xrightarrow{1/(1+s)} C_2X_5H_{(g)}$$

$$C_2X_{5(s)} + D \xrightarrow{s/(1+s)} C_2X_5D_{(g)}$$

(IV)–(1)

The values which give reasonable fit to the distribution observed on nickel are

$$p = 3, q = 2, r = 12 \text{ and } s = 1$$

Thus it is concluded that an adsorbed ethylene molecule has a 25% chance of desorbing and that an ethyl radical has a 92% chance of reverting to ethylene.

The results obtained for the reaction of ethylene with deuterium on a nickel wire (Turkevich et al., 1950) and on a nickel–kieselguhr catalyst (Wagner et al., 1952, 1953) show some inconsistencies when compared with results from nickel films. The wire catalysed olefin exchange and gave a distribution of products at 90 °C similar to that obtained at much lower temperatures on the film, but hydrogen exchange was in this case found to be extensive. On the other hand the nickel–kieselguhr catalyst showed little activity for olefin exchange or hydrogen exchange in the temperature range of −50 to 50 °C; the principal product being dideutero ethane.

Discrepancies of this kind destroy confidence in many of the considerations of the catalytic activity of a metal. It is apparent that in many cases the detailed mechanism is dependent on the catalyst history and surface structure. Attempts to correlate the results obtained at different temperatures on catalysts of different overall activity seem generally unsatisfactory.

The formulation of a reaction mechanism consistent with our limited knowledge must leave the nature of the reactive hydrogen species unspecified. The adsorbed olefins may be written as π-bonded to the metal surface but this need not preclude the possibility that an intermediate σ–diadsorbed species may be formed.

$$
\begin{array}{l}
\left.
\begin{array}{c}
\mathrm{H_2C\!-\!\!-\!CH_2} \\
\ \ |\quad\ \ | \\
\ \ \mathrm{M}\quad\ \mathrm{M} \\
\\
or \\
\\
\mathrm{H_2C\!\!\mp\!\!CH_2} \\
\quad\ \mathrm{M}
\end{array}
\right\}
\ \underset{\longrightarrow}{+\mathrm{D}}\
\begin{array}{c}
(B) \\
\mathrm{CH_2D} \\
\diagup \\
\mathrm{H_2C}\qquad\!-\!\mathrm{H} \\
\ \ | \\
\ \ \mathrm{M} \\
\ \ |\ +\mathrm{D} \\
\downarrow
\end{array}
\ \underset{\longrightarrow}{}\
\left.
\begin{array}{c}
\mathrm{H_2C\!-\!\!-\!CHD} \\
\ \ |\quad\ \ | \\
\ \ \mathrm{M}\quad\ \mathrm{M} \\
\\
or \\
\\
\mathrm{H_2C\!\!\mp\!\!CHD} \\
\quad\ \mathrm{M}
\end{array}
\right\}
\qquad \text{(IV)–(2)}
$$

$$(A) \qquad\qquad \mathrm{H_2DC\!-\!CH_2D} \qquad\qquad (C)$$

In Equation (IV)–(2), (B) is the half-hydrogenated adsorbed species with a single bond to the surface. The observation that a nickel film catalyses multiple olefin exchange shows that the processes (A) → (B) and (B) → (C) are rapid on this metal. In this case there will be half-hydrogenated species of varying degrees of deuteration and so the product is a variety of deuteroethanes.

(c) Other transition metals: Kemball (1956) has obtained results for the reaction of ethylene with deuterium on iron, rhodium and tungsten. The initial distributions of products are in Table 23.

The product distribution on iron is similar to that found for nickel showing that iron is also an effective catalyst for olefin exchange. This behaviour is to be contrasted with the inactivity of iron for ethane exchange (Anderson and Kemball, 1954a). The exchange of ethane on iron at low temperatures is limited by a high activation energy, whereas ethylene is apparently easily adsorbed and desorbed and hence undergoes exchange.

On rhodium the rates of olefin exchange and deuteration are almost equal. In this case there is also a significant amount of hydrogen-deuterium exchange.

On tungsten only 7% of deutero-ethylene is formed and the ethane distribution peaks sharply at ethane-d_2. This behaviour of tungsten arises because the chance of alkyl reversal is small and is consistent with the small amount of multiple exchange observed for ethane on tungsten films.

For the reaction of ethylene with deuterium on other transition metals we are dependent on the results obtained on supported catalysts. For a palladium/alumina catalyst between −36 ° and 67 °C Bond et al. (1962) found much olefin exchange but almost no hydrogen exchange. It is concluded that palladium also allows efficient alkyl reversal and relatively easy olefin desorption.

Contrasting results were found for a platinum catalyst (Bond, 1956; Bond et al., 1964). At low temperatures very little olefin exchange or hydrogen exchange were observed; however, a redistribution occurs leading to all of the possible deuteroethanes. It is concluded that while alkyl reversal must readily occur, olefin desorption from platinum is difficult.

Rather similar results were found for an iridium/alumina catalyst (Bond et al., 1964). The pattern of activity of the transition metals becomes clearer from the isomerization reactions of the higher olefins.

(d) Copper and Gold: Campbell and Emmett (1967) in their study of the reaction on alloy films, found that a copper film gave measurable hydrogenation of ethylene at 140–200 °C. The activation energy of 12 kcal mole^{-1} varied through zero to negative values at above 200 °C. The kinetics were shown to be first order in both hydrogen and ethylene pressures.

Bond (1966) suggests that on a copper catalyst the selective hydrogenation of diolefins and alkynes to mono-olefin occurs because the latter are weakly adsorbed. The reaction on copper differs from that on the transition metals, not only in the strength of olefin adsorption but also in the role of hydrogen. Numerous authors have considered that hydrogen does not dissociatively adsorb on copper at room temperature (Trapnell, 1953; Kington and Holmes, 1953). Pritchard (1965) has, however, argued that hydrogen is adsorbed by copper.

It seems likely that surface concentrations of both olefin and hydrogen are low on the surfaces of both copper and gold. Campbell and Emmett (1967) found that gold films had barely measurable catalytic activity for the ethylene hydrogenation at 486 °C. The activity of gold for hydrogenation is discussed further when considering the reaction of cyclohexene (section IV.A.2).

2. Reactions of the Higher Olefins

(a) C_3-C_4 olefins: The reactions of propene and the butenes with hydrogen and deuterium have not been studied on evaporated metal films; however, there have been experiments on wires, powders and supported catalysts. Hirota and Hironaka (1966) have investigated the hydrogenation of propene on powders of nickel, palladium and platinum. They conclude that the rate determining step is the addition of hydrogen to the half-hydrogenated species and that for nickel only, the lifetime of chemisorbed propene is shorter than the time interval for exchange. Veda et al. (1969) have extended this work to copper and rhodium and conclude that, while all these metals have non-dissociatively adsorbed propene, palladium also has an allyl type species and platinum a trans-propenyl species,

(D)

The reactions of the butenes over wires and supported catalysts have been reviewed in detail by Bond and Wells (1964). It has been shown that iron, nickel and palladium are again active in olefin exchange and also catalyse cis–trans isomerization. The mechanism of cis–trans isomerization can be formulated as

where the π-bonded structures could alternatively have been represented as σ-diadsorbed. The isomerization occurs via free rotation about the single carbon–carbon bond in the mono-adsorbed intermediate. A similar mechanism explains double bond shift if the alkyl radical dissociates hydrogen from the terminal carbon.

Evidence has been sought in the isomerization and double bond migration reactions for the participation of π-allylic species in the mechanism. The evidence is generally inconclusive.

On platinum, results from solution and gas phase reactions show that while hydrogenation rates are high, the rate of isomerization is generally low. This might have been expected as it was previously noted that platinum is a poor catalyst for olefin exchange due to difficulty in desorbing the olefin. The mechanism of isomerization must necessarily involve this same step.

The behaviour of iridium resembles that of platinum and it is concluded that olefin desorption from this metal is also difficult.

Rhodium is an effective hydrogenation catalyst but its isomerization behaviour is found to be temperature dependent. At >80 °C the probability of olefin desorption is apparently high and exchange and isomerization are as extensive as on nickel. At lower temperatures the behaviour of rhodium resembles that of platinum and the products are predominantly the saturated hydrocarbons.

(b) Cyclic Olefins: Erkelens *et al.* (1961) have investigated the reactions of cyclohexene with deuterium on films of nickel, iron, palladium, platinum and tungsten. The results are in Table 24. The

TABLE 24

Cyclohexanes from the deuteration of cyclohexene on platinum, tungsten, nickel, palladium and iron films

catalyst	Pt	Pt	W	Ni	Ni	Pd	Pd	Fe	Fe
temp (°C)	-35	0	-35	-35	0	-35	0	-35	0
d_0	5.7	5.5	2.8	21.3	18.5	3.4	4.9	18.7	12.3
d_1	23.3	17.2	20.7	22.6	20.0	12.8	12.3	20.5	14.9
d_2	33.9	24.6	52.4	19.3	19.9	20.9	21.6	17.3	14.2
d_3	19.5	16.0	13.1	14.0	14.3	17.9	17.0	13.0	11.9
d_4	10.5	10.0	5.0	9.6	10.0	14.9	13.6	9.3	9.2
d_5	4.1	6.3	2.9	6.0	6.5	11.2	10.4	6.2	7.1
d_6	1.6	4.9	1.7	3.5	4.1	7.8	7.7	4.2	5.4
d_7	0.8	3.9	0.4	1.9	2.6	4.9	5.1	2.7	4.4
d_8	0.3	3.5	0.5	0.9	1.5	2.7	3.7	1.9	3.9
d_9	0.3	2.9	0.3	0.6	0.8	1.8	2.3	1.7	3.9
d_{10}	0	2.4	0.2	0.3	0.4	1.0	1.2	1.6	4.1
d_{11}	0	1.8	0	0	0.5	0.6	0.2	1.5	4.7
d_{12}	0	1.0	0	0	0.9	0.1	0	1.4	4.0
M†	2.32	3.53	2.21	2.22	2.58	3.61	3.59	2.86	4.22

† M refers to the mean deuterium content of the saturated hydrocarbons as atoms per molecule.

Distribution taken when cyclohexene had been consumed.

experiments were extended to cyclopentene, cycloheptene, norbornylene and 4-methylcyclohexene on iron. At -35 °C the products were in the range cycloalkane-d_0 to cycloalkane-d_n where n is the number of carbon atoms. It is considered that the products arise from the inter-conversion of adsorbed cycloalkene and cycloalkyl radicals and that the shape of the molecule restricts the formation of more highly deuterated species. This effect was already noted in the exchange of cyclopentane (section III). Small amounts of more highly deuterated products are presumed to form via the deutero-olefins.

On iron at somewhat higher temperatures more highly deuterated cycloalkanes were formed and there was extensive hydrogen–deuterium exchange. The multiple exchange process leading to these more highly deuterated cycloalkanes has an activation energy 4–5 kcal mole^{-1} greater than that leading to the lightly deuterated products.

This multiple exchange process was not observed on iron films sintered at $\sim$200 °C before reaction. Sintered iron films also failed to effectively catalyse the disproportionation of cyclohexene to cyclohexane and benzene and it has been suggested that the sintering process may remove sites required for the formation of a π-allyl species, a possible intermediate in both the disproportionation and multiple exchange processes.

Hilaire (1969) has recently reported reactions of a number of cyclic olefins with deuterium on nickel films and palladium catalysts. He concludes that the multiple exchange reactions of various substituted cyclopentenes on palladium are explained by π-allyl intermediates but that nickel favours the formation of 1–2 adsorbed intermediates such as

Erkelens et al. (1963) have measured the reactions of cyclohexene with hydrogen and deuterium on gold films (Figure 5). With hydrogen at 196–342 °C cyclohexane and benzene were formed; the proportion of benzene increasing with temperature. The occurrence of both hydrogenation and dehydrogenation implies that species such as $C_6H_{11(s)}$ and dissociated species such as $C_6H_{9(s)}$ down to $C_6H_{6(s)}$ must co-exist on the catalyst surface. The ratio of benzene to cyclohexane is not an equilibrium ratio but is determined by the kinetics of formation.

This reaction is of particular interest as, while it is known that unsaturated hydrocarbons are weakly adsorbed on gold, the adsorption of hydrogen is not detected (Trapnell, 1953). The results of Pritchard and Tompkins (1960) for the adsorption of atomic hydrogen on gold at −183 °C suggest that the adsorption of hydrogen molecules on gold

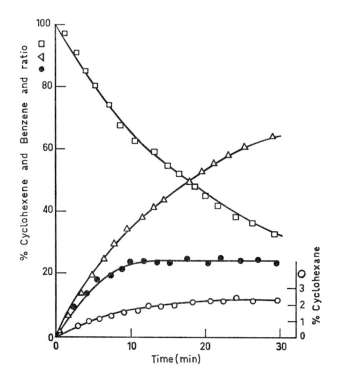

FIG. 5. The reaction of cyclohexane and hydrogen on 82 cm² gold at 342 °C: □, cyclo-hexene; △, benzene; ○, cyclohexane; ●, ratio of benzene:cyclohexane.

is endothermic and that chemisorbed hydrogen atoms on gold are metastable with respect to hydrogen gas at low temperatures. However, it is apparent that in the reaction of cyclohexene on gold the presence of hydrogen or deuterium influences the course of the reaction. Thus it is found that the exchange of cyclohexane occurs fairly readily, the formation of cyclohexane is first order in hydrogen and that the formation of benzene is inhibited by hydrogen.

The exchange product is predominantly cyclohexane-d_0. The d_1 and d_2 compounds are formed only after considerable olefin exchange has occurred. This shows that self-hydrogenation of cyclohexene is more important on gold than the addition of deuterium. It is suggested that the self-hydrogenation occurs via chemisorbed hydrogen atoms from the dissociative adsorption of the cyclohexene. The assumption that this hydrogen atom concentration is much lower than the hydrocarbon concentration, leads to the expectation that while these hydrogen atoms would be available for the self-hydrogenation, their collisions to form molecular hydrogen would be infrequent. It is concluded that gold is a relatively effective catalyst for the dehydrogenation of cyclohexene but a poor catalyst for hydrogenation because the frequency factor of the addition reaction is low owing to the low surface concentration of adsorbed hydrogen.

Chambers and Boudart (1966) have also investigated the selectivity of gold for the hydrogenation and dehydrogenation of cyclohexene. A gold powder reduced in hydrogen was found to give results in essential agreement with the film data of Erkelens *et al.* (1963). It was found that the selectivity could be controlled by the hydrogen pressure and that in the presence of small quantities of oxygen, benzene was the sole product. This suggests that the action of oxygen is to remove the small concentration of hydrogen atoms on the surface and so inhibit the hydrogenation reaction.

3. *Conclusions on Olefin Reactions*

The activation energies and rate expressions determined for olefin hydrogenation over metal films are in general agreement with those found on bulk and supported catalysts. However, the exchange and isomerization reactions, which are dependent on the detailed mechanism, show substantial dependence on the form of the catalyst. For this reason, generalizations which attribute specific activity to the nature of the metal should be qualified: the structure of the catalyst surface may in part determine the reaction products.

There is clear evidence in the comparison of the exchange reactions of ethylene with deuterium on nickel films with those on wire and supported nickel catalysts and in the reactions of cyclohexene on sintered iron films that surface structure can influence the distribution of products. Contamination of the surface by other adsorbed gases is a likely cause of variation in the surface properties of the catalyst. For this reason results on evaporated films are the most likely to give valid information on the catalytic activities of the clean metals.

The general conclusion is that iron, nickel and palladium are active in isomerization and olefin exchange reactions. Rhodium is less active and platinum and iridium almost inactive for these reactions. The strengths of adsorption of ethylene obviously vary on these metals, the strongest adsorption being indicated by those metals which do not catalyse olefin exchange.

The precise nature of the surface intermediates is still in dispute, but it seems clear that in, say, the reaction of ethylene, the surface contains adsorbed ethylene, ethyl radicals and hydrogen atoms. There is no

G

evidence in the reactions of the simplest olefins to decide whether π-bonded or σ-diadsorbed species are active in the reaction, and the evidence for the π-allyl intermediates postulated to explain the reactions of the higher olefins is usually equivocal. There is an obvious need for further work on the isomerization and exchange reactions of substituted olefins on metal films in order to clarify and confirm the data obtained on supported catalysts and from reactions in solution.

B. HYDROGENATION OF ALKYNES

The hydrogenation of acetylene has not been widely studied on evaporated metal films. Beeck (1950) has reported the hydrogenation of acetylene to ethane on films of transition metals. The activity of nickel was found to be low; less than one hundredth of the activity for ethylene hydrogenation. Rhodium and platinum were about ten times as active as nickel and palladium about twenty times as active. On all four metals the activation energy for acetylene hydrogenation was 6–7 kcal mole^{-1}. Iron and tungsten were found to be inactive at room temperature.

The generally low activity for acetylene hydrogenation is associated with the fact that acetylene is adsorbed much more strongly than ethylene. On a nickel film the initial heat of adsorption for acetylene is higher than for ethylene and shows only little fall with coverage (Bond, 1962). For ethylene the heat of adsorption falls as the coverage increases so that under reaction conditions hydrogen can compete for adsorption sites. Acetylene is evidently so strongly bound that most of the surface is effectively poisoned.

On powdered and supported metal catalysts the multiply unsaturated hydrocarbons hydrogenate to give both mono-olefin and alkane. The relative amounts of these products depend on the catalyst and the conditions. Furthermore, if the reactant molecule has four or more carbon atoms a distribution of isomeric olefins may be obtained.

The selective action of catalysts to perform this partial hydrogenation has been the subject of much study. While some of the specific actions of supported catalysts may be attributed to a pore structure which results in diffusion control of the reaction mechanism, it seems also likely that the action of supported catalysts is in this case due to a selective poisoning of the surface by the reactants.

High yields of ethylene are obtained from the hydrogenation of acetylene on metal catalysts which have much higher activities for ethylene hydrogenation in the absence of acetylene. The small amounts of ethane produced in such a selective hydrogenation apparently arise from one residence of the hydrocarbon on the surface. The chance of ethylene returning to the surface after desorption is low because it has to compete with the much more strongly adsorbed acetylene.

The selectivity of the transition metal catalysts for the hydrogenation of acetylene to ethylene may arise from the mechanism of the surface reaction or from the low equilibrium concentration of adsorbed ethylene under the reaction conditions. It is found that the selectivity, i.e. the ability to produce the olefin, increases with increasing temperature and decreases with increasing initial hydrogen pressure. Two effects may be attributed to an increase of temperature. If the activation energy for olefin desorption is greater than that for olefin hydrogenation on the surface, then increasing temperature will favour the desorption of the olefin product. Alternatively, increasing temperature may decrease the surface concentration of hydrogen, thereby inhibiting the hydrogenation of the adsorbed olefin. The increase of hydrogen pressure would be expected to increase the surface concentration of hydrogen resulting in the converse effect.

A further specific action of the catalyst may be observed in the hydrogenation of dialkylalkynes. In the hydrogenation of 2-butyne on transition metal catalysts the addition of two hydrogen atoms to the adsorbed 2-butyne results in the formation of mainly cis-2-butene. However, some trans-2-butene is usually observed together with 1-butene. A study of the product distributions of the reaction with deuterium shows that these products are formed directly from the adsorbed 2-butyne and not by the subsequent readsorption and isomerization of cis-2-butene (Bond and Wells, 1964).

Bond and Wells (1965) have suggested that the formation of trans-2-butene involves the addition of hydrogen to a free radical form of the half-hydrogenated state which is supposed to be in equilibrium with the normal form.

$$
\begin{array}{ccc}
\mathrm{CH_3 \quad CH_3} & & \mathrm{CH_3 \quad CH_3} \\
\diagdown \quad \diagup & & \diagdown \quad \overset{\bullet}{} \diagup \\
\mathrm{C{=}C} & \rightleftarrows & \mathrm{C{-}C} \\
| \quad \diagdown & & | \quad | \diagdown \\
\mathrm{M \quad\ H} & & \mathrm{M \ M \quad H}
\end{array}
\qquad \text{(IV)-(4)}
$$

Addition of hydrogen to this free radical would lead equally to cis- or trans-2-butene. A free radical intermediate of this kind was previously proposed for the catalytic polymerization of acetylene. However, it seems improbable that a free-radical such as this should be able to exist on a metal surface except in a reaction transition-state. The predominant

mechanism, leading to cis-2-butene may be represented as in reaction
(IV)–(5).

$$
H_3C—C\equiv C—CH_3 \;\rightleftharpoons\; \underset{\substack{| \\ M}}{\overset{\substack{CH_3 \\ \diagdown}}{C}}\!\!=\!\!\underset{\substack{\diagdown \\ H}}{\overset{\substack{CH_3 \\ \diagup}}{C}} \;\rightarrow\; \underset{\substack{\diagup \\ H}}{\overset{\substack{CH_3 \\ \diagdown}}{C}}\!\!=\!\!\underset{\substack{\diagdown \\ H}}{\overset{\substack{CH_3 \\ \diagup}}{C}} \qquad (IV)–(5)
$$

cis-2-butene

1-butene is thought to be formed by the initial isomerization of the
reactant to 1,2-butadiene, as in reaction (IV)–(6), followed by hydrogenation.

$$
\underset{\substack{| \\ M}}{\overset{\substack{CH_3 \\ \diagdown}}{C}}\!\!=\!\!\underset{\substack{\diagdown \\ H}}{\overset{\substack{CH_3 \\ \diagup}}{C}} \;\longrightarrow\; \underset{\substack{M \; M}}{H_2C\equiv C\equiv}\underset{\substack{\diagdown \\ H}}{\overset{\substack{CH_3 \\ \diagup}}{C}} \qquad (IV)–(6)
$$

From the small yields of 1-butene and the distribution of deutero-
butenes it is concluded that the formation of adsorbed 1,2-butadiene
is difficult.

In the reactions of acetylene with deuterium evidence was sought for
the production of deutero-acetylenes. Bond and Wells (1966a, b, c)
found that alumina-supported rhodium, palladium, iridium and platinum showed no acetylene exchange. More recently (Bond, et al., 1968)
it was found that alumina-supported ruthenium and osmium at 100–
200 °C gave appreciable amounts of exchanged acetylene and it is suggested that acetylene exchange and deuteration occur independently
(cf. benzene, section IV.C.).

The importance and variety of the reactions of the alkynes on supported catalysts warrant further study of these reactions on evaporated
metal films. Suitable reactions may be achieved by control of surface
structure during film preparation or by subsequent modification. Controlled poisoning has often been used to produce catalysts of desired
selectivity and there are attractive possibilities of modifying the
catalytic behaviour of clean metal surfaces towards the reactions of the
alkynes in this way. Such experiments on well characterized surfaces
could make significant contributions to our understanding of the role of
surface structure in controlling reaction mechanism.

C. HYDROGENATION OF AROMATIC COMPOUNDS

1. Reactions of Benzene with Hydrogen

There have been a number of studies of the hydrogenation of benzene over evaporated metal films. Beeck and Ritchie (1950) found that the reaction could be followed over evaporated films of nickel and iron at 20–60 °C. The reaction rate was found to be independent of benzene pressure and proportional to a positive fractional power of the hydrogen pressure which depended somewhat on temperature. The activation energy was 8.7 kcal mole^{-1} for both metals and nickel was found to be more than twice as active as iron.

In the vapour phase hydrogenation of benzene and toluene over evaporated films of nickel and tungsten some poisoning of the catalyst surface by hydrocarbon residues has been detected (Anderson, 1957b). The rate and extent of this poisoning increases with temperature and results in falling rates of hydrogenation. On poisoned nickel rates passed through a maximum at 90 °C. In some of the reactions cyclohexene was produced along with the major product cyclohexane. The hydrogenation of cyclohexene over nickel was found to be extremely rapid and strongly inhibited by benzene.

In this work the reaction was found to be first order in hydrogen pressure. Comparison of the rates with those observed by Beeck and Ritchie shows that the latter were working with considerably poisoned films. Their observed dependence of the rate on a fractional power of the hydrogen pressure was apparently a result of this. In order to avoid the effects of poisoning it is necessary to have hydrogen in considerable excess over the stoichiometric quantity for reaction.

Anderson and Kemball (1957) studied the reaction on a number of metal films and found that the order of activity is

$$W > Pt > Ni > Fe > Pd$$

This order is remarkably different from that found for the hydrogenation of ethylene in that palladium has moved from high activity to lowest. The high activity of tungsten is also to be contrasted with the inactivity of this metal for acetylene hydrogenation. The observed activation energy on platinum and palladium was ~9 kcal mole^{-1} and it is apparent that the variations in activity are due to variations in the pre-exponential factor. Silver films were found to be inactive for hydrogenation and in other work copper catalysts showed almost no activity (Greenhalgh and Polanyi, 1939). van der Plank and Sachtler (1968) in their study of the reactions of benzene with hydrogen and deuterium over copper-nickel alloy films obtained new results for pure nickel. The hydrogenation of benzene was found to be accompanied by poisoning and rate data were obtained only from the initial reaction on a film. For nickel at 150 °C the observed reaction rate was 1.5×10^{14} molecules $cm^{-2} sec^{-1}$ and the activation energy 12 kcal $mole^{-1}$. The reaction order with respect to hydrogen was found to be positive and to increase with temperature. In these experiments a high hydrogen/benzene ratio was used (>20) in order to minimize the poisoning effects. van der Plank and Sachtler believe that the formation of carbonaceous residues on the metal surface is the result of C–H bond cleavage and that this process is to some extent reversible, as was previously suggested by Anderson (1957b).

$$C_6H_{6(g)} \rightleftarrows C_6H_{(6-n)(s)} + nH_{(s)}$$

It is concluded that the positive and temperature dependent order with respect to hydrogen is in agreement with this assumption of a partial dissociation of C–H bonds in adsorbed benzene.

Madden and Kemball (1961) measured the hydrogenation of benzene over evaporated nickel films in a flow system. The results at 0–50 °C are in substantial agreement with the above studies made in static systems. Madden and Kemball find that the slow step in the reaction is the formation of $C_6H_{8(s)}$. The subsequent stages in the hydrogenation are apparently fast, in agreement with the observation that the hydrogenation of cyclohexene is 200 times faster than that of benzene at -34 °C and that the divergence of the rates increases with temperature.

The formulation of a reaction mechanism for the hydrogenation of benzene must await the presentation of results of the exchange reaction with deuterium. However, it is worth considering at this stage to what extent the measurements of the hydrogenation of benzene on supported catalysts are in agreement with the results obtained on metal films. Generally the results are in agreement except that some studies on films have been taken to higher temperatures (~ 200 °C) which results in hydrogenolysis to C_1–C_6 alkanes (see e.g. Kubicka, 1968). Some variations in order of activity, depending on the nature of the support, have been observed. On alumina the order is Rh > Ru > Pt > Pd (Amano and Parravano, 1957) but on silica the order is Pt > Rh > Ru > Pd > Co > Ni > Fe (Schuit and van Reijen, 1958).

The variations in activity are again found to amount to a variation in the pre-exponential factor. The activation energy found on supported catalysts is approximately constant in the range 7–11 kcal $mole^{-1}$ found on films.

2. Deuterium Exchange Reactions of Benzene

The early studies of the simultaneous exchange and deuteration of benzene were on metal foil catalysts (Horiuti et al., 1934; Farkas and

Farkas, 1937b; Greenhalgh and Polanyi, 1939). These experiments did not include detailed analyses of the distribution of deuterium atoms in the products.

The reactions of benzene with deuterium over evaporated metal films have been studied by Anderson and Kemball (1957). Results obtained for films of platinum, palladium and silver are in Table 25.

TABLE 25

Initial products from exchange of benzene with deuterium

Metal	Temp (°C)	Activation energy (kcal mole^{-1})	percentages of isotopic species						M
			d_1	d_2	d_3	d_4	d_5	d_6	
Pd	29.5	13	61.8	17.7	7.1	3.8	3.5	6.1	1.8
Pt	−43.5	18	77.6	13.0	2.8	2.3	2.0	2.3	1.5
Ag	373	6	71.2	17.0	3.5	2.1	2.5	3.7	1.4

Although the activation energies for exchange are higher than for hydrogenation, the exchange reaction proceeds faster and is measured at lower temperatures. Silver was found to be inactive for hydrogenation but catalysed the exchange with deuterium. A similar observation had previously been made for a copper powder catalyst (Greenhalgh and Polanyi, 1939).

Although the exchange and deuteration of benzene occur simultaneously, Anderson and Kemball presented evidence that the two processes are independent. The deuteration was zero order in benzene pressure and led to the production of cyclohexanes with six or more deuterium atoms. Very small amounts of cyclohexane-d_5 were found on platinum but only minute quantities of cyclohexanes of lower deuterium content were detected. The further exchange of cyclohexane was inhibited by benzene and the mean deuterium content of the cyclohexanes rose from the initial value of six throughout the experiment increasing as the extent of benzene exchange increased. These results suggest that the act of deuteration involves the addition of six deuterium atoms to each benzene molecule.

The behaviour of benzene is here in contrast to the deuteration of an olefin where the mechanism of exchange and deuteration have common steps and give rise to deuterated products containing from zero to the maximum possible numbers of deuterium atoms. The absence of cyclohexane with low deuterium content indicates that no such redistribution occurs in the deuteration of benzene. The deuteration of benzene apparently involves adsorption by the opening of one of the double bonds. This would lead to the reactivity of the two remaining double bonds of the molecule being enhanced because of the loss of resonance energy on adsorption. Anderson and Kemball suggested that the addition of deuterium atoms may occur in pairs either from the surface or from molecules from the gas phase since the addition of deuterium atoms singly would give rise to half-hydrogenated species which would be expected to bring about further exchange reactions of the kind observed for the deuteration of olefins.

The relative rates of exchange and deuteration vary with different metals. Exchange is much faster than deuteration on nickel and the relative importance of deuteration increases for other metals in the order iron, palladium and platinum. Deuteration may be faster than exchange on tungsten but the rate at -25 °C was found to be too fast to measure accurately. This variation of the relative rates of exchange and deuteration is evidence for two independent mechanisms. The distribution of exchange products shows that the formation of benzene-d_1 is favoured. Because this distribution of products resembles that found for the saturated hydrocarbons, Anderson and Kemball suggested that the resonance structure of the benzene molecule remains intact during the exchange reaction. They proposed an adsorbed phenyl group as an important intermediate in the reaction; however, more recent discussions have also recognised the importance of π-bonded benzene (Crawford and Kemball, 1962; Garnett and Sollich-Baumgartner, 1966).

Garnett and Sollich-Baumgartner discuss the limitations of the classical dissociative and non-dissociative mechanisms for the hydrogenation and exchange of benzene and investigate the mechanistic implications of π-bonded adsorbed states. It is considered that the π-adsorbed state will have a binding energy intermediate between that of van der Waal's adsorption and of dissociative chemisorption via a carbon–metal σ-bond. The suggested mechanism for the exchange of benzene with deuterium is illustrated for platinum in Figure 6.

The initial stage is the formation of a horizontally π-bonded species. A, represented as involving a nett charge transfer towards the metal. There are then three paths from this intermediate to the deutero benzene. One is by a dissociative π-state substitution mechanism. In this reaction the molecule rotates through 90° and changes to a vertically σ-bonded adsorbed species which may revert with exchange through a similar transition state to horizontally π-bonded benzene-d_1. An alternative path is via a non-dissociative π-state substitution mechanism. The transition state B is in this case π-bonded to the catalyst surface

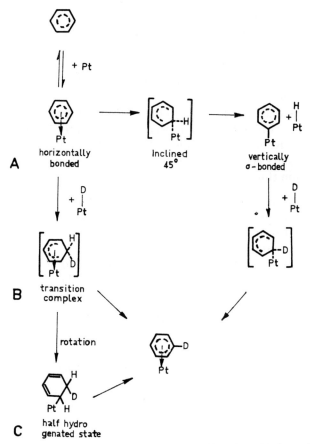

FIG. 6. A mechanism for catalytic exchange of benzene with deuterium.

and may react to give the adsorbed benzene-d_1, either directly by splitting off a chemisorbed hydrogen atom, or by rotation to form the σ-bonded half-hydrogenated state C.

The deuteration of benzene may occur via any of the species A, B or C. In the case of A, deuteration would involve reaction with a deuterium molecule in the van der Waals layer; for B and C reaction with an adsorbed deuterium atom would be expected.

There is no conclusive evidence for the validity of this reaction scheme from the experiments on benzene exchange, but there is considerable evidence for π-state adsorption in the study of competitive adsorption in liquid phase exchange and hydrogenation reactions of the homologous aromatic series. The evidence is presented and discussed in some detail by Garnett and Sollich-Baumgartner (1966).

3. Reactions of Alkylbenzenes

The reactions of a number of alkylbenzenes with deuterium on evaporated nickel films have been reported by Crawford and Kemball (1962). The exchange reactions of these molecules provide information on the catalytic activity for the making and breaking of different kinds of carbon–hydrogen bonds. The general treatment of the kinetics of exchange when the reactant molecule contains more than one type of exchangeable hydrogen has been discussed in Chapter 7, section IV.A.2. Crawford and Kemball were able to show that the hydrogen atoms in the alkylbenzenes could be grouped according to their ease of exchange with deuterium thus:

Group A—hydrogen atoms on carbon atoms α to the benzene ring and atoms on the ring not ortho to a ring substituent;

Group B—hydrogen atoms on the ring ortho to one but not two substituents;

Group C—hydrogen atoms on carbon atoms β and γ to the benzene ring, and (for m-xylene) the atom on the ring between the methyl groups.

Results for the initial rates of exchange are in Table 26.

TABLE 26

Exchange of alkylbenzenes with deuterium on nickel films

Compound	Numbers of hydrogen atoms in groups A B C	Temperature (°C)	Initial rates k_A k_B k_C			M
p-xylene	6 4 0	−10	104	4		1.8
o-xylene	8 2 0	−10	162	10		1.2
m-xylene	7 2 1	−10	161	12	0	1.4
toluene	6 2 0	−10	67	5		1.4
ethylbenzene	5 2 3	−10	92	3	<0.9	1.0
ethylbenzene	5 2 3	0	422	13	1.7	3.9
isopropylbenzene	4 2 6	0	90	11	4	1.7
n-propylbenzene	5 2 5	0	137	7	2	1.6

It is seen that the hydrogen atoms on carbon atoms α to the ring are in all cases the most reactive. It is suggested that this is a consequence of the low bond dissociation energies of these C–H bonds which are about 20–25 kcal mole^{-1} less than those for other aryl or alkyl C–H bonds. This high reactivity of the hydrogen atom in the methyl groups of p–xylene has also been noticed in the exchange with D_2O over a nickel powder catalyst (Hirota et al., 1961).

Crawford and Kemball propose that the exchange of the hydrogen atoms on the α-carbon atoms occurs through the formation of a species which is simultaneously bound to the surface by a π-bond holding the ring parallel to the surface and by a σ-bond from the alkyl substituent [Structure (E)].

$$
\text{(E)} \quad \text{(F)} \quad \text{(G)}
$$

Desorption of (G) leads to exchange, while interchange between (E) and (F) before desorption permits the multiple exchange indicated by the mean deuterium numbers in Table 26.

Exchange of the hydrogens on β and γ carbon atoms is considerably slower than for those on α carbon atoms. This difference is less pronounced at higher temperatures as it arises from a difference in activation energy. It is estimated that ($E_\alpha - E_\beta$) for iso-propyl benzene is 4 kcal mole^{-1}. The result for iso-propyl benzene shows that only one of the methyl groups undergoes exchange at a single adsorption of the molecule on the surface. This is consistent with a mechanism in which one of the methyl groups attached to the α-carbon atom interacts with the surface and the other is held at a distance by the tetrahedral arrangement of the bonds to the α-carbon atom. An analogous observation has been reported for the exchange of acetone on nickel (Kemball and Stoddart, 1957).

The results in Table 26 were obtained on nickel films deposited at 0 °C. Since the reactions occurred at this or lower temperatures, the film did not undergo thermal sintering after deposition. Crawford and Kemball also investigated reaction on nickel films that had been sintered at 200 °C for 30 minutes after deposition at 0 °C. Such a procedure introduces the possibility of contamination of the surface and a fall in activity due to a decrease in surface area. However, in this case the detailed distributions of products enabled some mechanistic conclusions to be drawn. It was found that the ring hydrogen atoms had substantially lower reactivity than those on the α-carbon atoms. This difference was

not found on unsintered films at somewhat lower temperatures. By
extrapolation of the temperature dependence it is concluded that the
behaviour cannot be due to the reaction temperature difference but is
connected with a change in the nature of the catalyst surface. This is an
indication that the mechanism of the exchange of ring atoms is different
from that for hydrogen atoms in the alkyl group. A mechanism for the
exchange of ring hydrogen atoms involving π-adsorbed states was discussed in section IV.C.2. Similar intermediates are probably involved in
reactions of cyclic olefins which were shown to be inhibited by the
sintering of iron films (Erkelens et al., 1961). Crawford and Kemball
point out that the low rate of exchange of the ring hydrogen atoms on
sintered films is related to a low frequency factor for the reaction. This
is consistent with the concept of reaction on specific surface sites, the
concentration of which is substantially reduced by sintering.

Harper and Kemball (1965) further investigated the exchange and
deuteration of p-xylene on films of palladium, platinum and tungsten.
The p-xylene molecule contains only two kinds of hydrogen atom, the
six in the methyl substituents and the four on the ring. Comparison
of its reactions on different metals is a useful test of the generality of the
observations on nickel.

The composition of the exchanged p-xylenes on these metals are in
Table 27. Reaction on palladium did not give rise to products contain-

TABLE 27

Composition of exchanged p-xylenes at ten per cent reaction

Catalyst	Temperature (°C)	percentages of isotopic species d_1 d_2 d_3 d_4 d_5 d_6 d_7 d_8	M
Palladium film	0	59.3 24.9 8.2 3.8 2.5 1.3 0 0	1.69
Platinum film	0	57.6 17.1 14.1 3.4 2.4 5.4 0 0	1.92
Tungsten film	−9.5	48.2 19.9 12.0 8.5 3.7 4.6 2.0 1.0	2.26

ing more than six deuterium atoms, indicating that only the hydrogens
in the methyl groups were reacting. The behaviour of palladium therefore resembles that of nickel. On platinum the ring atoms exchanged at
about 1/3 of the rate of the hydrogen atoms of the methyl groups. It is
apparent that a different type of reaction takes place on platinum.
The amounts of d_2 and d_3 products are similar and substantially greater
than d_4, d_5 and d_6, which suggests that on platinum exchange occurs
independently within the methyl groups, but within a given methyl

group there is a tendency for complete replacement of all three hydrogen atoms to occur.

On tungsten both kinds of hydrogen react at similar rates and the distribution of products is more complex. The enhanced amount of d_4 and the presence of d_7 and d_8 at an early stage of the reaction indicate that the exchange of ring hydrogens is contributing to the distribution. Harper and Kemball found it impossible to obtain individual rates for the two types of exchange because the p-xylene was rapidly deuterated. The dimethylcyclohexane formed had an average deuterium content >9 atoms/molecule, indicating that substantial amounts of exchange accompanied deuteration. Further exchange of the dimethylcyclohexane with deuterium took place on tungsten in contrast to the behaviour over the other two catalysts.

It is concluded that, with the possible exception of tungsten, side chain exchange is faster than either ring exchange or deuteration on the metals studied. Horrex and Moyes (1965) reported results for the reaction of toluene with deuterium on evaporated films of nickel, iron, palladium and platinum, and obtained results in essential agreement with those of Harper and Kemball. Their order for side chain exchange rates is Ni > Fe > Pd, Pt and for deuteration rates, Pt > Ni > Fe > Pd.

D. HYDROGENATION AND HYDROGENOLYSIS OF KETONES

The catalytic reaction of a ketone with hydrogen is found to lead either to addition of hydrogen across the carbon–oxygen double bond with the formation of a secondary alcohol or complete rupture between carbon and oxygen (hydrogenolysis) with the formation of a hydrocarbon and water. The nature of the products depends on the ketone, on the temperature and on the nature of the catalyst.

The hydrogenation of acetone over evaporated films of a number of metals leads mainly to 2-propanol. The order of catalyst reactivity determined by Stoddart and Kemball (1956) is Pt > Ni > Fe > W > Pd > Au. The activation energies are in Table 28. The low activation energies for gold and palladium are compensated by low frequency

TABLE 28

Hydrogenation of acetone to 2-propanol on films

Metal	Pt	Ni	Fe	W	Pd	Au
Temperature Range (°C)	−46	−45	0	0	25	130
	−22	24	98	88	116	203
Activation Energy (kcal mole^{-1})	5.0	8.8	12.5	9.1	5.8	5.7

factors. Small amounts of propane were observed in the initial stages of the reaction in addition to the major product of 2-propanol. This was most apparent on platinum, and decreased in the order Pt > W > Ni > Fe. Propane was not observed on Au or Pd. Because the production of propane was confined to the initial stage of the reaction, Stoddart and Kemball suggest that the hydrogenolysis reaction occurs by a separate mechanism on a limited number of active sites which are poisoned as the reaction proceeds.

The initial rate of production of 2-propanol was found to be halved when 2-propanol was added to the reaction mixture to a pressure equal to that of acetone. This suggests that 2-propanol competes with acetone for the surface and that the two are adsorbed with approximately equal strength. The overall kinetics were found to be zero order in both acetone and hydrogen pressures provided hydrogen was in excess. This indicates that the surface is simultaneously saturated with acetone and hydrogen. It is suggested that this condition is obtained not because hydrogen can displace the strongly adsorbed acetone, but because the size and geometry of the acetone molecule necessarily leaves sites accessible only to hydrogen. As the ratio of hydrogen pressure to acetone pressure is reduced, the coverage of the vacant sites between the adsorbed acetone molecules becomes dependent on the pressure of hydrogen.

The reaction of acetone with deuterium over evaporated metal films results in exchange which is generally much more rapid than addition (Kemball and Stoddart, 1957). The distributions of products are in Table 29. On nickel and iron, significant deuteration accompanied the exchange, and the accumulation of 2-propanol caused a decrease in the exchange rate. While it was concluded that in the hydrogenation of acetone, the product and reactant competed for the surface on approximately equal terms, the above result suggests that this is not true for

TABLE 29

Initial products from exchange of acetone with deuterium on films

Metal	Temp (°C)	Activation Energy (kcal mole^{-1})	d_1	d_2	d_3	d_4	d_5	d_6	M
Fe	22.5	9.9	94.2	3.6	2.2	0.0	0	0	1.08
W	69.5	10.0	92.2	6.2	1.3	0.4	0.1	0	1.11
Ag	230.8	14.6	90.5	8.5	1.1	0.0	0	0	1.11
Au	120.6	13.9	76.1	11.2	5.6	2.5	2.1	2.5	1.51
Pt	−35.5	3.4	64.6	16.1	17.6	0.7	0.5	0.5	1.58
Pd	−34.5	10.1	65.3	16.8	11.6	2.8	1.8	1.8	1.64
Ni	−23.0	7.3	52.8	11.2	17.5	6.7	4.6	7.1	2.20

exchange. The difference arises because the adsorption of the methyl groups is important in exchange whereas hydrogenation is concerned with adsorption at the carbon–oxygen bond. On tungsten, deuteration accompanied exchange, but in this case there was no fall in rate as the 2-propanol accumulated. Tungsten is known to adsorb hydrocarbons more strongly than iron and nickel and it is suggested that the methyl groups of the acetone are strongly adsorbed.

The predominance of the d_1 compound in the distribution of the deutero acetones shows that the most important process on all metals is the simple exchange involving only one methyl group. The initial adsorption then appears to involve the dissociative adsorption at one methyl group with the formation of an acetonyl radical and adsorbed hydrogen atom. However, the mean deterium number was generally found to be somewhat greater than unity, showing that some multiple exchange does occur. For tungsten this apparently occurs by a repeated second point adsorption of the molecule bringing each methyl group in turn in contact with the surface. Such a process involving a 1–3 adsorbed intermediate has been previously observed for the exchange of neopentane on tungsten (Kemball, 1954b). For the other metals, the sharp cut-off in the distribution at d_3 shows that multiple exchange is confined to within one methyl group.

The most extensive studies of the reactions of higher ketones on evaporated metal films have been with cyclopentanone and cyclohexanone (Kemball and Stoddart, 1958). It was found that the hydrogenation of the cyclic ketones to the cyclic alcohols is generally similar to the hydrogenation of acetone. The confirmation of the observation that ketone and alcohol are adsorbed with equal energies suggests that ketones and alcohols in general form the same adsorbed complex on the surface.

The order of reactivity of the metals and the activation energies for the hydrogenation of the cyclic ketones correspond with those for acetone. It is concluded that a common mechanism involving the desorption of the same type of intermediate is rate determining. Kemball and Stoddart suggest that the slow desorption is the breaking of the secondary carbon–metal bond of the half-hydrogenated ketone complex, (step 2)

$$
\begin{array}{ccccc}
\overset{\textstyle R}{\underset{\textstyle \diagdown}{R}}\!\!\diagdown & & \overset{\textstyle R}{\underset{\textstyle \diagdown}{R}}\!\!\diagdown & & \overset{\textstyle R}{\underset{\textstyle \diagdown}{R}}\!\!\diagdown \\
\text{C—O} & & \text{C—OH} & & \text{C—OH} \\
|\;\;| & \xrightarrow{\hspace{1cm}} & | & \xrightarrow{\hspace{1cm}} & | \\
\text{M   M} & (1) & \text{M} & (2) & \text{H}
\end{array}
$$

adsorbed ketone   half-hydrogenated   alcohol

Hydrogenolysis reactions were found to occur to a much greater extent with the cyclic ketones than with acetone. The main product of hydrogenolysis on all metals was found to be the saturated hydrocarbon, i.e. cyclopentane or cyclohexane from the corresponding ketones. Tungsten also produced the corresponding cyclic olefins. The order of activity of the metals for hydrogenolysis of cyclopentanone was found to be Pt ≫ Pd > Rh > W > Ni. This order is quite different from that for hydrogenation of the ketone which was the same as that given above for acetone.

The kinetics of the exchange of cyclopentanone with deuterium on tungsten and nickel are similar to those found for the exchange of acetone. The d_1 compound predominated and no exchange products above d_4 could be detected. On palladium rapid exchange of cyclopentanone to d_4 occurred but for cyclohexanone the exchange went beyond d_5. Kemball and Stoddart conclude that neither of the expected explanations based either on exchange of one side of the ring at a time or the differing reactivities of α and β hydrogens, can be proved by the available results. Cornet and Gault (1967) have investigated the reactions of 2-methylcyclopentanone on a number of catalysts including nickel films. They find that both cis- and trans-2-methylcyclopentanol are produced and that the isomerization does not proceed via a simple adsorbed ketone intermediate (above). From the distribution of deuterium in both alcohol and ketone they conclude that the mechanism of deuteration cannot depend only on the addition of two deuterium atoms to the adsorbed ketone. The mechanism proposed by Cornet and Gault involves a triadsorbed π-allylic species in which the carbonyl group and an adjacent carbon atom in the ring interact with the surface. It is considered that such a species could react with both adsorbed hydrogen from below and molecular hydrogen from above.

V. Hydrogenolysis and Skeletal Isomerization Reactions

By hydrogenolysis, we refer here to catalytic reactions which result in rupture of a [C–X] bond, where the atom X may be C, S, N or halogen: that is, the net reaction at this bond may be represented as

$$[C\!-\!X] + H_2 \to [CH + HX] \qquad\qquad (V)\text{--}(1)$$

where it is understood that C and X may be bonded to other atoms to satisfy ordinary valence requirements. On some catalysts, but particularly on platinum, hydrogenolysis of saturated hydrocarbons is accompanied by skeletal isomerization, of which the conversion of n-butane to isobutane may be cited as an example to make clear the meaning of the term. The two reactions are under some circumstances, mechanistically related.

A. REACTIONS OF ALIPHATIC AND ALICYCLIC HYDROCARBONS

In all of this discussion we shall be dealing with reactions occurring in the presence of excess gas phase hydrogen, unless otherwise stated.

1. *Isomerization reactions*

The skeletal isomerization of saturated hydrocarbons over evaporated films of platinum was first demonstrated by Anderson and Baker (1963) and Anderson and Avery (1963) and this work was confirmed and extended in considerable mechanistic detail by Anderson and Avery (1966) and by Gault and his collaborators (Barron *et al.*, 1963, 1966). This has clearly demonstrated that this type of skeletal isomerization reaction can proceed on the surface of platinum alone, and provides an alternative reaction path for isomerization to that which exists with dual-function catalysts of the platinum/alumina-silica type (Mills *et al.*, 1953; Sinfelt *et al.*, 1960). The reality of a dual-function isomerization mechanism has been amply demonstrated by Weisz (1962) and involves the following reaction path

$$
\begin{bmatrix} \text{aliphatic or} \\ \text{alicyclic} \end{bmatrix}_1 \xrightarrow[\longrightarrow]{\text{Pt site } (-H_2)} \text{olefin (1)} \xleftarrow{\begin{array}{c}\text{acidic}\\\text{site } (+H^+)\end{array}} \begin{array}{c}\text{carbonium}\\\text{ion (1)}\end{array}
$$

$$
\begin{bmatrix} \text{aliphatic or} \\ \text{alicyclic} \end{bmatrix}_2 \xrightarrow[\longrightarrow]{\text{Pt site } (+H_2)} \text{olefin (2)} \xleftarrow{\begin{array}{c}\text{acidic}\\\text{site } (-H^+)\end{array}} \begin{array}{c}\text{carbonium}\\\text{ion (2)}\end{array}
$$

$$(V)-(2)$$

Two important features distinguish isomerization reactions occurring only at a platinum surface from those involving a dual-function catalyst. In work with evaporated platinum films, no acidic sites are available, and in the definitive experiments care was taken to avoid contacting the reaction mixture with heated glass as a precaution against the (unlikely) activity of the glass as a source of acidic sites. More importantly, the reaction proceeds readily at a quaternary carbon atom where olefin formation is impossible.

As we shall see, experimental evidence shows the operation of two distinct mechanisms with metal-only isomerization reactions. The first is a simple C–C bond shift and probably occurs via a 1–3 adsorbed intermediate. The second proceeds through an adsorbed carbocyclic

H

intermediate. The size of the reactant molecule is an important factor influencing which mechanism operates. Molecules of insufficient size to give a C_5 carbocyclic intermediate isomerize by the bond shift mechanism. With larger molecules both processes are possible and they often occur co-operatively. We shall begin by discussing the two mechanisms separately.

2. Isomerization via a bond shift mechanism

Table 30 gives some results found by Anderson and Avery (1966) for reactions of neopentane and the butanes over polycrystalline evaporated films of platinum. The hydrogenolysis and isomerization reactions have activation energies and frequency factors that are very similar, and this strongly suggests that the mechanisms are closely related. The presence of an iso-structure in the molecule (neopentane, isobutane) favours the occurrence of isomerization.

In addition to this skeletal isomerization reaction, Anderson and Avery showed that in a suitable isotopically labelled hydrocarbon, a reaction of positional isomerization occurred. Thus, with n-butane-1-^{13}C as the starting material, the isomerization reaction products over platinum were 2-methyl (^{13}C)-propane (that is, peripherally labelled isobutane) and n-butane-2-^{13}C

$$CH_3 CH_2 CH_2{}^{13}CH_3 \begin{array}{c} \nearrow (CH_3)_2 CH\,{}^{13}CH_3 \\ \\ \searrow CH_3 CH_2{}^{13}CH_2 CH_3 \end{array} \qquad\text{(V)-(3)}$$

and the amount of n-butane-2-^{13}C was in a constant ratio to the amount of isobutane produced, this ratio being 25%. At the same time, no scrambling of the ^{13}C occurs, so the isomerization reactions are entirely intramolecular. In neither this, nor in the hydrogenolysis reaction does a recombination of discrete surface residues occur.

In a purely *formal* sense these isomerizations can be represented as group transfers thus

$$\text{(V)-(4)}$$

TABLE 30

Reactions of neopentane and the butanes over platinum films

	Activation energy (kcal mole^{-1})	$\log_{10} A$*	initial reaction products (mole percent)							percent of parent giving isomerization product
			CH_4	C_2H_6	C_3H_8	iso-C_4H_{10}	n-C_4H_{10}	iso-C_5H_{12}	n-C_5H_{12}	
neopentane (239–299 °C)	21 (<270 °C)	21.2 (<270 °C)	14	5	4	10	3	59	5	78
isobutane (256–299 °C)	21	21.0	24	6	20	—	50			68
n-butane (256–300 °C)	21	21.1	32	29	28	11	—			20

* A, frequency factor, molecules sec^{-1} cm^{-2}.

For this reaction to proceed via an adsorbed cyclic intermediate the latter would have to be a three membered carbocyclic ring. One would expect this to be of unreasonably high energy judging from the known energy of cyclopropane relative to propane and, furthermore, Anderson and Avery showed that in the reaction of methylcyclopropane itself over platinum, the distribution of reaction products was markedly different from that required to be consistent with the butane isomerization results. Anderson and Avery (1966, 1967a) have proposed a mechanism based upon a 1–3 adsorbed intermediate, which when formed from neopentane is

$$
\begin{array}{cc}
\text{CH}_3 & \text{CH}_3 \\
\diagdown & \diagup \\
& \text{C} \\
\diagup & \diagdown \\
\text{CH}_2 & \text{CH} \\
| & \parallel \\
\text{Pt} & \text{Pt}
\end{array}
\qquad (A)
$$

The mechanism was formulated as an extension to the general theory of carbonium ion, free radical, and carbanion rearrangements discussed by Zimmerman and Zweig (1961). The proposed reaction scheme is shown in (V)–(5)

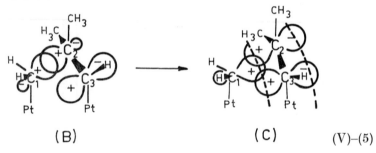

(B)   (C)   (V)–(5)

Species (B) in (V)–(5) corresponds to (A), and C_3 is double bonded to the surface: (C) is a bridged intermediate in which C_2 has been rehybridized to sp_2 and the π-system extends over C_2, C_3 and a surface atom. The likelihood of isomerization was assessed by Anderson and Avery (1967a) in terms of the energy difference between (B) and (C) using Zimmerman and Zweig's isomerization criterion that the energy of the bridged intermediate should be lower than that of the precursor. It was shown that isomerization was favoured where there is partial electron transfer from the adsorbed hydrocarbon residue to the surface metal atom. Values of 2.2 and 1.7 have been quoted for the electronegativities of an isolated platinum atom and a π-bonded carbon atom, and this difference is such as to favour isomerization by this mechanism. This treatment also included in the calculation the explicit effect of methyl group hyperconjugation, and the following decreasing order of isomerization activity was predicted: neopentane > isobutane > n-butane, which agrees with the experimental results.

3. *Butane isomerization on* (111) *and* (100) *platinum surfaces*

The relative rate of isobutane isomerization has been shown by Anderson and Avery (1966) to be markedly increased by using a (111) platinum film surface. On the other hand, this did not occur with n-butane, nor did it occur with either iso- or n-butane over a (100) platinum surface. A triangular array of adjacent sites on a (111) platinum surface can be readily fitted by an adsorbed isohydrocarbon, and this structure also fits to allow the carbon orbitals to be directed normally to the surface. On simple geometric grounds, this adsorbed structure is specific to the (111)/isohydrocarbon system. The extra residue reactivity resulting from this structure can be understood in terms of the mechanism discussed in the previous section. Thus, if double bonding with the surface occurs at two out of the three carbon atoms,

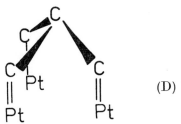

(D)

by comparison with the two-point adsorbed structure (e.g. A), structure (D) can provide for an increase in isomerization rate on purely statistical grounds by a factor of two. However, a more important factor may well be the extent to which (D) allows electron removal from the π-system thus lowering the energy of the bridged structure.

With supported catalysts, the proportion of (111) surface exposed can be expected to depend on the method of preparation. This has recently been explored by Boudart *et al.* (1968) who have correlated an increasing facility for neopentane isomerization with the increasing temperature of catalyst pretreatment (425–900 °C). This probably occurs because an increasing pretreatment temperature allows an increased proportion of the thermodynamically stable (111) faces to be formed (cf. section II Chapter 7). A conclusion which follows is that if the catalyst surface is such as to expose a reasonable proportion of sites upon which three-point adsorption can occur, these will dominate the isomerization by bond shift if the reactant molecule contains the required isostructure. Furthermore, if a reactant molecule is large enough to contain both an iso- and a n-structural unit, isomerization *by this mechanism* will favour the former.

This discussion has so far tacitly assumed that an adsorbed hydrocarbon molecule only undergoes a single bond shift process in one period of residence on the surface. However, it will be seen from Table 30 that about one-twelfth of the total initial isomerization product from neopentane was n-pentane, the balance being isopentane. Bearing in mind that since we are considering initial reaction products, the chance of n-pentane being formed by the re-adsorption of isopentane already produced is very small, the only reasonable conclusion is that the n-pentane is formed by two consecutive bond shifts during one residence period. In a similar way Boudart and Ptak (1970) has shown that in the isomerization of neopentane on a supported gold catalyst at 358 °C, all but 10% of the reacted parent appears as isomerization product, and of this, 33% is n-pentane and 67% isopentane. In general one would expect a process requiring two consecutive shifts to be of lesser importance, and this appears to be the case.

Platinum is by far the most active catalyst for skeletal isomerization. Anderson and Avery (1966) reported that palladium films had a small activity for n- and isobutane isomerization at 270–310 °C, but hydrogenolysis accounted for 95–98% of the total reaction. Boudart (1969) has found that unsupported gold powder is active for neopentane isomerization at 400 °C, and some 80% of the reaction product was isopentane. The activity of iridium films is limited entirely to hydrogenolysis. Boudart (1969) has reported neopentane isomerization with a silica supported iridium catalyst. It seems likely that this difference between the behaviour of iridium films and supported iridium catalysts is due both to the use of a flow reaction system with the latter which tends to minimize hydrogenolysis of isomerization products, and to contamination of the supported iridium surface which makes this less active for hydrogenolysis than a clean film.

4. *Isomerization of larger molecules: carbocyclic intermediates*

Some typical data for isomerization of some of the hexanes over platinum catalysts are given in Table 31. These reactions have been particularly studied by Gault and his collaborators. With supported platinum catalysts, the proportions of the products appear to depend on the nature of the catalyst and on the space velocity and degree of conversion, but even so, the significant products remain restricted to those indicated in Table 31 as having non-zero values. The products from the film catalysts refer to low parent conversions.

TABLE 31

C_6 reaction products from some hexanes over platinum catalysts

reactant and catalyst	initial C_6 reaction products (mole percent)								reference
	2-MP	3-MP	n-H	2,3-DMB, CHe,n-He	neo-H	MCP	CH	B	
2-methylpentane film, 281 °C	—	23.2	18.5	0	0.1	58.2	0	0	(1)
n-hexane, film, 274 °C	28.4	14.5	—	0	0	43.6	5.0	8.5	(2)
2-methylpentane, 10% Pt/Al$_2$O$_3$	—	59–79	18–30	0	0	2–11	0	0	(1)
3-methylpentane, 10% Pt/Al$_2$O$_3$	80–91	—	7–18	0	0	1–2	0	0	(1)

2-MP ≡ 2-methylpentane: 3-MP ≡ 3-methylpentane: n-H ≡ n-hexane: 2,3-DMB ≡ 2,3-dimethylbutane: neo-H ≡ neohexane: MCP ≡ methylcyclopentane: CH ≡ cyclohexane: B ≡ benzene: CHe ≡ cyclohexene: n-He ≡ n-hexene.
(1) Corolleur (1969); (2) Anderson and Macdonald (1969b).

A feature immediately apparent from Table 31 is the formation of very substantial amounts of cyclic products. Starting from n-hexane, closure to a C_5 (methylcyclopentane) or a C_6 (cyclohexane and benzene) ring is possible, but the former is strongly preferred. From 2- and 3-methylpentane a C_6 ring cannot be produced directly and is, in fact, almost totally absent. The formation of methylcyclopentane and cyclohexane clearly demonstrates the existence of adsorbed C_5 and C_6 carbocyclic species, and the further dehydrogenation of the C_6 species is no doubt responsible for the formation of benzene. The high energy of a C_4 ring presumably precludes the production of cyclobutanes in the reaction products, and for the same reason one may conclude, *a fortiori*, that adsorbed C_4 carbocyclic species are unlikely to be of major significance. In no case has olefin (e.g. n-hexene or cyclohexene) been detected as a reaction product when excess hydrogen is present in the reaction mixture.

It was previously reported by Barron *et al.* (1966) that the isomerization of n-hexane and 3-methylpentane over platinum films gave substantial amounts (25–35%) of 2,3-dimethylbutane, while over platinum/alumina catalysts only very small amounts of this product (1–2%) were claimed. However, the more recent results of Corolleur (1969) and

Anderson and McDonald (1969b) with film catalysts have failed to confirm the formation of this compound (cf. Table 31), and it remains at present unresolved whether this difference is due to an uncontrolled difference in catalyst behaviour (e.g. Anderson and Macdonald used UHV platinum films), or whether it results from an analytical artifact in the earlier work. We incline to a belief in the first of these alternatives.

It is instructive to compare the pattern of the hydrogenolysis products from methylcyclopentane with the corresponding isomerization products. Thus, Maire and Gault (1967) found for the hydrogenolysis of methylcyclopentane at 275–316 °C over a 10% platinum/alumina catalyst (which behaved in many respects in a similar fashion to film catalysts), the ratio of products

$$\frac{\text{n–hexane}}{\text{2–methylpentane}} \simeq 6\text{–}10$$

while from Table 31 we see that the isomerization of 3-methylpentane gave

$$\frac{\text{n–hexane}}{\text{2–methylpentane}} \simeq 9$$

From considerations such as those given above, Gault suggested that in the C_6 series, isomerization proceeds via an adsorbed C_5 carbocyclic intermediate. For instance, skeletal interconversions involving n-hexane, 2- and 3-methylpentanes may be written

However, the temperatures for the reactions of the hexanes (cf. Table 31) are similar to those required for the bond shift mechanism, and thus one would expect, in general, reactions via a bond shift and via a carbocyclic intermediate to occur together when the reactant geometry permits. This problem has recently been examined by Corolleur (1969) in the reaction of 2-methylpentane-2-^{13}C

$$CH_3-\overset{\overset{\displaystyle CH_3}{|}}{\underset{\underset{\displaystyle H}{|}}{^{13}C}}-CH_2-CH_2-CH_3$$

and the results of this reaction are set out in Table 32, which shows the

TABLE 32

Isotopic species in the hexanes from the reaction
of 2-methylpentane-2-^{13}C over platinum films

component	composition of each component (mole percent)		
2-methylpentane	3%	93% (parent)	4%
3-methylpentane	0–2%	57–61%	39–41%
n-hexane	20%	63–64%	17–18%

The dot represents the position of the ^{13}C atom.

proportions of the various isotopic species present in the "hexane" component of the reaction mixture at an early stage of the reaction.

A reaction path passing only through an adsorbed C$_5$ carbocyclic intermediate (cf. reaction (V)–(6)) cannot account for the formation of n-hexane-1-^{13}C, n-hexane-3-^{13}C, 3-methylpentane-1-^{13}C and 3-methylpentane-2-^{13}C which appear as important products. Furthermore, a single bond shift process can, from the listed products, only account for n-hexane-2-^{13}C and n-hexane-3-^{13}C. It is clear that (at least) two consecutive processes are required to account for the distribution of products, and the fact that we are speaking of initial reaction products means that these consecutive processes must occur in a single residence of a molecule on the surface, that is, without intervening desorption and readsorption, although the latter no doubt becomes increasingly important at higher conversions. If we restrict ourselves for the moment to no more than two consecutive processes, then in general an adsorbed reactant molecule has the choice of the reaction paths shown in Figure 7 where all of the processes occurring within the broken line take place

H*

within the adsorbed phase in a single residence period. The data in Table 32 are, in fact, not sufficient for a completely unambiguous decision to be made about the reaction path. However, the following points emerge. If the overall process is limited to two consecutive bond

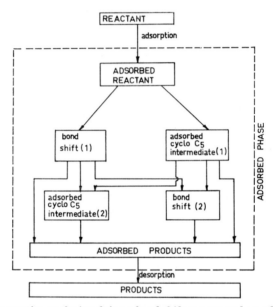

FIG. 7. Possible reaction paths involving a bond shift process and an adsorbed cyclo–C_5 intermediate for skeletal isomerization.

shifts, this would account for all of the products with the exception of 3-methylpentane-1-^{13}C. The presence of the latter is not certain since the mass spectral data only yield the sum of 3-methylpentane-1- and -2-^{13}C. However, the overall reaction path cannot be *limited* to two consecutive bond shifts since an adsorbed C_5 carbocyclic species is inferred from the formation of methylcyclopentane. Again, the overall reaction cannot be limited to two consecutive C_5 carbocyclic intermediates, since this will not explain the formation of five of the listed products, the most important of which are n-hexane-1- and -3-^{13}C. All of these are important omissions, and one can immediately say that while these particular paths may contribute they are, of themselves, insufficient. For an economy of hypothesis, all of the reaction products with the exception of 3-methyl (^{13}C)-pentane can be accounted for by assuming two operating paths which we may summarize as in Figure 8a. The proportion of 3-methyl (^{13}C)-pentane is quite small (0–2%, cf. Table 32), and if this is ignored, the reaction scheme in Figure 8a is adequate.

To account for the formation of this compound, Corolleur (1969) suggests the possibility of a third step within one path, as in Figure 8b. The reaction schemes outlined in Figures 7 and 8 are readily specialized in terms of a given reactant, and Figure 9 illustrates this with part of

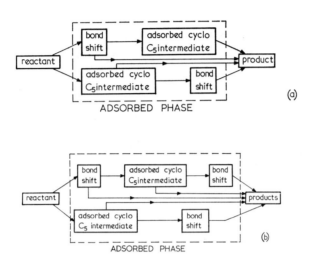

FIG. 8. (a), upper; Main reaction paths for skeletal isomerization of 2-methylpentane-2-^{13}C over platinum.
(b) lower; Complete reaction scheme for skeletal isomerization of 2-methylpentane-2-^{13}C over platinum.

the reaction path for the reaction of 2-methylpentane-2-[13]C: here, the desorption of methylcyclopentane is not shown explicitly.

The work of Anderson and Avery (1966) showed that in the isomerization of isopentane over platinum films, only a very small amount ($< 1\%$) of neopentane is produced, while the data in Table 31 show a corresponding absence of neohexane as a significant product in the hexane series. These observations are in agreement with the mechanisms discussed above, and in particular it may be noted that the bond shift mechanism as formulated in reaction (V)–(5) does not allow the formation of a quarternary carbon atom to occur. Nevertheless, Barron *et al.* (1966) have previously reported the formation of small but significant amounts of neohexane ($< 5\%$) from the reaction of n-hexane and 3-methylpentane over platinum films. We can only offer again the same comments about this apparent disagreement as have already been previously made with respect to 2,3-dimethylbutane. If necessary, a number of possibilities present themselves for the formation of products

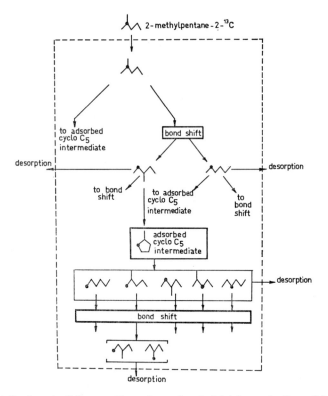

Fɪɢ. 9. Detail of part of the reaction scheme for skeletal isomerization of 2-methyl-pentane-2-^{13}C over platinum.

containing a quaternary carbon. The production of neohexane could be formulated via an absorbed C_4 carbocyclic intermediate from an immediate precursor with a 2-methylpentane skeleton. Or again, reaction scheme (V)–(5) could be modified to enable a quaternary carbon atom to be formed at C_3, if C_3 could be bound to the surface by a π-only bond. Alternatively, a modified bond shift mechanism could be formulated via an adsorbed C_3 carbocyclic intermediate.

Recent work by Gault (1969) on the reaction of 2-methyl (^{13}C)-3-methylbutane and 2,3-dimethylbutane-2-^{13}C has confirmed that an adsorbed C_4 carbocyclic reaction intermediate is of little importance, and this agrees with Corolleur's (1969) conclusion which was reached from a study of the isomerization of pentane-1-^{13}C over platinum film catalysts.

These distributions of initial products which we have discussed are controlled by kinetic factors. However, it is useful to take note of some of the relevant thermodynamic quantities which place a boundary on the reaction possibilities. The following discussion is confined to a temperature of 550 °K (277 °C) which is a typical reaction temperature. Free energies of formation ($\Delta G^0_{f\,550\,°K}$) interpolated from API Project 44 data are (kcal mole^{-1}): n-hexane, 35.7; 2-methylpentane, 35.0; 3-methylpentane, 35.6; neohexane, 35.2; 2,3-dimethylbutane, 36.1; methylcyclopentane, 39.3; cyclohexane, 40.9; benzene, 41.4; 1-hexene, 48.4. By using these data we may compute equilibrium constants and equilibrium partial pressures in the usual way, from which the following points emerge. The equilibrium constants for equilibria between the isomeric hexanes are such that, at equilibrium all isomers would be present at appreciable concentrations: in particular, the equilibrium constants for n-hexane $\rightleftarrows$ neohexane, and n-hexane $\rightleftarrows$ 2,3-dimethylbutane are 1.58 and 0.695 respectively, so the absence of these products from the reaction of n-hexane (cf. Table 31) is kinetic and not thermodynamic in origin.

Four dehydrogenation equilibria will be illustrated:

(a) n-hexane (g) $\rightleftarrows$ methylcyclopentane (g) $+ H_2$(g)

(b) n-hexane (g) $\rightleftarrows$ cyclohexane (g) $+ H_2$(g)

(c) n-hexane (g) $\rightleftarrows$ benzene (g) $+ 4H_2$(g)

(d) n-hexane (g) $\rightleftarrows$ 1-hexene (g) $+ H_2$(g)

The corresponding equilibrium constants are (a) 3.8×10^{-2} atmos, (b) 8.75×10^{-3} atmos, (c) 5.61×10^{-3} (atmos)^4, (d) 1.12×10^{-5} atmos. From this it follows that, starting with a typical reaction mixture containing 50 torr hydrogen and 5 torr n-hexane, the partial pressures in each case at equilibrium would be, (a) methylcyclopentane, 1.86 torr; (b) cyclohexane, 0.59 torr; (c) benzene, 4.99 torr; (d) 1-hexene, 8.5×10^{-4} torr. It should be noted that the presence of appreciable amounts of dehydrogenation product from reactions (a)–(c) is dependent on a relatively low hydrogen partial pressure. For instance, if the hydrogen pressure were 760 torr instead of 50 torr in the above example, the partial pressures at equilibrium would be (a) methylcyclopentane, 0.19 torr; (b) cyclohexane, 0.04 torr; (c) benzene, 0.03 torr; (d) 1-hexene 5.6×10^{-5} torr. For mechanistic studies it is clearly an advantage to work at lower pressures where a wider range of products becomes observable.

In general summary we may say that these isomerization reactions in the hexane and pentane series are dominated by paths to which the bond shift process and reaction by an adsorbed C_5 carbocyclic intermediate are of major importance. The relative contributions that these two processes make depends on the nature of the catalyst as well as hydrocarbon geometry. Thus, on platinum films and on 10% platinum/

alumina catalysts these processes are of comparable importance, but on a 0.2% platinum/alumina catalyst, reaction by an adsorbed C_5 carbocyclic intermediate is dominant (Coroleur 1969). Moreover, Gault (1969) has found that the proportions of the products from the reaction of 2-methylpentane-2-^{13}C and of n-hexane-2-^{13}C on platinum vary appreciably from film to film: in other words, the relative contributions from bond shift and carbocyclic mechanisms vary. Clearly the proportions of these mechanisms must depend on features of film structure which are not controlled during the normal method of preparation and use of polycrystalline platinum films. We anticipate here the discussion in section V.A.6 where it is pointed out than an ultra-thin platinum film strongly favours a carbocyclic intermediate. Now, with the current experimental technique a platinum film of normal thickness will always be accompanied by an ultra-thin region at the edge region: the area of this ultra-thin region which is heated to reaction temperature will certainly vary from experiment to experiment, since it will depend on matters like the positioning of the heating oven around the reaction vessel and the exact geometry of the evaporation filament. We suggest that this is probably the most important (but not the only) feature responsible for this sort of variability in reaction mechanism.

Finally we should remark on the likely importance of adsorbed C_6 carbocyclic intermediates in isomerization reactions with molecules larger than the hexane series. Moreover, a bond shift process would be capable of effecting ring enlargements and contractions: an illustration of this may well occur in a reaction studied by Kazanskii (1967) over a platinum/carbon catalyst, the conversion of 2,2,4-trimethylpentane to m- and p-xylenes; a process that occurs via 1,1,3-trimethylcyclopentane and which involves ring enlargement of the latter into the cyclohexanes.

5. Hydrogenolysis reactions

The simplest hydrocarbon hydrogenolysis reaction is that with ethane which can yield methane as the sole reaction product. On the other hand with larger molecules a range of reaction products is possible since the adsorbed reactant may fragment at more than one C–C bond and, furthermore, even if only one such bond is broken in the reaction of a given molecule, the reactant will often contain more than one stereochemically distinguishable type of C–C bond. Over film catalysts, the extremes of behaviour are typified by nickel and platinum. Figure 10 shows the dependence on temperature of the initial products from the reaction of some aliphatic hydrocarbons over nickel films; the formation of methane is the dominant reaction mode (190–270 °C), and

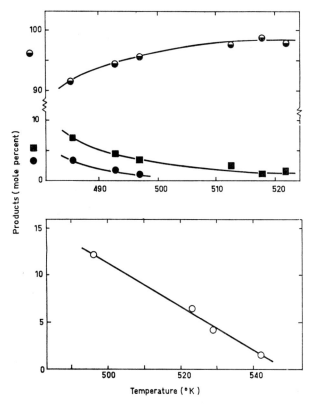

Fig. 10. Temperature dependence of composition of products from hydrogenolysis on nickel films. Upper: Products from neopentane: methane, ◓; ethane, ◼; propane, ●. Lower: Proportion of ethane in products from propane (balance of products methane).

the proportion of ethane increases with increasing temperature (Anderson and Baker, 1963). However, it is clear that this dominant methane production is to some extent dependent both on the nature of the hydrocarbon and on the type of catalyst. Thus, the reaction of methylcyclobutane occurs on nickel films at 180 °C, and although C_1–C_5 products are formed, the most important are the pentanes from ring opening, and this is no doubt due to the enhanced reactivity of the strained ring (Maire et al., 1965). Again, propane hydrogenolysis was examined over a nickel/kieselguhr catalyst by Morikawa et al. (1937) (at 138–184 °C), who found that a simple reaction to methane and ethane in equal proportions occurred. It seems very likely that on a clean surface such as a film, extensive dehydrogenative adsorption can readily occur, and this leads to extensive fragmentation. This is supported by some recent work by Kochloeft and Bazant (1968) who examined the hydrogenolysis of C_8–C_{10} aliphatics over a nickel catalyst containing 8% Al_2O_3: at 220 °C the predominate reaction was successive degradation to methane and a hydrocarbon containing one carbon less than the starting material, rather than extensive fragmentation.

Over platinum films (150–300°C) the main features of the initial product distributions from C_3–C_4 hydrocarbons may be accounted for by the assumption that in any particular molecule on the surface, only one C–C bond is broken, and that the chance of any given bond being broken is independent of whether it is 1°–2°, 2°–3° etc., as shown in Table 33 (Anderson and Avery, 1966).

TABLE 33

Hydrogenolysis products from C_3–C_4 aliphatic hydrocarbons over platinum films

		products (mole percent)		
		C_1	C_2	C_3
propane	initial obs.	~50	~50	
	calc.	50	50	
n-butane	initial obs.	36	33	31
	calc.	33	33	33
isobutane	initial obs.	48	12	40
	calc.	50	0	50

The reaction of various methyl substituted cycloalkanes has been studied by Maire et al. (1965) over both platinum films and platinum/alumina catalysts. The reactions are limited to C–C bond rupture within the ring (ring opening), and the various alternative reaction modes are

Again, the reactions can mostly be described by assuming that only one C–C bond ruptures in any given molecule. The temperature ranges for reaction are: (E), 50–120 °C; (F), 50–150 °C; (G), 160–260 °C; (H), 230–310 °C; (I), 320 °C; so of these compounds, it is only with the cyclopentanes where the reactivity for C–C bond rupture in the ring is similar to the reactivity of aliphatic molecules. With these cycloalkanes the product distributions are temperature dependent. For instance, Table 34 gives the temperature dependence of the ratio r from the re-

TABLE 34

Temperature dependence of r from reaction of
methylcyclobutane over platinum films

reaction temperature (°C)	51	72	82	105	120	150
r		2.06	1.80	1.66	1.44	1.30	1.0

action of methylcyclobutane over platinum films, where

$$r = \frac{\text{initial proportion of isopentane}}{\text{initial proportion of n-pentane}} = \frac{\text{chance of ring opening at (b)}}{\text{chance of ring opening at (a)}}$$

Thus, unlike the simple aliphatic hydrocarbons, with these cycloalkanes the relative chance of C–C bond rupture is influenced by energy factors, and the data of Table 34 correspond to the activition energy for opening at (b) being about 2–3 kcal mole^{-1} lower than at (a). There is a trend, therefore, for each C–C bond in the ring to open with approximately equal facility towards the upper end of the reaction temperature range. However, with supported catalysts the behaviour is to come extent dependent on the nature of the catalyst.

The reactions of C_3–C_4 aliphatics were also studied over films of tungsten, rhodium and palladium (Anderson and Baker, 1963); Anderson and Avery, 1966). The nature of the reaction fell between the extremes of nickel and platinum with tungsten tending towards the behaviour of nickel.

Many of these hydrogenolysis reactions on metal films are self-poisoned due to strongly adsorbed residues, and the reaction rates decrease with time. This has been noted with C_5–C_6 aliphatics and cyclo-alkanes on platinum and tungsten.

The kinetic parameters for hydrocarbon hydrogenolysis reactions are collected into Table 35 and here data for supported catalysts has been included for comparison with data obtained over evaporated films. The following main points emerge from these data. The activity of the supported catalysts can vary very widely for a given hydrocarbon (e.g. ethane), and this variability can be reflected in either the activation energy as with nickel/silica,alumina catalysts, or in the frequency factor, as with cobalt/silica catalysts. The range of activation energies

TABLE 35

Kinetic parameters for hydrogenolysis reactions

hydrocarbon	catalyst	E^*	$\log_{10}A$†	rate ∝ $p_{Hc}^x\, p_{H_2}^y$		temperature range (°C)	reference
				x	y		
C_2H_6	Ni film	58	35.8	—	—	254–273	(1)
	Ni/kieselguhr	52	—	0.7	−1.2	181–214	(2)
	Ni on various silica, alumina supports	28–42	—	0.9 to 1.0	−1.7 to −2.4	175–335	(3), (4), (5), (6)
	W film	27	26.2	—	—	173–183	(1)
	Pt film	57	34.2	~1.0	—	274–340	(7)
	Pt/silica	54	31.8	0.9	−2.5	344–385	(8)
	Pd film	50	31.9	~1.0	—	273–358	(7)
	Pd/silica	58	33.6	0.9	−2.5	377–392	(8)
	Ru/silica	32	28.1	0.8	−1.3	177–257	(8)
	Os/silica	35	—	0.6	−1.2	127–162	(9)
	Rh/silica	42	31.8	0.8	−2.2	192–242	(8)
	Ir/silica	36	28.7	0.7	−1.6	177–212	(8)
	Co on various silica supports	29–31	~25.5	1.0	−1 to 0	219–288	(4), 10), (11)
						288–330	(4)

Reactant	Film					E range	Ref
	W film	18	21.7	—	—	211–207	(1)
						180–190	(1)
n-$\mathrm{C_4H_{10}}$	Ni film	34	28.5	—	—	188–209	(1)
	W film	7	16.4	—	—	144–164	(1)
	Pt film	21	21.1	0.7	1.4	256–300	(7)
	Pd film	~38	~27	−0.3	~0	276–310	(7)
iso-$\mathrm{C_4H_{10}}$	Ni film	30	26.8	—	—	201–221	(1)
	Pt film	21	21.0	0.5	1.4	265–299	(7)
	(111) Pt film	19	19.5	—	—	294–305	(7)
	Pd film	21	21	−0.2	0.1	270–311	(7)
neo-$\mathrm{C_5H_{12}}$	Ni film	32	26.3	—	—	222–265	(1)
	W film	11	17.5	—	—	202–219	(1)
	Pt film	21	21.2	—	—	239–270	(7)
neo-$\mathrm{C_6H_{14}}$	Ni film	>25	—	—	—	181–200	(1)

*E, activation energy, kcal mole^{-1}; †A, frequency factor, molecules cm^{-2} sec^{-1}.

(1) Anderson and Baker (1963); (2) Kemball and Taylor (1948); (3) Taylor *et al.* (1965); (4) Sinfelt *et al.* (1965); (5) Taylor *et al.* (1964); (6) Yates *et al.* (1964); (7) Anderson and Avery (1966); (8) Sinfelt and Yates (1967); (9) Sinfelt and Yates (1968); (10) Yates *et al.* (1965); (11) Sinfelt and Taylor (1968).

over the various nickel catalysts is particularly large, there being a difference of about 30 kcal mole^{-1} between the value for the ethane reaction over nickel films and the lowest figure for nickel/silica catalysts. It is significant that this lowest activation energy is for a catalyst with the lowest proportion of metal (1%), and that the reaction over a 15% nickel/kieselguhr catalyst gave an activation energy approximating to that for a nickel film catalyst: thus this variability may be due in part to a variation in the electronic properties of the metal as a consequence of a variation in metal particle size. However, at least with nickel, this is unlikely to be the main factor and it is difficult to escape the conclusion that on most supported catalysts the reaction has been studied on a surface that is qualitatively different from that of a metal film. We are not here implying mere crystallographic differences of surface metal structure, but a difference in the chemical composition of the catalyst surface. We believe it is extremely important to note that whereas most metal films show appreciable and sometimes severe evidence for self-poisoning in hydrogenolysis reactions, the supported catalysts are much less affected in this way: this self-poisoning is a feature of the behaviour of clean metal surfaces, and the difference cannot be accounted for merely in terms of surface area. The conclusion follows that the initial surface on most supported catalysts is not clean metal in the sense that this term is applied to metal film catalysts, but we have, however, no adequate data with which further to characterize their structure.

Although ethane hydrogenolysis has been examined over a very wide range of catalysts, data for other hydrocarbons are largely confined to metal film catalysts, and reactions over films therefore provide some insight into the influence of hydrocarbon geometry on the reaction. We consider that the evidence discussed below leads to the following conclusions. Hydrogenolysis occurs from multiply adsorbed residues: in the case of ethane only the 1–2 adsorption mode is possible. However, with molecules larger than ethane, other reaction paths are possible: one is via adsorption into the 1–3 mode, and another involves adsorption as a π-allylic species. These two paths are both more facile than that for the reaction of ethane, and they offer parallel reaction routes for hydrogenolysis, of which the relative importance depends on hydrocarbon geometry and on the catalyst. The single most distinctive feature is that on all metals studied, the activation energy for ethane hydrogenolysis is much larger than for higher saturated hydrocarbons. In as much as the C–C bond energy in ethane ($\sim$83 kcal mole^{-1}) is not much different from that in the other lower aliphatic hydrocarbons (e.g. 85 kcal mole^{-1} in C_3H_8 and 78 kcal mole^{-1} for the central bond in n-butane, Cottrell,

1958), it is likely that this activation energy difference will be associated with a basically different hydrogenolysis mechanism for ethane compared to the other hydrocarbons. As pointed out previously by Anderson and Baker (1960) and by Anderson and Avery (1966), the hydrogenolysis data for neopentane are important because a 1–2 adsorption mode is impossible with this hydrocarbon. The data in Table 35 show that the activation energy for the neopentane reaction over nickel, tungsten and platinum is much lower than for ethane, but is comparable with values for C_3 and C_4 hydrocarbons. Furthermore, if hydrogenolysis could occur by adsorption at only a single carbon atom, it would be difficult to explain the effect of the different geometries of ethane and neopentane on the activation energy. It is thus difficult to avoid the conclusion that with C_3 and C_4 hydrocarbons, reaction via adsorption into the 1–3 mode which is mandatory for neopentane, can in principle make an important contribution. Thus, using neopentane as the archetype for this sort of process, we note that over platinum both hydrogenolysis and skeletal isomerization have the same activation energy, and we conclude that these reactions are mechanistically related in that the slow step in both involves the conversion of a 1–3 adsorbed species into a bridged intermediate, as shown in reaction (V)–(5) for neopentane. However, if the bridged intermediate is attacked by hydrogen, hydrogenolysis results rather than completion of the skeletal rearrangement. The detailed manner in which this hydrogen attack occurs is uncertain, but overall, the attack may be written

in which H has been added to C_A and C_B, and this is followed by further reaction with hydrogen to give product desorption. This process could occur by attack by H_2 either from the gas phase or physically adsorbed on an adjacent site. It is also possible to formulate (V)–(7) by a mechanism in which species (J) is attacked sequentially by two adjacent chemisorbed H atoms. In either case, if attack occurs from a third adjacent site, one would expect an adsorbed species such as (D) (*vide supra*) which is attached to three adjacent sites to be more resistant to hydrogenolysis for steric reasons, and to favour isomerization. This appears to be the case.

Although the mechanism is best characterized on platinum for which isomerization is an important reaction path, similar arguments based on hydrocarbon geometry apply to these reactions on other metals to which we suggest reaction (V)–(7) is also applicable. In these cases attack by hydrogen may make isomerization only of very minor importance (palladium and tungsten), or totally absent.

However, considerations of molecular geometry show that a reaction path additional to that via a 1–3 adsorption mode must be possible in many cases. Thus for instance, cyclopentane and cyclohexane undergo hydrogenolysis with a reactivity much greater than ethane and comparable to the other aliphatics. These cycloalkanes cannot adopt a 1–3 adsorption mode of the type discussed for neopentane, and Barron et al. (1966) suggested that adsorption as a π-allylic species is a precursor to ring opening. Reaction through such an intermediate will not result in ring opening adjacent to a gem-dimethyl substituent (although this could occur via a 1–3 adsorbed intermediate). In fact, 1,1-dimethylcyclopentane reacts over supported platinum as shown in Table 36 with zero chance of ring opening adjacent to the gem-dimethyl

TABLE 36

Hydrogenolytic ring opening modes for methylcyclopentanes
on platinum/carbon catalysts*

| cyclopentane derivative | percentage ring opening in indicated position | | | | |
	C_1-C_2	C_2-C_3	C_3-C_4	C_4-C_5	C_5-C_1
†Methyl-	5	32	26	32	5
1,1 demethyl-	0	37	26	37	0
1,2 dimethyl- (trans)	5	5	42.5	42.5	5
1,3 dimethyl- (trans)	14.2	14.2	14.2	45	14.2
1,2,3 trimethyl-	0	0	0	100	0

* 10–20% Pt/carbon; reaction at 300–320 °C, data from Newham's summary of the extensive work of Kazanskii and his collaborators (Newham, 1963); † A very similar ring opening scheme is found on platinum films (Anderson and Baker, 1963).

substituents. There is also clear evidence from Table 36 that there is a relatively lower chance of ring opening adjacent to a single methyl substituent, and the effect of the substituent is probably steric. There is a similar influence of methyl substituents on the position of opening in C_3 and C_4 rings over platinum, palladium and nickel catalysts: the reader is referred to the work of Maire *et al.* (1965) while Newham (1963) provides a summary of the older literature, particularly from Russian sources. However, the effects of alkyl substituents are not properly understood. For instance (cf. Newham, 1963) ring opening in n-propylcyclopentane is apparently abnormal, and supported catalysts can show abnormal behaviour depending on their history: it is possible that when this happens the reaction is not confined solely to the metal surface.

We have emphasized in this section the general importance of de-hydrogenative adsorption as a precursor to hydrogenolysis. Although both of the reaction paths which we have discussed above are illustrations of this, it is quite possible that other and more extensively dehydrogenated species may also contribute to the reaction. Clearly an ethane-type hydrogenolysis mechanism is, in principle, possible with higher aliphatic hydrocarbons provided the bond to be broken is not to a quaternary carbon atom. Extensively dehydrogenated residues of this type will favour C–C bond rupture, and since the residues are strongly bound to the surface, extensive fragmentation and methane production will result. This concept that extensively dehydrogenated residues are important in hydrogenolysis has been emphasized by a number of workers, among whom we may cite Cimino *et al.* (1954), Anderson and Baker (1963), Maire *et al.* (1965) and Freel and Galwey (1968). It is, for instance, useful here to note that hydrogenolysis reactions on rhodium and nickel films lead to extensive methane production, and with both these metals there is clear evidence from deuterium exchange data for extensively dehydrogenated species, involving particularly the loss of two hydrogens from a carbon atom, and this has been summarized in section III.

We may say in summary that hydrogenolysis reactions of aliphatic and alicyclic hydrocarbons on metals are complex processes in which at least three distinct types of reactions can occur, and these can act both competitively and consecutively to extents that depend both on the nature of the hydrocarbon and the metal. As an example of consecutive reactions we may take the reaction of neopentane over nickel films where the initial C–C bond rupture must proceed via a 1–3 adsorbed intermediate, but the fragments so initially produced may, before desorption, undergo extensive degradation probably via a 1–2 adsorbed intermediate, so that methane is the major initial product. On the other hand, with non-quaternary hydrocarbons larger than ethane, reactions by (i) a 1–3 adsorbed intermediate, (ii) a 1–2 adsorbed intermediate and (iii) an adsorbed π-allylic intermediate are all potentially competitive.

In a reaction as complex as this, the identification of the rate control-ling step is particularly hazardous. However, it does not seem unreason-able to accept that for reaction via a 1–3 adsorbed species, the rate is controlled by the formation of the bridged intermediate. For the other mechanisms, the most frequently made assumption is that the rate is controlled by the step in which the C–C bond is ruptured. While this would often seem to be a reasonable assumption, it may not always be valid. There are situations which make it necessary to consider the possibility that product desorption is rate controlling. For instance, it is known that on iron film catalysts, ethane exchange with deuterium does not occur before C–C bond rupture: if bond rupture were rate control-ling, all of the steps connecting the immediate precursor to bond rupture with gas phase ethane should be in equilibrium, and thus exchange should be observed. Again, with nickel films Anderson and Baker (1963) noted that the rate of methane production from ethane hydro-genolysis was lower than the corresponding rate of methane exchange with deuterium: on the assumption that the rate of methane exchange was desorption controlled, this would suggest that on this metal the hydrogenolysis reaction was controlled by the rate of methane desorp-tion. This conclusion has been supported by Freel and Galwey (1968). However, Anderson and Baker's argument is not valid if the rate of exchange is adsorption controlled, and this now seems the more likely. At the moment we are inclined to believe that C–C bond rupture is rate controlling on nickel, but the matter should still be considered as open to doubt. The desorption rate will, of course, be dependent on the size of the adsorbed residue in question. Thus, consider a reaction that leads to extensive molecular fragmentation in a single residence time of the re-actant on the surface. Since the adsorbed residues are extensively de-hydrogenated, one must expect a $C_n(n > 1)$ surface residue to be more strongly bound to the surface (at more than one carbon atom) than a C_1 surface residue. Hence it is easy to envisage a situation where only C_1 residues can desorb (as methane), and the reaction producing them could still be controlled by the rate of C–C bond rupture.

That the hydrogenolysis of ethane proceeds via a 1–2 adsorbed inter-mediate was suggested by Kemball and Taylor (1948) and this has since been elaborated by Cimino et al. (1954) who made the additional reasonable assumption that the 1–2 adsorbed intermediate is dehydro-genated to an extent that depends on the catalyst. Thus, writing initial ethane adsorption as

$$C_2H_6 \rightleftarrows C_2H_{a(s)} + \left(\frac{6-a}{2}\right)H_2 \qquad (V)–(8)$$

followed by

$$H_2 + C_2H_{a(s)} \rightarrow CH_{x(s)} + CH_{y(s)} \qquad (V)-(9)$$

and then followed by rapid reaction of $CH_{x(s)}$ and $CH_{y(s)}$ to give methane,

$$CH_{x(s)}, CH_{y(s)} \xrightarrow{\quad H_2 \quad} CH_4 \qquad (V)-(10)$$

and the assumption is made that (V)–(8) is in equilibrium, and (V)–(9) controls the rate. Since gas phase hydrogen will also be in adsorption equilibrium, we may write θ', the surface coverage by $C_2H_{a(s)}$ as

$$\theta' = \frac{a\,K_p\,p_{C_2H_6}\,p_{H_2}^{-(6-a)/2}}{1 + a\,K_p\,p_{C_2H_6}\,p_{H_2}^{-(6-a)/2} + bp_{H_2}^{\frac{1}{2}}} \qquad (V)-(11)$$

assuming Langmuir adsorption, and with K_p the equilibrium constant for (V)–(8).

Expression (V)–(11) has been written down in the manner outlined in Chapter 7. Further development of the analysis depends on the approximation to be made to equation (V)–(11). Cimino *et al.* assumed that the term $bp_{H_2}^{\frac{1}{2}}$ can be neglected in the denominator of (V)–(11), and that θ' can then be written approximately as follows, with $0 < x < 1$

$$\theta' \propto [p_{C_2H_6}\,p_{H_2}^{-(6-a)/2}]^x \qquad (V)-(12)$$

In fact, as Kemball (1966) has pointed out, the assumptions involved in the conversion of (V)–(11) to (V)–(12) are open to doubt: it is obvious from the experimental pressure dependence data that hydrogen is an inhibiting gas, and furthermore, when a is small K_p will be small. Thus it is unlikely to be a good approximation to neglect $bp_{H_2}^{\frac{1}{2}}$ in the denominator of (V)–(11). If one assumes on the other hand that the term $bp_{H_2}^{\frac{1}{2}}$ is much larger than the other terms in the denominator, we obtain

$$\theta' \propto p_{C_2H_6}\,p_{H_2}^{-(7-a)/2} \qquad (V)-(13)$$

Equation (V)–(13) is likely to be a more physically realistic approximation than (V)–(12), but it clearly cannot account for an ethane pressure dependence exponent other than unity. In an empirical way we may take this into account by assuming an intermediate strength of adsorption and writing

$$\theta' \propto [p_{C_2H_6}\,p_{H_2}^{-(7-a)/2}]^x \qquad (V)-(14)$$

with $0 < x < 1$, which then amounts to a variant of the form used by Cimino *et al.* but with a somewhat different basis. The reaction rate will then be proportional to $\theta' p_{H_2}$ if, as assumed by reaction (V)–(9), $C_2H_{a(s)}$ is attacked by a hydrogen molecule. If attack is by $H_{(s)}$, the rate will be proportional to $\theta' \theta_H$ with θ_H proportional to a low power of hydrogen pressure (zero power in the approximation that (V)–(13) is valid). In principle, the parameters x and a can be evaluated by comparison with the experimental pressure dependence exponents. However, because of the uncertainties in the model, there is little profit in attempting to evaluate a in the hope that this will give an *accurate* estimate of the state of dehydrogenation of the 1–2 adsorbed intermediate. The most important general conclusion is that, irrespective of the details assumed, this reaction intermediate is, in most cases, extensively dehydrogenated. For instance, assuming the rate $\propto \theta' p_{H_2}$, for ethane hydrogenolysis on nickel/kieselguhr with the observed rate proportional to $p_{C_2H_4}^{0.7} \ p_{H_2}^{-1.2}$, the application of equation (V)–(12) gives a = 0, and (V)–(14) gives a = 1.

6. Reactions over ultra-thin platinum films

Anderson and Macdonald (1969b) have studied the reaction of n-hexane at 275 °C over ultra-thin platinum films. These were prepared by deposition on to mica at 300 °C under U.H.V. conditions to an average film thickness of the order of a monolayer of metal. Although the film could be detected analytically, its structure could not be resolved by electron microscopy, and it is presumed to consist of a dispersion of extremely small atom clusters, probably accompanied by some isolated metal atoms. By comparison with the behaviour of platinum films of "normal" thickness (that is, say, > 1000 Å), reaction of n-hexane over ultra-thin platinum films resulted in the formation of a much larger amount of methylcyclopentane in the initial reaction products (~70% compared with ~40%). This suggests that on ultra-thin films a much greater proportion of the overall reaction path occurred via a carbocyclic intermediate. This behaviour is quite analogous to that found by Corolleur (1969) for the reaction of n-hexane on 0.2% and 10% platinum/alumina catalysts, where the proportion of methylcyclopentane was much greater over the former. It is reasonable to propose that the carbocyclic reaction path is favoured by catalyst sites which consist of metal atoms with a low coordination to other metal atoms; for instance, an atom at an exposed corner of a small crystallite or an isolated metal atom on the substrate. (We also refer to the discussion in section V.A.4.) It is clear that ultra-thin metal films can be used as model systems to which the properties of supported technical catalysts may be compared.

B. REACTIONS OF MOLECULES WITH C–N, C–O, C–S AND C–Hal FUNCTIONS

1. *Reactions at the C–N bond*

The reactions of primary, secondary and tertiary aliphatic amines have been studied over a range of metal film catalysts in the presence of hydrogen by Kemball and Moss (1956, 1958) and by Anderson and Clark (1966). There is a basic distinction in behaviour between primary and secondary amines on the one hand, and tertiary amines on the other. We shall for the moment restrict attention to the methyl compounds. Methylamine and dimethylamine undergo two main reactions, (i) simple hydrogenolysis of the C–N bond, and (ii) a disproportionation between two amine molecules. However, these reactions are accompanied by more complex but less extensive processes which include carbon (or nitrogen) incorporation into the catalyst, and the production of compounds in which C–C bonds have been formed such as ethylenimine, acetonitrile, and C_2–C_4 hydrocarbons. Table 37 illustrates the complex pattern of products from methylamine.

We shall deal first with the main reactions where, again taking methylamine by way of example, the overall hydrogenolysis and disproportionation reactions may be represented by

$$CH_3 NH_2 + H_2 \rightarrow CH_4 + NH_3 \qquad \text{(V)–(15)}$$

$$2CH_3 NH_2 \rightarrow (CH_3)_2 NH + NH_3 \qquad \text{(V)–(16)}$$

with completely analogous reactions with dimethylamine. Hydrogenolysis and disproportionation were shown to have approximately the same Arrhenius parameters, and these are collected into Table 38 for the three amines. In general terms the behaviour of ethylamine is similar to methylamine: the main difference of any importance is that a little direct fission of the C–C bond can occur, but rhodium is the only metal of those examined on which C–C bond rupture approaches C–N bond rupture in importance. The reactions of ethylamine occur rather more readily than with methylamine, and this difference is due to higher frequency factors for the ethylamine reactions. This is due in turn to the absence of self-poisoning and carbon incorporation with ethylamine, which is the result of the very much lower surface concentration of C_1 units on the catalyst surface, as these require C–C bond rupture for their formation.

Quite substantial amounts of acetonitrile are produced by dehydrogenation of ethylamine. The corresponding dehydrogenation of methylamine would yield hydrogen cyanide, but this process is thermodynamically disallowed at the present reaction temperatures. However,

TABLE 37

Products from the reaction of CH_4NH_2 + H_2 over various metals

	Extent of reaction %	Composition of reaction product (mole percent)					Hydrocarbons				Carbon/nitrogen imbalance a
		NH_3	$(CH_3)_2NH$	$(CH_3)_3N$	CH_3CN	$\overset{CH_2-CH_2}{NH}$	C_1	C_2	C_3	C_4	
Pt	12	54	24	1.5	0	0	20	0.5^b	0	0	7.5% C loss
Pd	20	66	26	5.5	0	0	2	0.5^b	0	0	27% C loss
Ni	31	92	3.5	0	0	0	3	1.5^c	trace		86% C loss
W	12	58	16	1	7	11	2	3^d	1^d	1^d	7.5% C loss
Co	30	72	9	1	6	2	1	6^d	2^b	1^b	33% C loss
V	26	48	18	2	16	8	4	2	2	0	12% N loss
Cu	12	52	34	1	2	5	11	4	1	0	0

a Due to carbon (or nitrogen) incorporation into catalyst; b Saturate only; c Ratio of olefin to saturate 5:1; d Ratio of olefin to saturate 3:1.

TABLE 38

Activation energies and frequency factors for reactions of mono-, di-
and trimethylamines over metal films

	catalyst	E (kcal mole^{-1})	log$_{10}$A (A in molec sec^{-1}cm^{-2})
methylamine	palladium	21.5	23.5
	platinum	19.8	22.3
	nickel	15.1	19.5
	iron	23.1	23.3
	cobalt	9.7	17.1
	tungsten	16.5	18.4
	vanadium	12.9	18.1
	copper	18.3	20.5
dimethylamine	palladium	21.4	25.3
	platinum	10.5	18.6
trimethylamine	palladium	21	22.7

Anderson and Clark (1965) have shown that the catalytic reactions of
hydrogen with methylamine and with hydrogen cyanide are closely
related, and the surface residues in the two reactions are almost
certainly the same.

Two mechanisms have been proposed for the disproportionation
reaction; Kemball and Moss suggested that, in the reaction of methylamine for instance, hydrogenolysis results in the existence on the surface of C_1 residues which may recombine with adsorbed methylamine
residues. Writing $[C]_s$, $[C–N]_s$ etc. to represent the skeletons of adsorbed
residues without precise specification of the degree of hydrogenation or
the mode of binding to the surface, this mechanism may be represented
in outline as

$$CH_3 NH_2 \rightarrow [C–N]_s \rightarrow [C]_s + [N]_s \qquad (V)–(17)$$

$$[C–N]_s + [C]_s \rightarrow (CH_3)_2 NH \qquad (V)–(18)$$

However, Anderson and Clark (1966) pointed out objections to this
proposal. If reaction (V)–(18) is possible, in which a C–N bond is formed,
it is very difficult to see why (V)–(17) should not be reversible: however,
the latter has been shown to be not the case by the use of C^{13} and N^{15}
labelled methylamine. Secondly, the disproportionation reaction is
close to second order in amine pressure (cf. Table 39), and this is also
difficult to reconcile with reactions (V)–(17) and (18).

A mechanism for disproportionation which avoids these difficulties

TABLE 39

Pressure dependence exponents for rates of hydrogenolysis and
disproportionation reactions of amines on films

	CH_3NH_2 on Ni (210 °C)		CH_3NH_2 on Pd (181 °C)		$C_2H_5NH_2$ on Pt (123 °C)	
	exponent for		exponent for		exponent for	
	p_{H_2}	$p_{CH_3NH_2}$	p_{H_2}	$p_{CH_3NH_2}$	p_{H_2}	$p_{C_2H_5NH_2}$
Hydrogenolysis	0.4	−0.1	—	0.1	−0.7	0
Disproportionation	−0.2	1.7	—	2.0	—	2.0

was suggested by Anderson and Clark, involving a bimolecular surface
reaction between adsorbed $[C–N]_s$ residues,

$$[C\!\!-\!\!N]_s + [C\!\!-\!\!N]_s \rightarrow [C\!\!-\!\!N\!\!-\!\!C\!\!-\!\!N]_s \rightarrow \begin{cases} [C\!\!-\!\!N\!\!-\!\!C]_s \overset{(H)}{\rightarrow} (CH_3)_2NH \\ \overset{+}{[N]_s} \overset{(H)}{\rightarrow} NH_3 \end{cases}$$

$$(V)\text{--}(19)$$

The surface can accommodate a range of residues of the general type
$[C–N]_s$ of varying degrees of hydrogenation. That responsible for
reaction (V)–(19) must be weakly enough adsorbed for the surface concentration to be directly proportional to amine pressure, and it must
have a structure which can accommodate an analogous reaction with
dimethylamine: Anderson and Clark suggested a π-adsorbed species
$H_2C \displaystyle\mathop{\rlap{\textstyle\Vert}}_{M} NH$. Reaction scheme (V)–(19) is a simplification in that it
ignores the formation of products in which C–C bonds have been created.
The most important of these, namely acetonitrile, ethylenimine and C_2
hydrocarbons (cf. Table 37), very probably proceed from $[C–N–C]_s$

$$[C\!\!-\!\!N\!\!-\!\!C]_s \rightarrow \begin{bmatrix} C \\ \Big\| \diagdown \\ \diagup N \\ C \end{bmatrix}_s \begin{matrix} \text{(H) ethylenimine} \\ \text{(H)} \\ \rightarrow \text{acetonitrile} \\ \text{(H)} \\ \rightarrow C_2 \text{ hydrocarbons} \end{matrix} \qquad (V)\text{--}(20)$$

via a triangular intermediate which gives the desired products depending where the ring opens. It should be noted that these products are
also obtained from the reaction of dimethylamine, adsorption of which
gives $[C–N–C]_s$ directly, and more particularly the mechanism is confirmed by the large amounts of $C_2H_5NH_2$, CH_3 CN and C_2 hydrocarbons
produced from reaction of ethylenimine itself.

By a comparison of the catalytic reactions of hydrogen cyanide and methylamine, Anderson and Clark concluded that residues singly bonded to the surface

$$\underset{\text{M\ \ M}}{\underset{|\ \ \ |}{\text{H}_2\text{C}-\text{NH}}} \qquad \underset{\text{M}}{\underset{|}{\underset{\text{NH}}{\underset{|}{\text{CH}_3}}}} \qquad \underset{\text{M}}{\underset{|}{\underset{\text{CH}_2}{\underset{|}{\text{NH}_2}}}}$$

are not directly responsible for C–N bond rupture. This conclusion agrees with the methylamine exchange results discussed in a previous section showing that these singly bonded residues are involved in exchange under conditions where C–N bond rupture does not occur.

Both Kemball and Moss, and Anderson and Clark suggest that C–N bond rupture occurs from some state of $[\text{C–N}]_s$ (but certainly not the one suggested above as responsible for disproportionation), and that this is extensively dehydrogenated and multiply bonded to the surface. It will be seen from the data of Table 38 that on palladium films mono-, di- and trimethylamine undergo C–N bond rupture with approximately equal activation energies. One concludes that, at least on this metal, the mode of bonding of the nitrogen to the surface is not very important,

and presumably $\overset{\displaystyle\diagdown\ |\ \diagup}{\underset{\text{M}}{\underset{\downarrow}{\text{N}}}}$ will suffice. However, this conclusion may well

not be general to other metals, and more extensive data are needed to illuminate this problem and also to allow an assessment of the possible importance of 1–3 adsorption to the reaction.

2. Reactions at the C–O bond

There are very few data for hydrogenolysis of the C–O bond. An important feature of the reactions of aliphatic alcohols and ethers is the formation of carbon monoxide. With methyl alcohol this occurs fairly readily by simple dehydrogenation, and this has been observed over a number of evaporated metal films at about 170 °C (Anderson and Kemball, 1955). With higher alcohols and ethers, carbon monoxide formation requires C–C bond rupture, and this has been observed from ethyl alcohol and from diethyl ether over platinum films in the range 200–300 °C. However, C–O bond hydrogenolysis has been observed on platinum catalysts in both an alcohol and an ether. Farkas and Farkas (1939) observed the reaction of isopropyl alcohol with hydrogen over a platinum foil catalyst between 0° and 80 °C, and Whan and Kemball (1965) observed the hydrogenolysis of diethyl ether over an evaporated platinum catalyst between 30 and 150 °C giving ethane and ethyl alcohol as the primary products: the reaction was rapidly poisoned by strongly adsorbed residues resulting from C–C bond rupture, probably adsorbed carbon monoxide. However, over tungsten films ethers react only by dehydration with no C–O rupture (Imai *et al.*, 1968). Dehydration is favoured and C–C rupture suppressed by the heavy oxygen coverage of the surface. In all cases, the path for C–O rupture probably passes via a 1-2 diadsorbed intermediate.

3. *Reactions at the C–S bond*

Because of rapid incorporation of sulphur, hydrogenolysis of mercaptans and sulphides can be studied only with difficulty at a metal surface. However, these reactions have been examined over molybdenum and tungsten disulphide catalysts by Wilson and Kemball (1964) and Kieran and Kemball (1965).

Methyl mercaptan reacts in the presence of hydrogen over molybdenum disulphide in the temperature range 230 to 270 °C by two reactions: (i) hydrogenolysis to give methane and hydrogen sulphide, and (ii) disproportionation to give dimethyl sulphide and hydrogen sulphide. Reaction (i) takes place over tungsten disulphide only above 230 °C, but reaction (ii) occurs on this catalyst between 130 and 230 °C. With ethyl mercaptan, both of these reactions occur, but in addition there is a third process yielding hydrogen sulphide and ethylene: neither catalyst shows any activity below 300 °C for C–C bond rupture, which is thus relatively unimportant. The reactions of dimethyl sulphide and diethyl sulphide are analogous: the former gives only hydrogenolysis to methane and hydrogen sulphide, while the latter gives, in addition to simple hydrogenolysis to ethane and hydrogen sulphide, a decomposition reaction to ethylene and hydrogen sulphide.

The C–S bond hydrogenolysis reactions are zero order in mercaptan pressure and clearly proceed from strongly adsorbed intermediates, probably adsorbed in the 1–2 mode. Although the disproportionation reactions formally resemble those discussed in a previous section for amines, they differ in that they depend on mercaptan pressure to the 0.5–1 power. Kieran and Kemball suggest the most likely mechanism to be

$$C_2H_5SH_{(s)} + C_2H_{5(s)} \rightarrow (C_2H_5)_2S + H_{(s)} \qquad \text{(V)–(21)}$$

in which the mercaptan molecule is presumably adsorbed by a co-ordinative bond to the surface through the sulphur and the concentration of which depends on pressure, while it is reasonable to assume that the surface concentration of $C_2H_{5(s)}$ which is produced by the hydrogenolysis reaction will be not greatly dependent on mercaptan pressure.

4. *Reactions at the C–Hal bond*

The catalytic hydrogenolysis of ethyl chloride and of ethyl bromide has been studied by Campbell and Kemball (1961) on evaporated films of palladium, platinum and nickel and, in less detail, on films of iron, tungsten and rhodium. The reaction was the formation of ethane and hydrogen halide. Campbell and Kemball (1963) have studied the hydrogenolysis of t-butyl chloride over evaporated films of platinum and palladium. By comparing the results of Campbell and Kemball with Bond's (1962) discussion of alkyl halide reactivity, one may conclude that alkyl halides stand in the following decreasing order of reactivity for C–Hal bond hydrogenolysis in chloride and fluoride systems

$$\text{t-butyl} > \text{iso-propyl} > \text{n-propyl} > \text{ethyl} > \text{methyl}$$

and the halogen reactivity is

$$\text{Cl} > \text{Br} > \text{F}$$

The reactions are inhibited by the hydrogen halide product. No general statement can be made concerning the identity of the rate controlling step: thus, Campbell and Kemball suggest that on platinum and palladium films the rate of hydrogenolysis of t-butyl chloride is controlled by the rate of isobutane desorption, but that with ethyl chloride the initial dissociative adsorption is rate controlling. The pathway for C–Hal bond rupture is uncertain: however, by analogy with the previous discussion we suggest a 1–2 adsorbed intermediate from, for instance, ethyl chloride, involving a co-ordinative bond between the chlorine and the metal. With t-butyl chloride 1–3 adsorption would be required, a structure already postulated by Campbell and Kemball as involved in deuterium exchange.

VI. OXIDATION AND DEHYDROGENATION REACTIONS

The reactions discussed in previous sections have involved hydrogen, usually as a reactant present in excess, and most of the metals of catalytic interest in those reactions were the transition metals. In reactions involving oxygen many of these metals are converted to metal oxide and it is the latter which functions as a catalyst: the activity of nickel oxide for hydrocarbon oxidation is a case in point.

In this section we restrict attention to those metals which are sufficiently resistant to extensive oxidation under reaction conditions for the metal itself to be thus considered as the catalyst. In particular, silver, gold, rhodium, platinum and palladium have been studied in this way. A variety of reactions have been investigated on these metals, but

I

the two most important reactions which have been the subject of repeated and detailed investigation, are the oxidation of ethylene and the
dehydrogenation of formic acid. Hydrocarbon dehydrogenation reactions are not discussed in this section since many of the important ideas
have been covered implicitly in earlier sections.

A. CATALYTIC OXIDATION OF ETHYLENE

The oxidation of ethylene over evaporated films of platinum and
palladium has been studied by Kemball and Patterson (1962) and Patterson and Kemball (1963). The main products of reaction are carbon
dioxide and water, but traces of acetic acid and acetic anhydride were
detected by mass spectrometry. The temperature ranges were 5 to
100 °C for platinum and 50 to 140 °C for palladium. The reaction
kinetics were found to be zero order in oxygen pressure and first order
in ethylene pressure, except at high ethylene pressures when the reaction is zero order in ethylene. The higher activity of platinum is
attributed to the lower activation energy (11.7 kcale mole^{-1}) for
platinum as compared to 14.3 kcal mole^{-1} for palladium. Other ex-
ploratory experiments on gold, rhodium and tungsten show that the
order of activity is

$$Pt > Pd > Rh \gg Au > W$$

Kemball and Patterson correlate the decreasing order of activity of the
transition metals with the increasing order of the heats of oxygen adsorption. It is considered that oxygen is much more strongly adsorbed
than ethylene and that this accounts for the kinetic independence of
oxygen pressure. The results show that the small amounts of acetic
acid and acetic anhydride are produced by a side reaction and that
they do not occur as intermediates in the oxidation to carbon dioxide.

The most important catalyst for ethylene oxidation is silver. Under
carefully controlled conditions ethylene is selectively oxidized to ethylene oxide over silver catalysts of various types. Higher olefins are not
selectively oxidized and form only carbon dioxide and water when reacted over silver. The main features of the commercial process have been
reviewed by Voge and Adams (1967). Usually a flow reactor with a silver
catalyst on an inert carrier is used. The catalyst is usually moderated to
improve selectivity by the addition of electronegative elements such as
As, Sb, Bi, S or the halogens. A large excess of air or oxygen is found
necessary and the maximum selectivity for the production of ethylene
oxide is typically about 75%; the remaining reaction leading to carbon
dioxide. It has been shown (Voge and Atkins, 1952) that most of the
CO_2 is formed directly from the oxidation of ethylene and not by the

subsequent oxidation of the ethylene oxide. This is contrary to the result of Twigg (1946) who found that ethylene oxide could isomerize to acetaldehyde and undergo subsequent oxidation to CO_2. The latter conclusion may be valid for static systems but in the usual flow system it appears that the oxidation of ethylene oxide is unimportant.

The variable reactivities and selectivities of silver catalysts for the oxidation of ethylene have prompted investigation of the importance of surface structure in the behaviour of a silver catalyst. Kummer (1956) investigated the (110), (111) and (211) faces of silver single crystals. Only small differences in reactivity and selectivity could be detected. Wilson et al. (1959) evaporated silver films in vacuum and in the presence of nitrogen and hydrogen. The latter films, condensed on glass at 20 °C, were examined by grazing incidence electron diffraction. The results at that time were interpreted as indicating preferred orientation in the film surface. It is now known that films produced in this way are rough and the preferred orientation observed is not representative of the actual surface exposed to the gas (cf. Chapter 7, section II). Wilson et al., found that the films of different initial structure had recrystallized to the usual polycrystalline structure after catalysing the oxidation of ethylene at 250–280 °C. It seems that under reaction conditions unstable surface structures are easily modified and that careful examination of silver surfaces after reaction is an essential for a valid structural study.

Kummer and Wilson et al. showed that the selectivity of silver surfaces in the absence of any moderator was in the range 30–50%. Since only small amounts of moderators are needed substantially to increase the selectivity, and since many of the methods of producing silver catalysts do not preclude chlorine, sulphur or other moderating impurities, it is apparent that many studies of the reaction have been undertaken on catalysts already unintentionally moderated. This may in part explain the variable information available on the kinetics of the reaction.

In the presence of excess oxygen the rate of production of ethylene oxide is to a first approximation first order in ethylene pressure. Since ethylene is not appreciably adsorbed on bare silver the dependence of the reaction rate on ethylene pressure has been taken to indicate that the reaction occurs between gas phase ethylene and adsorbed oxygen. However, under reaction conditions the silver is at least partly covered by adsorbed oxygen and it is now considered likely that ethylene is weakly adsorbed by interaction with this adsorbed oxygen. Belousov and Rubanik (1963) have demonstrated that in the oxidation of a mixture of ethylene and propylene the presence of propylene inhibits the rate of oxidation of ethylene indicating that the two olefins compete for the surface.

Another point of controversy in the mechanism is whether the ethylene reacts with adsorbed oxygen atoms or molecules. The interaction of oxygen with silver has been studied in detail and it is readily shown that under reaction conditions bulk Ag_2O is thermodynamically unstable.

Twigg (1946) demonstrated that the coverage (θ) of oxygen under reaction conditions is of the order 0.1–0.4 depending on the temperature and gas pressure. This information does not disclose what form of the adsorbed oxygen participates in the reaction. It has been suggested that oxygen adsorption is rate determining. Although the rate of oxygen adsorption is fast on clean silver the rate decreases markedly with coverage and at $\theta = 0.7$ the rate of ethylene oxidation at 200 °C is over 100 times as fast as that of oxygen adsorption (Voge, 1957).

Gravimetric studies of the adsorption of oxygen on silver powders (Czanderna, 1964) have been interpreted as indicating adsorbed O_2^- on the surface at coverages above $\theta = 0.5$. In other work attempts have been made to identify the presence of adsorbed oxygen in the form of a surface peroxide. Vol and Shishakov (1962) have identified AgO_2 on the surface of a silver film by electron diffraction. It is shown that the AgO_2 is converted to silver by reaction with ethylene.

Direct measurement of oxygen uptake on evaporated silver films (Bagg and Bruce, 1963) has shown that the adsorption of oxygen at 200 °C is strongly dependent on the pressure in the range $10^{-3} - 10^{-1}$ torr. Under these conditions the amount of oxygen adsorbed corresponded to about one oxygen atom per 3.5 surface silver atoms. Boreskov and Khasin (1964) investigated the isotopic exchange of oxygen on evaporated silver films and found that the total adsorbed oxygen amounted to more than three monolayers. It is clear that the coverage and nature of the adsorbed species is dependent on oxygen pressure and that under reaction conditions some oxygen will be adsorbed as molecules which can take part in the oxidation to ethylene oxide.

It is significant that studies of the oxidation of ethylene carried out on silver films evaporated in ultrahigh vacuum and with pressures of both ethylene and oxygen less than 0.2 torr showed CO_2 and water as the only products (Mercer, 1969). With higher pressures of ethylene (10 torr) and oxygen (25 torr) Moss and Thomas (1967) found that evaporated silver films gave an ethylene oxide yield of approximately 20 per cent. Murray (1950) showed that a silver catalyst gave yields of 45 to 60 per cent for reactant pressures of 100–700 torr and that the yield was reduced by diminishing either total or oxygen pressure. All of these observations are consistent with a mechanism involving adsorbed oxygen molecules.

Mechanisms for the reaction involving adsorbed molecular oxygen have been proposed by Margolis (1963) and Voge and Adams (1967). The mechanism proposed by Voge and Adams involves the interaction of adsorbed O_2 with adsorbed ethylene to form ethylene oxide. The coupled reactions are

$$S + O_{2(g)} \rightarrow O_{2(s)}$$

$$O_{2(s)} + C_2H_{4(s)} \rightarrow C_2H_4O_{(g)} + O_{(s)}$$

$$4O_{(s)} + C_2H_{4(s)} \rightarrow 2CO_{(g)} + 2H_2O_{(g)} + 4S$$

$$2CO_{(g)} + O_{2(s)} \rightarrow 2CO_{2(g)} + S$$

where, as before, S represents a surface site. Such a scheme is consistent with the observation that the selectivity of the reaction is approximately constant over a range of conversions and over a wide range of temperatures. Since $O_{2(s)}$ leads to ethylene oxide it is suggested that the action of a moderator is to reduce the concentration of $O_{(s)}$ relative to $O_{2(s)}$. It is also suggested that the unique catalytic behaviour of silver in this reaction derives from this formation of $O_{2(s)}$ with a quasi-peroxide structure.

B. DEHYDROGENATION OF FORMIC ACID

The catalytic decomposition of formic acid on most metals leads almost exclusively to carbon dioxide and hydrogen. Many oxide surfaces favour the alternative mode of decomposition to yield water and carbon monoxide, although some dehydrogenation usually also occurs. Although the reaction is not specific to any one type of catalyst, repeated studies have been undertaken using this reaction as a test reaction with the object of elucidating the role of surface structure in catalytic activity. As a test reaction, formic acid decomposition has the advantage of experimental convenience. Furthermore, the dehydrogenation reaction can be readily studied in a regime where it is zero order in formic acid pressure, so the measured activation energy and frequency factor do not contain contributions from the adsorption equilibrium. Nevertheless, it should be remembered that any correlations found between the reaction kinetics and surface structure are specific for this particular reaction, and extrapolations to other reactions are extremely hazardous. A general review of the reaction has been given by Mars et al. (1963).

Because only the dehydrogenation reaction occurs on metals, the interpretation of the effect of catalyst structure has been based on analysis of the temperature dependence of reaction rate, and the alternative reaction path of dehydration can be ignored. On most metals it is found that the dehydrogenation reaction is zero order in formic acid pressure except at low pressures where the kinetics are first order. This is interpreted as indicating that at the higher pressures the whole of the reactive surface is covered by adsorbed intermediates.

Infrared spectroscopic studies of formic acid adsorption on a number of supported metal catalysts have identified the formate ion as an important chemisorbed entity (Sachtler and Fahrenfort, 1961; Fukuda et al., 1968). Jaeger (1967) focuses attention on whether interaction of formic acid with the metal is accompanied by the removal of metal atoms from the crystal lattice, that is, whether the interaction yields bulk formate at the surface, or is confined to the formation of chemisorbed formate without metal surface rearrangement. Sachtler and Fahrenfort have analysed rate data for the decomposition of formic acid on a number of metals, and by comparison with the behaviour of the bulk formates, have concluded that for those metals which form stable formates, for instance nickel and tungsten, the overall reaction rate is controlled by the rate of decomposition of the metal formate.

The study of formic acid dehydrogenation on metals undergoing bulk formate formation cannot be expected to yield any information on the effect of metal surface structure, and this is confirmed by the results of Rienacker and Volter (1959) showing that the activation energy on copper was independent of the surface plane exposed. The present discussion will be confined to an examination of some of the recent results obtained on silver for which the high instability of silver formate (Keller and Korosy, 1948) suggests that surface formate would be expected to form without removing metal atoms from the lattice (Jaeger, 1967).

Bagg et al. (1963) have examined the influence of surface structure and defects on the catalytic activity of evaporated silver films. The work stemmed from the earlier results of Sosnovsky (1955) on bulk single crystals of silver which suggested that different crystal faces had different activation energies for the zero order dehydrogenation of formic acid. The silver crystals used for this work were cleaned by argon ion bombardment. Different pretreatments resulted in different dislocation densities and Sosnovsky et al. (1958, 1959) suggested that active sites for the reaction are dislocations emerging at the surface. Bagg et al. studied four types of evaporated silver film; (1) deposited on glass at room temperature, (2) on mica at room temperature, (3) on mica at 280 °C and (4) on mica at 280 °C subsequently annealed at 400 °C. The different film structures were determined by electron microscopy and diffraction. The rate data for the catalytic decomposition of formic acid in the zero order region was resolved into the activation energy and frequency factor, and a compensation effect was found to operate. The activation energy was found not to depend on the length of grain boundaries, stacking faults and coherent twin boundaries. The concentration of dislocations in these films was approximately constant so that the suggested role of dislocations could not be confirmed. The most important surface factor in the reaction was found to be the type of crystal plane exposed; the lowest activation energy was found for a (111) surface.

More recently Jaeger (1967) has extended the work to silver films epitaxed on rock salt and showing preferred (100) orientation. By varying the thickness of these films and films evaporated onto mica, the density of dislocations was varied from 10^8 to 10^{10} per cm^2. The results are summarized in Table 40. Jaeger concludes that variations of the concentration of crystal defects have no influence on the rate of the reaction. The variation of the observed activation energy between 18 and 32 kcal mole^{-1} depends on the orientation of the surface. The lowest activation energy occurs on the (111) surface and it appears that the activation energy for the reaction depends on the stability of the chemisorbed species which in turn depends on the degree of unsaturation of surface metal atoms. From the estimates of the flat (111) and (100) areas in each surface, the activation energies for the separate crystal planes were derived by the procedure described in Chapter 7 section IV.B.1. Jaeger gives the values (in kcal mole^{-1}), (111) 16–17, (100) 24, polycrystalline 27, atomically rough 31.5.

Further insight into the mechanism of the decomposition of formic acid on silver and an explanation of the compensation effect found in Jaeger's results can be gained from the recent radiochemical study of Lawson (1968a). Carbon-14 labelled formic acid was adsorbed on polycrystalline and (111) epitaxed films of silver. By estimating the proportion of (111) area in the films, the coverage of formic acid could be calculated for the atomically smooth (111) plane alone and for the atomically rough surface. The coverage on the rough surface was found to be that expected for close-packed adsorption of freely rotating ions. On the (111) plane the coverage was found to be low. Lawson (1968b) has proposed that the adsorbed intermediate in formic acid decomposition and isotopic carbon exchange is a formic acid molecule bound to a chemisorbed formate ion and that this structure is preferentially formed on the (111) plane.

Lawson shows that the compensation effects observed in the dehydrogenation on silver are most likely associated with the different entropy

TABLE 40

Average catalytic constants and structural features of different types of silver catalysts for formic acid decomposition

Catalyst	Concentrations of Defects				Fraction of flat (111) or (100) in surface (%)	Activation Energy (kcal mole⁻¹)	Frequency Factor (A) log₁₀A (A in molecules cm⁻² sec⁻¹)
	Grain boundaries (cm⁻¹)	Stacking fault and coherent twin boundaries (cm⁻¹)	Noncoherent twin boundaries (cm⁻¹)	Isolated whole dislocations (cm⁻¹)			
1 (111) 1000 Å (thin)	0	5×10^3	1×10^4	10^9-10^{10}	75–85	26.8	26.8
1a Annealed	0	3×10^3	0	10^9-10^{10}	95	20.8	23.5
2 (111) 10000 Å (thick)	0	4×10^2	1×10^3	10^8	80	25.8	26.1
2a Annealed	0	4×10^2	1×10^3	10^8	95	20.8	23.4
3 (100) 1000 Å	0	2×10^4	0	10^{10}	<100	26.8	26.4
3a Annealed	0	6×10^3	0	10^9	<100	27.1	26.4
4 (100) + (111) 1000 Å	10^2-10^3	2×10^4	0	10^{10}	<100	26.6	26.2
4a Annealed	10^2-10^3	6×10^3	0	10^9	<100	22.8	24.4
5 Polycrystal on glass	3×10^4	7×10^4	0	$\sim10^9$	15	27.3	26.6
5a Annealed	3×10^4	3×10^4	0	$?\sim10^9$	8–15	27.1	26.8
6 Bulk single crystal random	0	0	0	10^7	0	31.5	28.5
6a Thermally etched (111)	0	0	0	10^7	50–75	23.8	24.8
7 Bulk single crystal random	0	0	0	10^7	0	31.5	28.5
7a Thermally etched (100)	0	0	0	10^7	50–75	29.9	27.5

Catalysts 1–5 are films

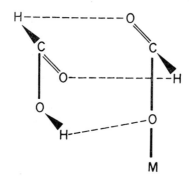

changes accompanying different modes of adsorption rather than with changes of the numbers of active sites.

VII. Prognosis for Future Work

It will be evident from the material in this Chapter that, at a phenomological level, a lot of information is now available about chemical reactions at the surface of metal film catalysts and, again at this level, a lot is understood about the behaviour of many important classes of compounds. However, our understanding of the influence of the metal surface on the reaction details remains rudimentary, and a theory with any reasonable degree of predictive power is still lacking. Metal film catalysts are not "practical" catalysts, and we only use them for the insight which such work may give towards the construction of a general theory of catalytic reaction kinetics. This being so, we believe that it is relatively unprofitable to proliferate further work which uses "as prepared" film catalysts (generally of a polycrystalline structure) without attempting to use the modern tools which are now available for the study of surface structure, and therefore without making a serious attempt to correlate reaction details with surface structure. We believe that current work with metal film catalysts is, in fact, moving towards a greater emphasis on detailed surface structure as a critical reaction variable.

References

Addy, J. and Bond, G. C. (1957). *Trans. Faraday Soc.* **53**, 368, 383, 388.
Alekseev, V. I. and Shvartsman, L. A. (1962). *Izv. Akad. Nauk. SSSR, Otd. Tekn. Nauk., Met. i Toplivo* No. 6, 171.
Alekseev, V. I. and Shvartsman, L. A. (1963). *Izv. Akad. Nauk. SSSR, Otd. Tekhn. Nauk., Met. i Toplivo* No. 1, 91.
Amano, A. and Parravano, G. (1957). *Adv. Catalysis* **10**, 242.
Anderson, J. and Estrup, P. J. (1968). *Surface Sci.* **9**, 463.
Anderson, J. R. and Kemball, C. (1954a). *Proc. R. Soc.* **A223**, 361.

I*

Anderson, J. R. and Kemball, C. (1954b). *Proc. R. Soc.* **A226**, 472.

Anderson, J. R. and Kemball, C. (1955). *Trans. Faraday Soc.* **51**, 966.

Anderson, J. R. (1957a). *Rev. Pure Appl. Chem.* **7**, 165.

Anderson, J. R. (1957b). *Aust. J. Chem.* **10**, 409.

Anderson, J. R. and Kemball, C. (1957). *Adv. Catalysis* **9**, 51.

Anderson, J. R. and Baker, B. G. (1960). *Nature Lond.* **187**, 937.

Anderson, J. R. and Avery, N. R. (1963). *J. Catal.* **2**, 542.

Anderson, J. R. and Baker, B. G. (1963). *Proc. R. Soc.* **A271**, 402.

Anderson, J. R. and Clark, N. J. (1965). *In* "Proc. 3rd Internat. Congr. Catalysis" (W. M. H. Sachtler, G. C. A. Schuit and P. Zwietering, eds.), North Holland, Amsterdam, p. 1048.

Anderson, J. R. and Avery, N. R. (1966). *J. Catal.* **5**, 446.

Anderson, J. R. and Clark, N. J. (1966). *J. Catal.* **5**, 250.

Anderson, J. R. and Avery, N. R. (1967a). *J. Catal.* **7**, 315.

Anderson, J. R. and Avery, N. R. (1967b). *J. Catal.* **8**, 48.

Anderson, J. R. and McConkey, B. H. (1967). *J. Catal.* **9**, 263.

Anderson, J. R. and McConkey, B. H. (1968). *J. Catal.* **11**, 54.

Anderson, J. R. and Macdonald, R. J. (1969a). *J. Catal.* **13**, 345.

Anderson, J. R. and Macdonald, R. J. (1969b). *J. Catal.* (in press.)

Arthur, J. R. and Hansen, R. S. (1962). *J. Chem. Phys.* **36**, 2062.

Bagg, J. and Bruce, L. A. (1963). *J. Catal.* **2**, 93.

Bagg, J., Jaeger, H. and Sanders, J. V. (1963). *J. Catal.* **2**, 449.

Balandin, A. A. (1958). *Adv. Catalysis* **10**, 96.

Baker, B. G. and Hoveling, A. W. (1969). *In* "22nd International Congress of Pure and Applied Chemistry," Sydney.

Barrer, R. M. (1936). *Trans. Faraday Soc.* **32**, 490.

Barron, Y., Maire, G., Cornet, D. and Gault, F. G. (1963). *J. Catal.* **2**, 152.

Barron, Y., Maire, G., Muller, J. M. and Gault, F. G. (1966). *J. Catal.* **5**, 428.

Beeck, O., Smith, A. E. and Wheeler, A. (1940). *Proc. R. Soc.* **A177**, 64.

Beeck, O. (1945). *Rev. Mod. Phys.* **17**, 61.

Beeck, O. (1950). *Discuss. Faraday Soc.* **8**, 118.

Beeck, O. and Ritchie, A. W. (1950). *Discuss. Faraday Soc.* **8**, 159.

Belousov, V. M. and Rubanik, M. Y. (1963). *Kinet. Catal.* **4**, 783.

Bénard, J., Oudar, J. and Cabané-Brouty, F. (1965). *Surface Sci.* **3**, 359.

Blyholder, G. and Neff, L. D. (1963). *J. Catal.* **2**, 138.

Bloch, J. (1963). *Z. Phys. Chem. (N.F.)* **39**, 169.

Bokhoven, C., van Heerden, R., Westrik, R. and Zwietering, P. (1955). *In* "Catalysis" (P. H. Emmett, ed.) Vol. 3, Reinhold, New York, p. 265.

Bond, G. C. (1956). *Trans. Faraday Soc.* **52**, 1235.

Bond, G. C. (1962). "Catalysis by Metals". Academic Press, London.

Bond, G. C., Webb, G., Wells, P. B. and Winterbottom, J. M. (1962). *J. Catal.* **1**, 74.

Bond, G. C., Phillipson, J. J., Wells, P. B. and Winterbottom, J. M. (1964). *Trans. Faraday Soc.* **60**, 1847.

Bond, G. C. and Wells, P. B. (1964). *Adv. Catalysis* **15**, 91.

Bond, G. C. and Wells, P. B. (1965). *J. Catal.* **4**, 211.

Bond, G. C. (1966). *Discuss. Faraday Soc.* **41**, 200, and following discussion.

Bond, G. C. and Wells, P. B. (1966a). *J. Catal.* **5**, 65.

Bond, G. C. and Wells, P. B. (1966b). *J. Catal.* **5**, 419.

Bond, G. C. and Wells, P. B. (1966c). *J. Catal.* **6**, 397.

Bond, G. C., Webb, G. and Wells, P. B. (1968). *J. Catal.* **12**, 157.
Boonstra, A. H. and van Ruler, J. (1966). *Surface Sci.* **4**, 141.
Boreskov, G. K. and Vassilevitch, A. A. (1961). *In* "Actes du Deuxième Congrès International de Catalyse". Editions Technip, Paris, p. 1095.
Boreskov, G. K. and Khasin, A. V. (1964). *Kinet. Catal.* **5**, 846.
Boudart, M., Aldag, A. W., Ptak, L. D. and Benson, J. E. (1968). *J. Catal.* **11**, 35.
Boudart, M. and Ptak, L. D. (1970). *J. Catal.* **16**, 90.
Burwell, R. L. and Briggs, W. S. (1952). *J. Am. Chem. Soc.* **74**, 5096.
Burwell, R. L. and Tuxworth, R. H. (1956). *J. Phys. Chem.* **60**, 1043.
Burwell, R. L., Shim, B. K. C. and Rowlinson, H. C. (1957). *J. Am. Chem. Soc.* **79**, 5142.
Burwell, R. L. and Pearson, R. G. (1966). *J. Phys. Chem.* **70**, 300.
Campbell, J. S. and Kemball, C. (1961). *Trans. Faraday Soc.* **57**, 809.
Campbell, J. S. and Kemball, C. (1963). *Trans. Faraday Soc.* **59**, 2583.
Campbell, J. S. and Emmett, P. H. (1967). *J. Catal.* **7**, 252.
Chambers, R. P. and Boudart, M. (1966). *J. Catal.* **5**, 517.
Cimino, A., Boudart, M. and Taylor, H. S. (1954). *J. Phys. Chem.* **58**, 796.
Clarke, J. K. and Kemball, C. (1959). *Trans. Faraday Soc.* **55**, 1.
Coekelbergs, R., Frennet, A. and Gosselain, P. (1959). *J. Chim. Phys.* **56**, 967.
Cornet, D. and Gault, F. G. (1967). *J. Catal.* **7**, 140.
Cottrell, T. L. (1958). "The Strength of Chemical Bonds". Butterworths, London.
Crawford, E., Roberts, M. W. and Kemball, C. (1962). *Trans. Faraday Soc.* **58**, 1761.
Crawford, E. and Kemball, C. (1962). *Trans. Faraday Soc.* **58**, 2452.
Corolleur, C. (1969). Ph.D. Thesis, University of Caen.
Czanderna, A. W. (1964). *J. Phys. Chem.* **68**, 2765.
Dawson, P. T. and Hansen, R. S. (1966). *J. Chem. Phys.* **45**, 3148.
Den Besten, I. E. and Selwood, P. W. (1962). *J. Catal.* **1**, 93.
Dobychin, S. L., Torshina, V. V. and Smolina, G. N. (1964). *Kinet. Catal.* **5**, 581.
Duell, M. J., Davis, B. J. and Moss, R. L. (1966). *Discuss. Faraday Soc.* **41**, 43.
Dwyer, F. G., Eagleston, J., Wei, J. and Zahner, J. C. (1968). *Proc. R. Soc.* **A302**, 253.
Ehrlich, G. (1961). Trans. 8th Vac. Symp. and 2nd Internat. Cong., Pergamon, London, p. 126.
Ehrlich, G. (1965). *In* "Proc. 3rd Internat. Congr. Catalysis" (W. H. M. Sachtler, G. C. A. Schuit and P. Zwietering, eds.) North Holland, Amsterdam, p. 133.
Eischens, R. P. and Pliskin, W. A. (1958). *Adv. Catalysis* **10**, 1.
Eischens, R. P. and Pliskin, W. A. (1961). *In* "Actes du Deuxième Congrès International de Catalyse". Editions Technip, Paris, p. 789.
Eley, D. D. (1941). *Proc. R. Soc.* **A178**, 452.
Eley, D. D. (1948). *Adv. Catalysis* **1**, 157.
Eley, D. D. (1950). *Discuss. Faraday Soc.* **8**, 34.
Eley, D. D. and Norton, P. R. (1966). *Discuss. Faraday Soc.* **41**, 135.
Erkelens, J., Galwey, A. K. and Kemball, C. (1961). *Proc. R. Soc.* **A260**, 273.
Erkelens, J., Kemball, C. and Galwey, A. K. (1963). *Trans. Faraday Soc.* **59**, 1181.
Farkas, A. (1936a). *Trans. Faraday Soc.* **32**, 416.
Farkas, A. (1936b). *Trans. Faraday Soc.* **32**, 922.
Farkas, A. and Farkas, L. (1937a). *Trans. Faraday Soc.* **33**, 678.
Farkas, A. and Farkas, L. (1937b). *Trans. Faraday Soc.* **33**, 827.
Farkas, A. and Farkas, L. (1939). *J. Am. Chem. Soc.* **61**, 1336.
Flanagan, T. B. and Rabinovitch, B. S. (1956). *J. Phys. Chem.* **60**, 724, 730.

Foss, T. G. and Eyring, H. (1958). *J. Phys. Chem.* **62**, 103.

Frankenburger, W. and Hodler, A. (1932). *Trans. Faraday Soc.* **28**, 229.

Freel, J. and Galwey, A. K. (1968). *J. Catal.* **10**, 277.

Fukuda, K., Nagashima, S., Noto, Y., Oniski, T. and Tamaru, K. (1968). *Trans. Faraday Soc.* **64**, 522.

Galwey, A. K. and Kemball, C. (1959). *Trans. Faraday Soc.* **55**, 1959.

Galwey, A. K. and Kemball, C. (1961). *In* "Actes du Deuxième Congrès International de Catalyse". Editions Technip, Paris, p. 1063.

Galwey, A. K. (1962). *J. Catal.* **1**, 227.

Garnett, J. L. and Sollich-Baumgartner, W. A. (1966). *Adv. Catalysis* **16**, 95.

Gault, F. G. and Kemball, C. (1961). *Trans. Faraday Soc.* **57**, 1781.

Gault, F. G., Rooney, J. J. and Kemball, C. (1962). *J. Catal.* **1**, 255.

Gault, F. G. (1969). Private communication, University of Caen.

Gleiser, M. and Chipman, J. (1962). *J. Phys. Chem.* **66**, 1539.

Good, R. H. and Muller, E. W. (1956). *In* "Handbuch der Physik". Band 21, Springer-Verlag, Berlin, p. 176.

Greenhalgh, R. K. and Polanyi, M. (1939). *Trans. Faraday Soc.* **35**, 520.

Gundry, P. M., Haber, J. and Tompkins, F. C. (1962). *J. Catal.* **1**, 363.

Gutmann, J. R. (1955). *J. Phys. Chem.* **59**, 478.

Hall, W. K. and Emmett, P. H. (1959). *J. Phys. Chem.* **63**, 1102.

Halsey, G. D. (1963). *J. Phys. Chem.* **67**, 2038.

Harper, R. J. and Kemball, C. (1965). *In* "Proc. 3rd Internat. Congr. Catalysis". (W. M. H. Sachtler, G. C. A. Schuit and P. Zwietering, eds.) North Holland, Amsterdam, p. 114.

Hayward, D. O. and Trapnell, B. M. W. (1964). "Chemisorption". Butterworths, London.

Hillaire, L., Maire, G. and Gault, F. G. (1967). *Bull. Soc. Chim. Fr.* 886.

Hilaire, L. (1969). Ph.D. Thesis, University of Caen.

Hirota, K., Kuwata, K., Otaki, T. and Asai, S. (1961). *In* "Actes du Deuxième Congrès International de Catalyse". Editions Technip, Paris, p. 809.

Hirota, K. and Hironaka, Y. (1966). *Bull. Chem. Soc. Japan* **39**, 2638.

Horiuti, J., Ogden, G. and Polanyi, M. (1934). *Trans. Faraday Soc.* **30**, 663.

Horiuti, J. and Polanyi, M. (1934). *Mem. Proc. Manchester Lit. Phil. Soc.* **78**, 47.

Horrex, C. and Moyes, R. B. (1965). *In* "Proc. 3rd Internat. Congr. Catalysis" (W. M. H. Sachtler, G. C. A. Schuit and P. Zwietering, eds.) North Holland, Amsterdam, p. 1156.

Imai, H., Kemball, C. and Whan, D. A. (1968). *Proc. R. Soc.* **A302**, 385.

Imai, H. and Kemball, C. (1968). *Proc. R. Soc.* **A302**, 399.

Jaeger, H. (1967). *J. Catal.* **9**, 237.

Jenkins, G. I. and Rideal, E. K. (1955). *J. Chem. Soc.* 2490, 2496.

Juza, R. (1966). *Adv. Inorg. Radiochem.* **9**, 81.

Kauder, L. N. and Taylor, T. I. (1951). *Science, N.Y.* **113**, 231.

Kazanskii, B. A. (1967). *Kinet. Catal.* **8**, 841.

Keller, A. and Korosy, F. (1948). *Nature, Lond.* **162**, 580.

Kemball, C. and Taylor, H. S. (1948). *J. Am. Chem. Soc.* **70**, 345.

Kemball, C. (1950). *Adv. Catalysis* **2**, 233.

Kemball, C. (1951). *Proc. R. Soc.* **A207**, 539.

Kemball, C. (1952a). *Proc. R. Soc.* **A214**, 413.

Kemball, C. (1952b). *Trans. Faraday Soc.* **48**, 254.

Kemball, C. (1953). *Proc. R. Soc.* **A217**, 376.

Kemball, C. (1954a). *Proc. R. Soc.* **A223**, 377.

Kemball, C. (1954b). *Trans. Faraday Soc.* **50**, 1344.

Kemball, C. and Wolf, F. J. (1955). *Trans. Faraday Soc.* **51**, 1111.

Kemball, C. (1956). *J. Chem. Soc.* 735.

Kemball, C. and Moss, R. L. (1956). *Proc. R. Soc.* **A238**, 107.

Kemball, C. and Stoddart, C. T. H. (1957). *Proc. R. Soc.* **A241**, 208.

Kemball, C. and Moss, R. L. (1958). *Proc. R. Soc.* **A244**, 398.

Kemball, C. and Stoddart, C. T. H. (1958). *Proc. R. Soc.* **A246**, 521.

Kemball, C. (1959). *Adv. Catalysis* **11**, 223.

Kemball, C. and Woodward, I. (1959). *Trans. Faraday Soc.* **56**, 138.

Kemball, C. and Patterson, W. R. (1962). *Proc. R. Soc.* **A270**, 219.

Kemball, C. (1966). *Discuss. Faraday Soc.* **41**, 190.

Kieran, P. and Kemball, C. (1965). *J. Catal.* **4**, 380.

Kington, G. L. and Holmes, J. M. (1953). *Trans. Faraday Soc.* **49**, 417.

Kochloeft, K. and Bazant, V. (1968). *J. Catal.* **10**, 140.

Kubicka, H. (1968). *J. Catal.* **12**, 223.

Kummer, J. T. (1956). *J. Phys. Chem.* **60**, 666.

Lawson, A. (1968a). *J. Catal.* **11**, 283.

Lawson, A. (1968b). *J. Catal.* **11**, 295.

Logan, S. R., Moss, R. L. and Kemball, C. (1958). *Trans. Faraday Soc.* **54**, 922.

Logan, S. R. and Kemball, C. (1960). *Trans. Faraday Soc.* **56**, 144.

Madden, W. F. and Kemball, C. (1961). *J. Chem. Soc.* 302.

Mah, A. D. (1960). U.S. Bureau Mines Rept. Invest. No. 5529.

Maire, G., Plouidy, G., Prudhomme, J. C. and Gault, F. G. (1965). *J. Catal.* **4**, 556.

Maire, G. and Gault, F. G. (1967). *Bull. Soc. Chim. Fr.* 894.

Maire, G. (1969). Unpublished work from Flinders University.

Margolis, L. Y. (1963). *Adv. Catalysis* **14**, 429.

Mars, P., Scholten, J. J. F. and Zwietering, P. (1963). *Adv. Catalysis* **14**, 35.

Masuda, M. (1965). *J. Res. Inst. Catalysis Hokkaido Univ.* **12**, 67.

Matsuzaki, I. and Tada, A. (1969). *J. Catal.* **13**, 215.

Maxted, E. B. (1951). *Adv. Catalysis* **3**, 129.

McCarroll, B. (1969). Unpublished work from General Electric Research and Development Center.

McElligott, P. E., Roberts, R. W. and Engel, T. W. (1967). *J. Chem. Phys.* **47**, 1537.

McKee, D. W. (1968). *Trans. Faraday Soc.* **64**, 2200.

Mercer, P. D. (1969). Unpublished work from C.S.I.R.O. Division of Tribophysics.

Meyer, E. F. and Kemball, C. (1965). *J. Catal.* **4**, 711.

Mignolet, J. C. P. (1950). *Discuss. Faraday Soc.* **8**, 105.

Mikovsky, R. J., Boudart, M. and Taylor, H. S. (1954). *J. Am. Chem. Soc.* **76**, 3814.

Mills, G. A., Heinemann, H., Milliken, T. H. and Oblad, A. G. (1953). *Ind. Eng. Chem.* **45**, 134.

Miyahara, K. (1956). *J. Res. Inst. Catalysis Hokkaido Univ.* **4**, 143.

Morikawa, K., Trenner, N. R. and Taylor, H. S. (1937). *J. Am. Chem. Soc.* **59**, 1103.

Morrison, R. A. and Krieger, K. A. (1968). *J. Catal.* **12**, 25.

Moss, R. L. and Thomas, D. H. (1967). *J. Catal.* **8**, 162.

Murray, K. E. (1950). *Aus. J. Sci. Res.* **A3**, 443.

208   J. R. ANDERSON AND B. G. BAKER

Newham, J. (1963). *Chem. Rev.* **63**, 123.
Ozaki, A., Taylor, H. S. and Boudart, M. (1960). *Proc. R. Soc.* **A258**, 47.
Patterson, W. R. and Kemball, C. (1963). *J. Catal.* **2**, 465.
Pauling, L. (1960). "The Nature of the Chemical Bond". 3rd ed., Cornell University Press.
Pearson, J. and Ende, V. J. C. (1953). *J. Iron Steel Inst.* **175**, III, 52.
van der Plank, P. and Sachtler, W. M. H. (1968). *J. Catal.* **12**, 35.
Pritchard, J. and Tompkins, F. C. (1960). *Trans. Faraday Soc.* **56**, 540.
Pritchard, J. (1965). *Nature, Lond.* **205**, 1208.
Richardson, F. D. (1953). *J. Iron Steel Inst.* **175**, III, 33.
Rienacker, G. and Volter, J. (1959). *Z. Anorg. Allg. Chem.* **302**, 299.
Ritchie, B. and Wheeler, R. (1966). *J. Phys. Chem.* **70**, 173.
Roberts, E., Emeleus, H. S. and Briscoe, E. V. A. (1939). *J. Chem. Soc.* 41.
Roberts, R. W. (1962a). *Trans. Faraday Soc.* **58**, 1159.
Roberts, R. W. (1962b). *J. Phys. Chem.* **66**, 1742.
Roberts, R. W. (1962c). *Nature, Lond.* **195**, 1094.
Roberts, R. W. (1963). *J. Phys. Chem.* **67**, 2035.
Roberts, M. W. and Ross, J. R. H. (1966). *Trans. Faraday Soc.* **62**, 2301.
Rooney, J. J., Gault, F. G. and Kemball, C. (1960). *Proc. Chem. Soc.* 407.
Rooney, J. J. (1963). *J. Catal.* **2**, 53.
Rooney, J. J. and Webb, G. (1964). *J. Catal.* **3**, 488.
Roth, J. A., Geller, B. and Burwell, R. L. (1968). *J. Res. Inst. Catalysis Hokkaido Univ.* **16**, 221.
Rowlinson, H. C., Burwell, R. L. and Tuxworth, R. H. (1955). *J. Phys. Chem.* **59**, 225.
Rye, R. R. (1967). Ph.D. Thesis, Iowa State University.
Sachtler, W. M. H. and Fahrenfort, J. (1961). *In* "Actes du Deuxième Congrès International de Catalyse". Editions Technip, Paris, p. 831.
Saleh, J. M., Kemball, C. and Roberts, M. W. (1961). *Trans. Faraday Soc.* **57**, 1771.
Saleh, J. M., Roberts, M. W. and Kemball, C. (1963). *J. Catal.* **2**, 189.
Saleh, J. M., Wells, B. R. and Roberts, M. W. (1964). *Trans. Faraday Soc.* **60**, 1865.
Schrage, K. and Burwell, R. L. (1966). *J. Am. Chem. Soc.* **88**, 4549, 4555.
Schuit, G. C. A. and van Reijen, L. L. (1958). *Adv. Catalysis* **10**, 242.
Schuit, G. C. A., van Reijen, L. L. and Sachtler, W. M. H. (1961). *In* "Actes du Deuxième Congrès International de Catalyse". Edition Technip, Paris, p. 893.
Selwood, P. W. (1957). *J. Am. Chem. Soc.* **79**, 4637.
Siegel, S. (1966). *Adv. Catalysis* **16**, 124.
Sinfelt, J. H., Hurwitz, H. and Rohrer, J. C. (1960). *J. Phys. Chem.* **64**, 892.
Sinfelt, J. H., Taylor, W. F. and Yates, D. J. C. (1965). *J. Phys. Chem.* **69**, 95.
Sinfelt, J. H. and Yates, D. J. C. (1967). *J. Catal.* **8**, 82.
Sinfelt, J. H. and Taylor, W. F. (1968). *Trans. Faraday Soc.* **64**, 3086.
Sinfelt, J. H. and Yates, D. J. C. (1968). *J. Catal.* **10**, 362.
Singleton, J., Roberts, E. and Winter, E. (1951). *Trans. Faraday Soc.* **47**, 1318.
Skinner, H. A. (1964). *Advances Organomet. Chem.* **2**, 49.
Solbakken, A. and Emmett, P. H. (1969). *J. Am. Chem. Soc.* **91**, 31.
Sosnovsky, H. M. C. (1955). *J. Chem. Phys.* **23**, 1486.
Sosnovsky, H. M. C., Ogilvie, G. J. and Gillam, E. (1958). *Nature, Lond.* **182**, 523.
Sosnovsky, H. M. C. (1959). *J. Phys. Chem. Solids* **10**, 304.

Stephens, S. J. (1959). *J. Phys. Chem.* **63**, 512.
Stoddart, C. T. H. and Kemball, C. (1956). *J. Colloid Sci.* **11**, 532.
Suhrmann, R. (1957). *Adv. Catalysis* **9**, 88.
Suhrmann, R., Heras, J. M., Heras, L. V. and Wedler, G. (1964). *Ber. Bunsenges. Phys. Chem.* **68**, 511.
Suhrmann, R., Heras, J. M., Heras, L. V. and Wedler, G. (1968). *Ber. Bunsenges. Phys. Chem.* **72**, 854.
Tamaru, K. (1959). *Trans. Faraday Soc.* **55**, 824, 1191.
Tamaru, K. (1961). *Trans. Faraday Soc.* **57**, 1410.
Tamaru, K. (1964). *Trans. Faraday Soc.* **60**, 765.
Taylor, H. S. and Jungers, J. C. (1935). *J. Am. Chem. Soc.* **57**, 679.
Taylor, T. I. (1957). *In* "Catalysis" (P. H. Emmett, ed.) Vol. V, Reinhold, New York, p. 257.
Taylor, W. F., Yates, D. J. C. and Sinfelt, J. H. (1964). *J. Phys. Chem.* **68**, 2962.
Taylor, W. F., Sinfelt, J. H. and Yates, D. J. C. (1965). *J. Phys. Chem.* **69**, 3857.
Taylor, G. F., Thomson, S. J. and Webb, G. (1968). *J. Catal.* **12**, 191.
Temkin, M. I. and Pyzhev, V. (1940). *Acta Phys.-chem.* (*URSS*) **12**, 327.
Thompson, S. O., Turkevich, J. and Irsa, A. P. (1951). *J. Am. Chem. Soc.* **73**, 5213.
Trapnell, B. M. W. (1951). *Proc. R. Soc.* **A206**, 39.
Trapnell, B. M. W. (1952). *Trans. Faraday Soc.* **48**, 160.
Trapnell, B. M. W. (1953). *Proc. R. Soc.* **A218**, 566.
Trapnell, B. M. W. (1956). *Trans. Faraday Soc.* **52**, 1618.
Turkevich, J., Bonner, F., Schissler, D. O. and Irsa, A. P. (1950). *Discuss. Faraday Soc.* **8**, 352.
Twigg, G. F. (1939). *Trans. Faraday Soc.* **35**, 934.
Twigg, G. H. and Rideal, E. K. (1939). *Proc. R. Soc.* **A171**, 55.
Twigg, G. H. (1946). *Proc. R. Soc.* **A188**, 92, 105, 123.
Twigg, G. H. (1950). *Discuss. Faraday Soc.* **8**, 152.
Veda, T., Hara, J., Hirota, K., Teratani, S. and Yoshida, N. (1969). *Z. phys. Chem.* (*NF*) **64**, 64.
Voge, H. H. and Atkins, L. T. (1952). Unpublished work, reported in Voge and Adams (1967).
Voge, H. H. (1957). *In* "Proc. 2nd Internat. Congr. Surface Activity" (J. H. Schulman, ed.) Vol. 2, Butterworths, London, p. 337.
Voge, H. H. and Adams, C. R. (1967). *Adv. Catalysis* **17**, 151.
Vol, Y. T. and Shishakov, N. A. (1962). *Izv. Akad. Nauk. S.S.S.R., Otd. Khim. Nauk.* 586.
Wagner, C. D., Wilson, J. N., Otvos, J. W. and Stevenson, D. P. (1952). *J. Chem. Phys.* **20**, 338.
Wagner, C. D., Wilson, J. N., Otvos, J. W. and Stevenson, D. P. (1953). *Ind. Eng. Chem.* **45**, 1480.
Wahba, M. and Kemball, C. (1953). *Trans. Faraday Soc.* **49**, 1351.
Wanninger, L. A. and Smith, J. M. (1960). *Chem. Weekblad* **57**, 273.
Weber, J. and Laidler, K. J. (1951). *J. Chem. Phys.* **19**, 1089.
Weisz, P. B. (1962). *Adv. Catalysis* **13**, 137.
Whan, D. A. and Kemball, C. (1965). *Trans. Faraday Soc.* **61**, 294.
Wilson, J. N., Voge, H. H., Stevenson, D. P., Smith, A. E. and Atkins, L. T. (1959). *J. Phys. Chem.* **63**, 463.
Wilson, R. L., Kemball, C. and Galwey, A. K. (1962). *Trans. Faraday Soc.* **58**, 583.

Wilson, R. L. and Kemball, C. (1964). *J. Catal.* **3**, 426.

Wood, B. J. and Wise, H. (1966). *J. Catal.* **5**, 135.

Wright, A. N. (1968). "Reactions of Clean Cadmium Surfaces"; Abstracts of Papers, 155th ACS National Meeting, April 1968, No. 5123.

Wright, P. G., Ashmore, P. G. and Kemball, C. (1958). *Trans. Faraday Soc.* **54**, 1692.

Yates, D. J. C., Taylor, W. F. and Sinfelt, J. H. (1964). *J. Am. Chem. Soc.* **86**, 2996.

Yates, D. J. C., Sinfelt, J. H. and Taylor, W. F. (1965). *Trans. Faraday Soc.* **61**, 2044.

Zimmerman, H. E. and Zweig, A. (1961). *J. Am. Chem. Soc.* **83**, 1196.

Chapter 9

Properties and Reactions of Alloy Films

D. R. ROSSINGTON

State University of New York
College of Ceramics at Alfred University,
Alfred, New York, U.S.A.

I. Introduction

In studying adsorption and reaction processes at metal surfaces there is usually a choice possible as to the physical nature of the substrate—powders, foils, filaments or evaporated films. The surfaces of powders, foils and filaments are contaminated with chemisorbed gases during their formation, and it is therefore necessary to clean such contaminated surfaces before they may be used in any adsorptive study.

Unfortunately many metals of interest in the catalytic and chemisorption field will melt at temperatures far below those necessary for the desorption of oxygen or nitrogen and consequently for these materials other cleaning methods are necessary. Reduction by atomic hydrogen, produced in an electrodeless discharge, has been effective in removing contaminants from the surface of low melting metals (Couper and Eley, 1950). This treatment results in the dissolution and chemisorption of hydrogen by the metal and it must therefore be followed by

211

vigorous outgassing procedures. There is also the inherent disadvantage of the very low surface area of a filament (often 0.5 to 1 cm² at the maximum) which renders it extremely sensitive to contamination by minute quantities of impurities.

For the above reasons the most widely used method nowadays for obtaining atomically clean metal surfaces is the evaporated film technique, in which a solid metal is outgassed in vacuo and then heated to a sufficient temperature to cause evaporation and subsequent condensation on the inner surface of a containing vessel.

With the continuous improvement of ultra-high vacuum techniques films may be evaporated which will remain uncontaminated for periods of up to several hours, thereby opening up whole new areas of experiments which are possible on such films. During evaporation the pressure above the film should not rise above 10^{-9} torr in order for the surface to remain clean. To some extent every film will act as a pumping system by gettering action, although this, of course, reduces the area of clean surface available for experimentation. By the evaporated film technique clean metal films, of thicknesses ranging from a few hundred to several thousand angstroms may be produced, with surface areas of the order of hundreds of square centimetres. Techniques for the preparation of these films and their properties are described elsewhere in this book.

For many years the catalytic properties of alloys have been of great interest to surface scientists. The electronic properties of the transition metals vary discontinuously when proceeding across the periodic table since electrons are added in increments of one. However, if an alloy of adjacent elements can be formed, a continuous variation of electronic structure may be obtained. Alloys formed between Group VIII and Group IB elements are of particular concern because of various theories of catalytic activity which are based upon the electronic constitution of the alloys. It is with such alloys that the present chapter will be largely concerned. A brief account of industrially important alloys, such as bronzes, brasses and nichrome, has been presented by Holland (1956).

Many aspects of Group VIII–Group IB alloys can be well described in terms of the simple band theory described by Mott and Jones (1936). Figure 1 shows a representation of this work for metals of the fourth and fifth periods. It is seen that the d and s bands overlap and that the broader $4s$ and $5s$ bands have energy levels which are lower in magnitude than the respective $3d$ and $4d$ bands. Therefore as the valency electrons are progressively added across these periods, they first go into the s band for the Group IB metals and then begin to fill the d band. The vertical lines in figure 1 indicate the Fermi surfaces of the indicated metals. For a material to be a conductor there must be unfilled energy

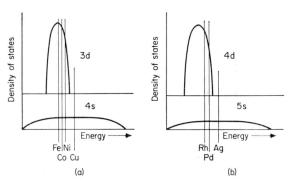

FIG. 1. Approximate representation of the band structure for some elements of Periods 4 and 5, the vertical lines representing Fermi surfaces.

levels directly above this surface, or, in the case of a semiconductor, there must be only a relatively narrow gap between the Fermi surface and the next available energy level. Thus, the area in a particular band to the right of the line indicating the Fermi surface in Figure 1 may be considered as being proportional to the number of vacancies or "holes" in that band. For example, nickel and palladium have approximately 0·6 "holes" or unpaired electrons, per atom in their d bands. Copper, silver and gold have filled d bands, but one "hole" per atom in their s bands.

As a Group IB element is alloyed with a Group VIII element, the valency s electrons from the former fill the holes in the d band of the latter. From the number of d band vacancies it is possible to calculate that an alloy of approximately 60 to 65 atomic % Group IB metal should contain sufficient electrons to completely fill the d band vacancies of the Group VIII metal. Evidence that these percentages are correct is supplied by the physical properties of the alloys. For palladium-silver alloys the paramagnetic susceptibility progressively decreases as silver is added to palladium until it reaches zero at approximately 60% silver (Pugh and Ryan, 1958). Similar behaviour is found for gold-palladium alloys (Vogt, 1932).

Several publications have studied the effect of the filling of d band vacancies on the catalytic activity of metals, some of the earliest being those of Couper and Eley (1950), Dowden (1950) and Reynolds (1950). Couper and Eley measured the activation energy of the paraortho hydrogen conversion on a series of gold-palladium alloy filaments and found, as shown in Figure 2, a marked rise near the composition at which there were just sufficient s electrons from the gold to fill the vacancies in the palladium d band. Dowden and Reynolds followed

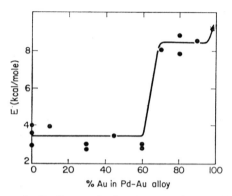

FIG. 2. Activation energies for the para hydrogen conversion on Pd–Au alloy wires. (Reproduced with permission from Couper and Eley (1950). *Discuss. Faraday Soc.* **8**, 172.)

reactions such as the hydrogenation of benzene, hydrogenation of styrene and various decomposition reactions over nickel-copper alloys in the form of powders and foils. Their results were related to magnetic properties of the substrates. Rienacker and Vormum (1956) studied the para hydrogen conversion over copper-nickel alloy foils. Their results were essentially similar to those of Couper and Eley except that in this case it is seen from Figure 3 the sharp rise in activation energy did not

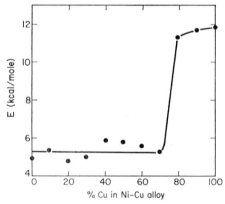

FIG. 3. Activation energies for the para hydrogen conversion on Ni–Cu foils. Redrawn from Rienacker and Vormum (1956).

occur until an alloy composition of approximately 70% copper was reached. In other studies on similar alloy systems the change in activity for reactions involving hydrogen did not occur exactly at the composition at which the d band is filled. These results suggest that the d band vacancy model represents an over simplification of the true picture.

Coles (1952) has studied the magnetic and thermo-electric properties of palladium-silver and nickel-copper alloys. Whilst the palladium-silver alloys showed the behaviour to be expected from the band theory of metals, the nickel-copper alloys showed consistent deviations, thought to be due to the differing proportions of the atomic volume occupied by the ions in copper and silver. The possibility of d–s promotion of electrons in the Group IB metal and the donation of s electrons from the d band by adsorbed hydrogen should also be considered.

In order to account for the almost constant activation energies of the alloys until the d band is filled, Couper and Eley (1950) suggested a mechanism for vacancy diffusion to the catalyst surface. Such vacancies could be created by the excitation of electrons from the necessary levels up to the Fermi surface. There is evidence for some correlation between activation energy for reactions on alloys and the height of the Fermi surface (Couper and Metcalfe, 1966), which will be discussed later.

Palladium-silver and palladium-gold alloy systems are well suited for a study of electronic factors in catalysis. They form a complete series of solid solutions having a face-centred cubic structure and the lattice constants varying by only a small factor from one end of the series to the other. Figure 4 shows the lattice constant for a complete series of palladium-silver alloys (Coles, 1956), varying by only 5% from 100% palladium to 100% silver.

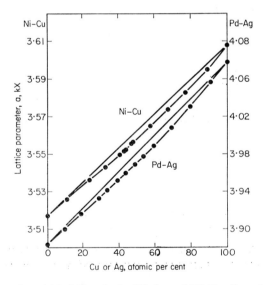

FIG. 4. Lattice spacings (a) in kX units for Pd–Ag and Ni–Cu alloys. (Reproduced with permission from Coles (1956). *J. Inst. Metals* **84**, 346.)

The departure from Vegard's law (deviation from linearity) in Figure 4 is shown more clearly in Figure 5. It is seen that the change in slope of the lattice spacing curve occurs at the composition of approximately 62% silver, where the d band of the palladium is just filled. Figures 4 and 5 also give data for the nickel-copper system and in this case there is no sharp change of slope at 60% copper, although there is one at approximately 35% copper, which is the composition at which the Curie temperature of the alloys reaches room temperature. This is further evidence that, in general, the electronic structure of nickel-copper alloys is more complicated than that of palladium-silver alloys.

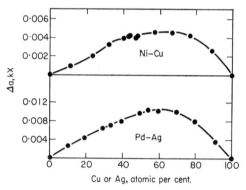

Fig. 5. Deviations (Δa) in kX units from Vegard's Law for Pd–Ag and Ni–Cu alloys. (Reproduced with permission from Coles (1956). *J. Inst. Metals* **84**, 346.)

The increase in lattice spacing in going from pure palladium to pure silver represents a weakening of the cohesive forces, caused by a decrease in contribution made by d electrons (Hume-Rothery and Coles, 1954). This point is further developed by Coles (1956).

However, even with such apparently ideal systems as palladium-silver alloys, care must be taken as their thermodynamic properties can be complex. At 1000 °K the formation of a random solid solution of palladium and silver is accompanied by a net *reduction* in entropy, (Pratt, 1960). Also palladium-rich alloys will tend to absorb hydrogen which can greatly affect reaction mechanisms.

All of the aforementioned considerations indicate the great potential of studies on alloy surfaces in order to elucidate reaction mechanisms and theories of catalytic activity. Practically all of the work cited above has been carried out on foils or filaments and it is an obvious extension of this work to study reactions on evaporated alloy film surfaces. There is a twofold reason for this: one is in order to obtain the advantages of working with evaporated film techniques mentioned earlier, the other is in order to study any possible fundamental difference between the surface structure of drawn filaments, rolled foils, and evaporated films.

II. Preparation

The ultimate goal in the preparation of alloy films for catalytic or chemisorption studies is the achievement of a reproducible method for the formation of completely homogeneous alloy films of known and controllable composition, over a reasonably large area in a reaction vessel.

It is only during the past 10 years or so that alloy thin films have been prepared under conditions which make them suitable for catalytic studies. Prior to that time most of the evaporated alloy films were made for a study of physical properties. One of the earliest investigations of alloy films was a brief study by Edwards (1933) on the optical properties of aluminium-magnesium alloy mirrors. These were formed by the simultaneous heating of separate pieces of the component metals. It was recognized that the higher vapour pressure of magnesium would result in a large concentration of this element in the alloy, but analyses were not reproducible, there being as much as a 50% variation in the magnesium reported in different analyses of the same alloy sample. More recently Belser (1960) had studied the alloying behaviour of a wide variety of thin films, obtained from some twenty different metal pairs. The films were prepared by the simultaneous or successive evaporation from independent sources or by a sputtering technique. They were deposited over the relatively limited area of a microscope slide, with thicknesses ranging from 100 to 1000 Å, largely for the physical properties of the alloys.

There are three distinct methods for the preparation of alloy films:

(1) Simultaneous evaporation of the two component metals from separate sources.

(2) Successive or alternate evaporation of the two component metals from separate sources.

(3) Evaporation from a single alloy source.

Each of these methods has particular advantages and disadvantages which will be discussed in detail in this section. All of them require subsequent heat treatment of the film to allow for interdiffusion of the atoms and homogenization of the alloy. All have been used by various workers in this field.

A fourth method for the preparation of alloy films, consisting of the flash evaporation of a finely divided powder of the component metals has been described by Harris and Siegel (1948). This technique requires the instantaneous evaporation of the powder, which is dropped from a continuously moving band on to a heater. The mechanical requirements for the moving parts of this apparatus would make it very difficult for use under high vacuum conditions.

Due to the different vapour pressures of metals their rates of evaporation at the same temperature will vary greatly and to some extent the best method for the formation of alloy films will depend upon the components. The partial vapour pressure of an alloy component may be approximately determined by the application of Raoult's law (Dushman, 1951) although the range of concentration over which this law is valid will vary greatly from one alloy system to another. In general it will give reasonably accurate results where the mole fraction of one component is greater than 0.8 (Holland, 1956). At intermediate concentrations the deviation from ideality is greater and there is no simple method for the determination of the vapour pressure of a component of a binary alloy.

A. SIMULTANEOUS EVAPORATION FROM SEPARATE SOURCES

Using this method the two component metals may be individually heated to different temperatures so that the relative vapour pressures and rates of evaporation may be controlled, thereby determining to some extent the composition of the final film. A typical procedure is that described by Moss and Thomas (1964, 1967a). Films of palladium-silver alloy were formed on the inner surface of spherical reaction vessels. The apparatus was attached to a conventional vacuum system and was baked out at 450 °C and 10^{-6} torr, and the silver and palladium outgassed at temperatures close to their evaporation points before actual evaporation was allowed to take place. Silver was evaporated from a loop of 0.5 mm diameter wire heated directly by passing the appropriate current through it. Palladium was evaporated from an 0.2 mm diameter wire wrapped around a tungsten heating filament. The deposition rate was kept essentially constant by only allowing a small percentage of the component metals to be vapourized. The weight of the film produced could be determined from the initial and final weight of the source materials. Specimens of the films for characterization purposes were obtained by placing a small piece of thin glass, akin to a microscope slide, in the reaction vessel, or by stripping portions of the film from the broken reaction vessel.

Even when simultaneous evaporation takes place regions of the film rich in one component will develop unless the condensing atoms can diffuse a sufficient distance to produce a homogeneous alloy. In fact, no matter which method is chosen for the preparation of evaporated alloy films, some form of heat treatment of the substrate, to facilitate such homogenization, is necessary.

From the work of Jost (1933) and Nachtrieb *et al.* (1957) the self diffusion of the components in a palladium-silver alloy may be determined.

The diffusion coefficient of palladium, D_{Pd}, in an 80% Ag–20% Pd alloy is given by

$$D_{Pd} = D_{Pd}^{\circ} \exp\left(-E_{Pd}/RT\right)$$

where

$$D_{Pd}^{\circ} = 6{\cdot}4 \times 10^{-6}\,\text{cm}^2\,\text{sec}^{-1}$$

$$E_{Pd} = 20{\cdot}2\,\text{kcal mole}^{-1}$$

For a range of palladium-silver alloys the diffusion coefficient of silver, D_{Ag}, is given by

$$D_{Ag} = D_{Ag}^{\circ} \exp\left(-8{\cdot}2\,X\right) \exp\left(-E_{Ag}/RT\right)$$

where

$$D_{Ag}^{\circ} = 0{\cdot}270\,\text{cm}^2\,\text{sec}^{-1}$$

$$E_{Ag} = 43{\cdot}7\,\text{kcal mole}^{-1}$$

$$X = \text{atomic fraction of palladium}$$

$\log D_{Ag}$ is a linear function of reduced temperature T_m/T where T_m is the melting point of the alloy taken from the phase diagram for the system (Ruer, 1906). The equation for this linear relationship is

$$D_{Ag} = 0{\cdot}27 \exp\left(-17{\cdot}75\,T_m/T\right)$$

and the function is shown in Figure 6.

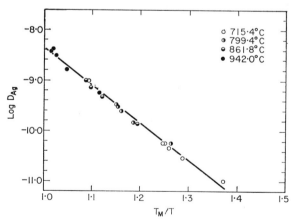

FIG. 6. Diffusion coefficient of silver as a function of reduced temperature. (Reproduced with permission from Nachtrieb, Petit and Wahrenberg (1957). *J. Chem. Phys.* **26**, 106.)

From these figures Moss and Thomas have calculated the times necessary for adequate interdiffusion of palladium and silver atoms at various
temperatures. In a time interval of one hour at temperatures of 100 °C,
200 °C, 300 °C and 400 °C, palladium atoms will diffuse distances of
26 Å, 460 Å, 3000 Å and 11,000 Å respectively. In one hour at corresponding temperatures silver atoms will only diffuse distances of 3 ×
10^{-4} Å, 0·15 Å, 9 Å and 150 Å respectively. Thus silver atoms will only
diffuse a reasonable distance at relatively high temperatures and a substrate temperature of at least 400 °C is necessary for the complete
homogenization of a palladium-silver alloy film.

The positioning of the component evaporation sources is critical to
the achievement of a homogeneous film over an extended area. Figure 7

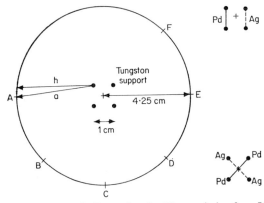

Fig. 7. Plan of evaporation vessel. (Reproduced with permission from Moss and Thomas
(1964). *Trans. Faraday Soc.* **60**, 1110.)

shows a plan of the evaporation vessel used by Moss and Thomas. With
this arrangement a particular film was found to have compositions of
70, 69, 73, 68 and 69 weight percent at positions B, C, D, E, and F
respectively, compared to an average value of 72 weight percent calculated from the weight loss of the sources. In later work on palladium-
silver alloy films, Moss and Thomas (1967) continued to use the techniques of slow simultaneous evaporation from separate sources, the
films being laid down at a rate of 15 Å thickness per minute.

Mader *et al.* (1967) used a method of forming copper-silver and cobalt-
gold alloy films in which the composition of the film could be monitored
and controlled while it was being laid down on a substrate. The two
component metals were simultaneously evaporated from resistance
heated alumina crucibles and "vapour quenched", i.e., condensed on a
substrate held at liquid nitrogen temperature. The deposition rate of
each component was controlled by a Bayard-Alpert ionization gauge in

the vapour stream of the vaporant, and the signal from this gauge, when compared to that from a gauge measuring background pressure only, was used to control the evaporation current.

Goldsztraub and Michel (1951) formed silver-magnesium alloys by the simultaneous evaporation of magnesium from a tungsten heating coil and silver from a molybdenum heating coil. The films were evaporated on to collodian films placed in different positions around the two sources, thus obtaining alloys of varying compositions from a single evaporation procedure. There was no heat treatment to homogenize the films. Electron diffraction was used to characterize the films, which gave essentially the same results as X-ray diffraction, with the exception of γ phase with a narrow region of stability, which was not observed in the weak X-ray patterns.

B. SUCCESSIVE EVAPORATION FROM SEPARATE SOURCES

The technique of successive evaporation has been more widely used than other methods and most of the early workers in the production of alloy films used this method. Burr et al. (1944) used this method for the formation of copper-zinc alloy films, as did Koenig (1944) for a similar system, and Belser (1960) for a wide range of alloys from 20 metal pairs.

This method was also used in the first preparation of alloy films suitable for catalytic investigations, by Gharpurey and Emmett (1961). Copper-nickel alloy films were evaporated from separate sources, the nickel directly from a 0.02 inch diameter wire and the copper from a 0.01 inch diameter wire wrapped around a tungsten heating loop. The plane of the nickel loop was normal to that of the tungsten loop. After evacuation of the system a pressure of 50 torr hydrogen was admitted and the sources reduced by being kept at a dull red heat for 10 minutes. The reaction vessel was then heated to 500 °C for one hour after which time it was evacuated and left pumping overnight at 500 °C. The source filaments were then degassed by being again heated to a dull red heat.

The appropriate quantity of nickel was evaporated by passing a constant current of 6.0 amp for the required period of time through the wire. All of the copper was evaporated from the tungsten wire. After evaporation was complete the films were homogenized by heating to 300 °C in 50 torr hydrogen overnight. The criterion for homogenization was that the resulting catalytic properties of the films of similar composition were identical whether the nickel or the copper was evaporated first. No analyses were carried out on the films, their compositions being determined from the weights of the component metals evaporated. In a more recent publication Campbell and Emmett (1967) report that X-ray diffraction work has been carried out in order to determine the degree of homogenization of the films. Their work with gold-nickel films showed little or no alloying of the metals, evaporated simultaneously, when the films were held in 50 torr hydrogen at 500 °C for 24 hours. This is in contrast with the results of Belser (1960) who found gold-nickel alloy formation at 190 °C.

The technique of successive evaporation from separate sources was used by Sachtler *et al.* (1965a, 1965b, 1965c) in the preparation of copper-nickel films for an extensive study of surface properties under ultra-high vacuum conditions. These authors gave careful consideration to the alternative methods for preparation of alloy films. Evaporation from an alloy wire was unsuccessful due to the large difference in vapour pressure of the components. Complete outgassing of the alloy wire was not possible since the highest temperatures were reached at the end of the evaporation process, and the amount of gas then developed was significant under the ultra-high vacuum conditions. They also found a strong tendency for the formation of a ternary alloy with the supporting filament, (tungsten, tantalum or rhenium) resulting in the filament burning out and the atoms of the supporting metal being incorporated in the resulting film.

The technique of simultaneous evaporation was also rejected on the grounds of insufficient outgassing of the component metals and alloying between the nickel metal and the tungsten or tantalum supporting cup material.

For these reasons, and the fact that subsequent heat treatment of the films for homogenization was necessary in any case, the method of successive evaporation was used. The tendency for nickel to alloy with a supporting metal was avoided by evaporating directly from a nickel wire. Copper was evaporated from a tantalum support. The copper-nickel films were evaporated directly under ultra-high vacuum conditions in the apparatus in which subsequent measurements of the work function were carried out. The apparatus used by Sachtler and Dorgelo is shown in Figure 8. The fact that the two sources of nickel and copper were geometrically separated means that the nickel/copper ratio varied over the lateral dimensions of the evaporated film. Use of this fact was made in the later measurements. The whole system was outgassed at 400 °C and 10^{-10} torr for 16 hours and the metal elements subjected to high-frequency eddy current heating. During the initial period of evaporation of the component metal the pressure was below 10^{-9} torr. The films were homogenized by heating to 200 °C.

Moss and Thomas (1964) found that the successive evaporation of palladium and silver on to a substrate at 0 °C followed by heat treatment at 530 °C produced homogeneous films. It was found that 3 hours treatment at this temperature was necessary for complete

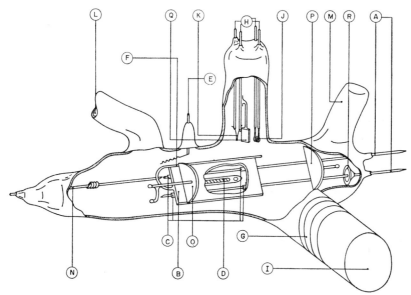

FIG. 8. Sliding-cathode phototube. (A) tungsten rods, (B) movable glass plate, substrate for film (photocathode), (C) glass sleeves, each a sliding fit on (A) ((B) and (C) are given an electrically conducting coating), (D) glass encapsulated iron slug, (E) anode connection, (F) strip for collecting film specimen for X-ray examination, (G) graded seal to quartz window (I), (H) evaporation source leads, (J) and (K) evaporation sources, (L) and (M) tubulation to vacuum system, (N) tungsten rod onto which (B) may be moved during filament degassing, (O), (P), (Q) and (R) evaporation screens. (Reproduced with permission from Sachtler and Dorgelo (1965b). *J. Catal.* 4, 654.)

homogenization of the film. Figure 9 shows the development of alloy with time; (a) shows part of the diffraction pattern for the component metal immediately after deposition at 0 °C. After heating for 1 hour at 530 °C (Figure 9b) the start of alloy formation is indicated, showing a silver-rich composition compared with the expected composition. A further hour's heating gave a palladium-rich alloy (Figure 9c) until after 3 hours at 530 °C (Figure 9d) the diffraction peaks were approximately symmetrical about the positions calculated from the bulk lattice constants of the expected alloy composition. The order of deposition of palladium or silver was found to have a marked effect on the alloying which occurred during the deposition of the metals. With palladium deposited first, the diffraction pattern obtained from the dual layer, before any heating, showed little or no alloying. When silver was deposited first there was considerable alloying upon the subsequent deposition of palladium. As well as the rates of diffusion of the alloy components other factors, such as the temperature of the primary deposit being raised by the latent heat of condensation of the secondary deposit (Fujiki, 1959), may play a part in the initial alloying process

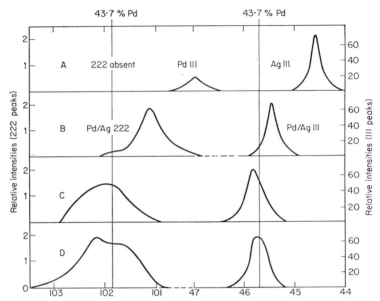

FIG. 9. X-ray diffractometer patterns (111 K$_{\alpha 1}$ and 222 K$_{\alpha 2}$ only) from dual Pd and Ag layer. (A) as deposited at 0 °C. After heating at 530 °C (B) for 1 hour, (C) for 2 hours, (D) for 3 hours. Expected positions for peaks for a 43.7% Pd–Ag alloy are shown. (Reproduced with permission from Moss and Thomas (1964). *Trans. Faraday Soc.* **60**, 1110.)

and the importance of the order of deposition. Moss and Thomas (1964) also found that the evaporation of the palladium from a high temperature tungsten heating loop subjected the silver deposit to radiant heat and they considered this to be the major factor in the importance of order of deposition in their work.

In general Moss and Thomas indicate that they favoured the simultaneous evaporation technique for the formation of palladium-silver alloys as they found that evaporation on to a substrate at 400 °C provided a convenient method of obtaining homogeneous films, whereas with successive evaporation of the components on to a substrate at 0 °C, subsequent heat treatment at 530 °C was necessary for homogenization. In later publications (1967a, 1967b) they also used the simultaneous evaporation method.

C. EVAPORATION FROM AN ALLOY SOURCE

Due to the differences in vapour pressure of metals the evaporation of an alloy source will result in fractionation taking place, the component with the higher vapour pressure being preferentially evaporated. The composition of films, therefore, evaporated from an alloy source, bears no simple relation to the source composition, and when this technique is employed the weight loss of the source alloy gives no indication of the evaporated film composition. Thus an independent method of analysis, for example X-ray flourescence spectroscopy, is essential in order to determine the film composition.

The success of this method depends upon the relative vapour pressures of the two components. The composition of the vapour from the alloy wire is governed approximately by Raoult's law and therefore if the vapour pressures of the components are extremely different at the evaporation temperature the component with the higher vapour pressure will be deposited first, following by the remaining component. The fractionation which takes place results in a layered type of film similar to that obtained by the successive evaporation technique. This is essentially the case Sachtler and Dorgelo (1965b) found for nickel-copper films and is one of the reasons why they decided to use successive evaporation.

In those cases where the difference in vapour pressures is not too extreme, evaporation from an alloy source provides a convenient method of producing alloy films. The advantages of this method are simplicity of deposition and the fact that the components are evaporated from the same point within the reaction vessel. The component with the higher vapour pressure will volatilize preferentially and the composition of the film will, therefore, be richer in that material than the parent alloy. However, this is not detrimental provided that the film composition can be determined by another method. It is important to note that even when evaporation from an alloy source is suitable, heat treatment to homogenize the film is a necessary procedure following evaporation. In most cases it is preferable to evaporate the alloy wire from some heating element, such as a tungsten or tantalum filament. It is therefore necessary to ensure that no low-melting ternary alloy formation can take place between the binary alloy parent source and the heating filament. This was one reason which prevented the evaporation of a nickel-copper alloy source from a supporting filament as mentioned previously (Sachtler and Dorgelo, 1965b).

Rossington and Runk (1967) found the alloy evaporation technique very suitable for the preparation of palladium-silver films of compositions varying from approximately 50% Pd–50% Ag to 20% Pd–80% Ag over areas of the order of 120 cm^2 and with variations in the composition of the order of a maximum of 3% over the active area of the films. The films were evaporated inside cylindrical reaction vessels as shown in Figure 10. The alloy wire to be evaporated was wound uniformly around the tungsten filament with the aid of a lathe and speed reduction gear.

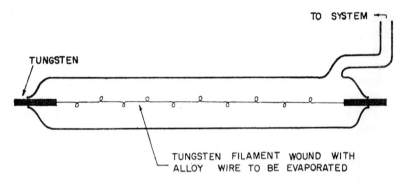

Fig. 10. Reaction vessel used for evaporation from alloy wires. (Reproduced with permission from Rossington and Runk (1967). *J. Catal.* **7**, 365.)

TABLE 1

Characterization data for 80% Ag – 20% Pd film.

Preparation

Length of wire	19 cm
Diameter of wire	0.25 mm
Evaporating current	4.5 amp
Evaporating time	3 min
Substrate temperature	673°K
Annealing temperature	673°K
Annealing time	1 hr
Apparent area	120 cm²
Total reaction volume	500 cm³
Weight deposited	17 mg
Film thickness	1300 Å

Composition and Homogeneity Analysis

Section of vessel	X-ray diffraction results			X-ray fluorescence results	
	Corrected $2\theta_{(111)}$	a_0	% Ag	$\dfrac{\text{Ag K}\alpha}{\text{Ag K}\alpha + \text{Pd K}\alpha}$	% Ag
A	38.57	4.043	82	0.77	82
B	38.56	4.043	82	0.81	85
C	38.55	4.041	81	0.77	82
D	38.56	4.043	82	0.79	83
E	38.56	4.043	82	0.80	85
Average composition			82% ± 1%		83% ± 2%

(Reproduced with permission from Rossington and Runk (1967). *J. Catal.* **7**, 365.)

To obtain homogeneous alloy films, the following evaporation procedure was used: After overnight baking out periods at about 500 °C,
the vessel temperature was brought down to 400 °C and the alloy wire
was evaporated. Evaporation currents from 3.5 to 5.5 amps and
evaporation times of 3 min were used. The deposited films were then
annealed for 1 hr at 400 °C before being cooled to room temperature.
After evaporation of the films the reaction vessels were broken open and
pieces of film taken from five sections along the entire length of the
vessels and removed for analysis. These are the sections designated A,
B, C, D and E in Tables 1 to 4.

It was found that the composition of the films could be reproducibly
determined by careful control of the evaporating current passed through
the tungsten filament. For instance, the evaporation of 50% Pd–50% Ag
alloy wire caused by passing a current of 4.5 amp through the tungsten

TABLE 2

Characterization data for 70% Ag – 30% Pd film.

Preparation

Length of wire	16 cm
Diameter of wire	0.25 mm
Evaporating current	5.5 amp
Evaporating time	3 min
Substrate temperature	673 °K
Annealing temperature	673 °K
Annealing time	1 hr
Apparent area	120 cm²
Total reaction volume	500 cm²
Weight deposited	22 mg
Film thickness	1700 Å

Composition and Homogeneity Analysis

Section of vessel	X-ray diffraction results			X-ray fluorescence results	
	Corrected $2\theta_{(111)}$	a_0	% Ag	$\dfrac{\text{Ag K}\alpha}{\text{Ag K}\alpha + \text{Pd K}\alpha}$	% Ag
A	38.85	4.011	67	0.70	74
B	38.75	4.021	72	0.68	72
C	38.83	4.014	68	0.69	73
D	38.84	4.012	67	0.69	73
E	38.95	4.006	65	0.69	73
Average composition			68% ± 3%		73% ± 1%

(Reproduced with permission from Rossington and Runk (1967). *J. Catal.* **7**, 365.)

K

TABLE 3

Characterization data for 50% Ag – 50% Pd film I.

PREPARATION AND COMPOSITION DATA
FOR THE 50% SILVER–50% PALLADIUM FILM I

Preparation

Length of wire	60 cm
Diameter of wire	0.1 mm
Evaporating current	4.0 amp
Evaporating time	3 min
Substrate temperature	673 °K
Annealing temperature	673 °K
Annealing time	1 hr
Apparent area	120 cm²
Total reaction volume	500 cm³
Weight deposited	33 mg
Film thickness	2500 Å

Composition and Homogeneity Analysis

Section of vessel	X-ray diffraction results			X-ray fluorescence results	
	Corrected $2\theta_{(111)}$	a_0	% Ag	$\dfrac{\text{Ag K}\alpha}{\text{Ag K}\alpha + \text{Pd K}\alpha}$	% Ag
A	39.15	3.982	52	0.49	52
B	39.20	3.977	50	0.49	52
C	39.20	3.977	50	0.48	51
D	39.21	3.976	49	0.50	53
E	39.19	3.978	50	0.50	53
Average composition			50% ± 2%		52% ± 1%

(Reproduced with permission from Rossington and Runk (1967). *J. Catal.* **7**, 365.)

wire for 3 minutes resulted in a film of composition 82% ± 2% Ag. Evaporation of the same wire by passing a current of 5.5 amp for 3 minutes resulted in a film of composition of approximately 70% ± 2% Ag.

Tables 1 to 4 give complete experimental details for the preparation and composition of four silver-palladium films. The 80% Ag and 70% Ag films were obtained by the evaporation of 50% Ag–50% Pd alloy wire, and the 50% Ag films were obtained by the evaporation of an alloy wire of composition 30% Ag–70% Pd. The tables give an indication of the degree of uniformity of composition of a film along the entire length of a reaction vessel, and also the degree of reproducibility between the analysis techniques used. These techniques are discussed in the following section concerned with the characterization of alloy films.

TABLE 4

Characterization data for 50% Ag – 50% Pd film II.

Preparation

Length of wire	60 cm
Diameter of wire	0.1 mm
Evaporating current	4.0 amp
Evaporating time	3 min
Substrate temperature	673 °K
Annealing temperature	673 °K
Annealing time	1 hr
Apparent area	120 cm^2
Total reaction volume	500 cm^2
Weight deposited	29 mg
Film thickness	2200 Å

Composition and Homogeneity Analysis

Section of vessel	X-ray diffraction results			X-ray fluorescence results	
	Corrected $2\theta_{(111)}$	a_0	% Ag	$\frac{Ag\ K\alpha}{Ag\ K\alpha\ +\ Pd\ K\alpha}$	% Ag
A	39.22	3.975	48	0.49	52
B	39.23	3.974	48	0.48	51
C	39.21	3.976	49	0.50	53
D	39.25	3.972	47	0.49	52
E	39.23	3.974	48	0.48	51
Average composition			48% ± 1%		52% ± 2%

(Reproduced with permission from Rossington and Runk (1967). *J. Catal.* **7**, 365.)

III. CHARACTERIZATION AND STRUCTURE

In characterizing alloy films there are generally two distinct criteria to be observed. Firstly the existence of a single phase must be determined and secondly the composition of this phase. In those cases where evaporation from separate sources takes place the average composition of the film can be determined by the weight loss of the sources. Where a separate method for composition analysis is required X-ray fluorescence techniques have been found to yield fruitful results. Electron microscopy has also been used to obtain information concerning the structure of evaporated alloy films.

A. X-RAY DIFFRACTION

Standard X-ray diffraction techniques (e.g. Barrett, 1952) may be used for alloy film work. In general the lattice constants for alloy films show only small deviations from those for the bulk alloy. Figure 4 shows these constants for a series of Pd–Ag films, compared to the values for a bulk alloy. Moss and Thomas (1964, 1967a) have discussed the broadening of X-ray lines of films compared to those of bulk metals due to the effect of the small size of crystallites forming the film and the possible occurrence of a range of lattice parameters due to incomplete alloy formation.

The line broadening, B_C, due to crystallite size is given by

$$B_C = \frac{K\lambda}{L \cos \theta} \qquad \text{(III)–(1)}$$

where

$$
\begin{aligned}
K &= \text{Scherrer constant} \\
\lambda &= \text{radiation wavelength} \\
L &= \text{crystallite size normal to substrate} \\
\theta &= \text{Bragg angle}
\end{aligned}
$$

The broadening due to a variation in lattice parameter caused by incomplete alloying is given by

$$B_a = 2 \tan \theta \frac{da}{a} \qquad \text{(III)–(2)}$$

where da is the range of lattice parameter a.

Moss and Thomas (1964, 1967a) measured the breadths of the 111 and 222 diffraction lines of Pd–Ag films and obtained values for (da/a). They showed that a simple treatment of analysis (Jones, 1938) was sufficient, by assuming a triangular distribution of lattice parameters and that a line broadened by instrumental and crystallite size effects has the Cuachy form, then

$$\left(\frac{da}{a}\right)^2 = \frac{3}{\pi^2} \frac{B_2{}^2 \cos^2 \theta_2 - B_1{}^2 \cos^2 \theta_1}{\sin^2 \theta_2 - \sin^2 \theta_1} \qquad \text{(III)–(3)}$$

where the subscripts refer to the 111 and 222 lines having an observed breadth, B.

Typical values of (da/a) for Pd–Ag films as given by Moss and Thomas are

% Pd in film	da/a
5.5	1.4×10^{-3}
9.0	1.7×10^{-3}
20.0	2.3×10^{-3}
35.0	6.3×10^{-3}

Differences between lattice parameters of bulk silver and silver films of 1000 Å thickness have been reported as being approximately 0.05% (Moss et al., 1963).

The development of superficial homogeneity in films, for example during heat treatment, may be conveniently followed by X-ray diffraction studies. Figure 11a (Rossington and Runk, 1967), shows a pattern for a film evaporated from a 30% Ag–70% Pd alloy wire and deposited at room temperature, and figure 11b the pattern after annealing for 1 hour at 400 °C. It may be clearly seen that as deposited initially the film contained an alloy phase together with free palladium. The pattern

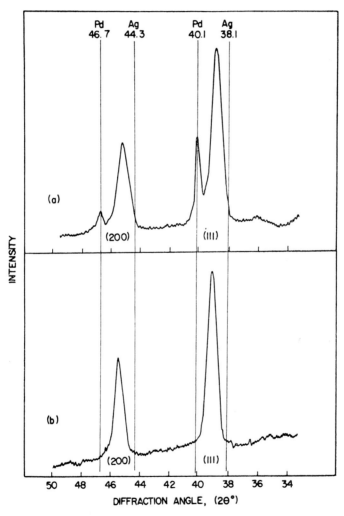

FIG. 11. X-ray diffraction pattern [(111) and (200) peaks only]. (a) Film deposited on substrate at room temperature. (b) Film deposited on substrate at 400 °C. (Reproduced with permission from Rossington and Runk (1967). *J. Catal.* **7**, 365.)

in Figure 11b, however, shows well defined peaks indicating a homogeneous film.

For measuring the diffraction patterns of Pd–Ag films Rossington and Runk (1967) used a copper target tube with a nickel filter, operating at 50kV and 16 ma. Beam and detector slits of 3° and 0.05° respectively were used, with a goniometer speed of 0.2 degrees/cm. To correct for low angle systematic errors of the diffractometer, spectroscopically pure silver powder was run as a standard with each set of measurements. Random errors were reduced by taking the average of at least three measurements for the recorded 2θ values. From the relative intensities of the 111, 200, 220, 311 and 222 peaks of the face-centred cubic alloys it was established that the alloy films were polycrystalline with no preferred crystalite orientation. The lattice parameters were calculated from the 111 spacing. Comparisons were made with lattice parameters calculated from the other planes and the differences were found to be less than experimental errors. Experimental variations caused less than 1% error in the final alloy composition calculated from X-ray diffraction.

Sachtler and Dorgelo (1965b) used X-ray diffraction to analyse their Cu–Ni films. The photographs were calibrated by reference to the diffraction lines of α-quartz which was applied as a thin layer over the alloy sample surface.

B. X-RAY FLUORESCENCE

In those cases where an analysis method complementary to X-ray diffraction is necessary, X-ray fluorescence techniques are particularly suitable. However, it should be emphasized that the data obtained by this method are purely quantitative. By itself X-ray fluorescence neither determines the homogeneity of a film, nor, indeed, the existence of a single alloy phase. Therefore this technique must be used in conjunction with X-ray diffraction.

The basic property of matter which makes possible chemical analysis by X-ray fluorescence spectroscopy is its ability to absorb electromagnetic energy in the X-ray region of the spectrum and re-emit this energy in the form of secondary X-rays. An incident quantum of X-radiation ionizes an atom by removal of an electron from the K or L shell. Subsequently an electron from the L, M or N shell falls into the vacant orbital in the K or L shell emitting a quantum of secondary X-radiation with an energy corresponding to the energy difference between its initial and final states. This energy is less than that of the original ionizing quantum, since the original electron was removed to infinity, and therefore the secondary X-ray quantum has a longer wavelength than the ionizing quantum. Since the energy difference between any two shells or subshells is a function of the atomic number, each element will produce a spectrum of secondary X-rays characteristic of that particular element, permitting its identification in a sample of ir-radiated material.

The frequency of the secondary X-rays varies as the square of $(Z{-}1)$ when Z is the atomic number, and the small number of energy levels in the innermost shells, plus particular restrictions imposed by selection rules, leads to X-ray spectra which are simple compared to the optical spectra of the same element. The notation used is as follows: the first letter, e.g. K, designates the shell into which the electron passes when the secondary quantum is emitted; the second (Greek) letter and its subscript refer to the energy level previously occupied.

An example of this notation is shown in Figure 12. $K_{\alpha1}$ designates the quantum produced when an electron is transferred from the L_{III} to the K state. The $k_{\alpha1}$ lines are strongest for each element, but when the $K_{\alpha1}$ and $K_{\alpha2}$ lines are not resolved they are observed as a single, intense and slightly asymmetric line designated as just the K_{α} line. This K_{α} line is the one most commonly used in analytical work.

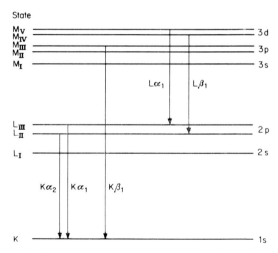

Fig. 12. Energy level diagram (not to scale) showing origin of the five strongest X-ray lines.

This technique is particularly applicable to the analysis of Pd–Ag films, by comparing the relative intensities of the AgK_{α} and PdK_{α} fluorescent X-ray emissions, which occur at 2θ angles of 23.87 and 25.00 degrees, respectively. A calibration curve is necessary and Figure 13

shows one as determined by Rossington and Runk (1967). Palladium-
silver alloy wires of known composition were used for the calibration.
The deviation from linearity in Figure 13 is a real effect and is caused by

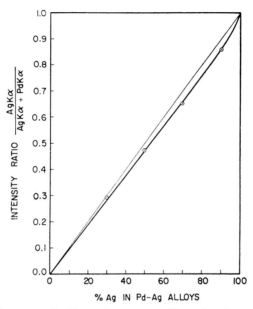

FIG. 13. Calibration curve for X-ray fluorescence. (Reproduced with permission from
Rossington and Runk (1967). *J. Catal.* **7**, 365.)

an enhancement effect of silver on palladium. The energy absorbed by
palladium from the AgK emission is stored and then re-emitted with the
Pd Kα emission.

For determination of alloy film composition a tungsten target was
used, operating at 65 kV and 55 ma. The goniometer was operated at a
speed of 0.5 degrees/minute. An average of at least three peak intensities were converted into percent silver content using the calibration
curve. Tables 1 to 4 show the degree of homogeneity of these films at
various positions along the reaction vessel and give a comparison with
the X-ray diffraction analysis data.

By the nature of the two techniques, the critical sample thickness is
much greater for fluorescence than for diffraction. This may be shown
by an examination of an inhomogeneous film deposited at room temperature. Differences in the diffraction patterns of both sides of such a
film revealed the expected concentration gradients of silver and palladium (Rossington and Runk, 1967). The original layers to be condensed contained an alloy phase rich in silver and often free silver, while

the final layers to be condensed contained an alloy phase less rich in silver and pure palladium, as shown in Figure 11a. No differences were observed in the K spectra of both sides of these inhomogeneous films, indicating that X-ray fluorescence yielded the bulk composition. For films deposited and annealed at 400 °C, it was concluded that complete homogeneity was obtained when the alloy compositions determined by both techniques were in agreement.

C. ELECTRON MICROSCOPY

Electron microscopy can yield valuable information as to the state of crystallite size and aggregation of alloy films. Figure 14 shows transmission micrographs of Pd–Ag films used for catalysing the oxidation of ethylene (Moss and Thomas, 1967a). It may be observed that the structure varies consistently with composition, the silver-rich films showing coalescence of the crystallites, the palladium-rich films containing a compact mass of small crystallites. These authors also showed that the structure of these films was due to the method of film formation and not due to any subsequent catalytic reaction.

Most investigations have assumed a homogeneous composition through the thickness of an evaporated film, but Sachtler et al. (1965b, 1965c) present evidence that in the case of Cu–Ni films phase separation can occur, the composition of the surface being different to that of the interior. In general, Sachtler and Dorgelo (1965b) found three ranges of alloy composition where different phases were formed. Alloys containing approximately 50% silver and 50% nickel were split into two phases, a copper-rich alloy containing 10% nickel and a second phase of almost pure nickel. On the copper-rich side of the film only the copper was detected and on the nickel side, both phases. These films were formed by evaporation from separate sources. From measurements of the work function of these Cu–Ni films, which is discussed later, (Figure 17), it was concluded the copper-rich alloy in the two-phase system is located at the surface and the nickel-rich alloy below the surface. Sachtler and Jongpier (1965c) have developed a model for the location of phases in these films and suggest that a two-phase system rather than the ideal solid solution may be the thermodynamic equilibrium situation for Cu–Ni alloys of random composition at temperatures in the region of 200 °C.

For an alloy of composition Cu_xNi_{1-x} the free energy at the temperature of formation T is given by

$$\Delta G_{(x)} = \Delta H_{(x)} - T\Delta S_{(x)}$$

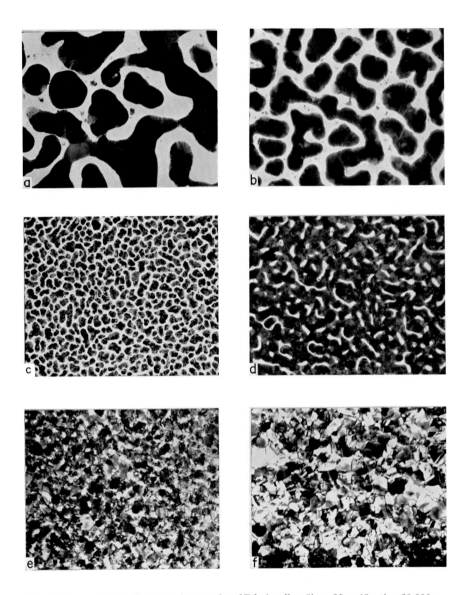

FIG. 14. Transmission electron micrographs of Pd–Ag alloy films. Magnification 20,000×. (a) 12.6% Pd, (b) 36.7% Pd, (c) 51% Pd, (d) 76% Pd, (e) 94% Pd, (f) 19.4% Pd. Films (a), (b), (d), (e) deposited and annealed at 400 °C, catalysed C_2H_4 oxidation at 240 °C. Film (c) prepared similarly, not used for reaction. Film (f) deposited at 400 °C, not annealed, catalysed C_2H_4 oxidation at 240 °C. (Reproduced with permission from Moss and Thomas (1967a). *J. Catal.* **8**, 151. British Crown Copyright.)

For an ideal solution

$$\Delta H_{ideal} = 0$$
$$\Delta S_{ideal(x)} = - R \; x \ln x + (1 - x) \ln (1 - x)$$

and deviations from ideality are described in terms of excess functions

$$\Delta H_{excess(x)} = \Delta H_{real(x)}$$
$$\Delta S_{excess(x)} = \Delta S_{real(x)} - \Delta S_{ideal(x)}$$

Sachtler and Jongepier (1965c) have used values of these excess thermodynamic fractions given by Rapp and Maak (1962) and Vecher and Gerasimov (1963) to calculate ΔG values at 200 °C, assuming no temperature dependence of $\Delta S_{(x)}$ and $\Delta H_{(x)}$. These values are shown in figure 15 where it may be seen that there are minima at $x = 0.82$ and at

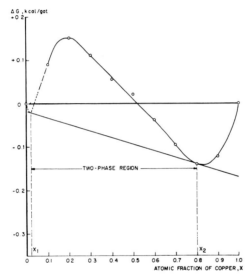

FIG. 15. $\Delta G_{200 \, °C}$ for copper-nickel alloys. (Reproduced with permission from Sachtler and Jongepier (1965c). *J. Catal.* **4**, 665.)

$x \ll 0.1$. For alloy compositions between these values there exist two phases, one containing 18% Ni and 82% Cu, the other almost pure nickel. The free energy diagram may be divided into distinct regions according to composition and Sachtler and Jongepier present the following model for the phases in each region.

$$\text{Range 1: } x_2 \leq x \leq 1$$

A homogeneous alloy, rich in copper, with the surface concentration equal to the bulk concentration.

Range 2: $x_1 < x < x_2$

A two phase system. Each crystallite consists of a centre of almost pure nickel ($x = x_1$) surrounded by an alloy of composition x_2. The composition of the outer layer is constant, only its relative thickness will depend upon the overall composition.

Range 2a: $x_1 < x < (x_1 + \Delta x)$

A two phase system with phase boundaries at the surface. Crystallites of almost pure nickel ($x = x_1$) covered by *patches* of alloy ($x = x_2$).

Range 3: $0 < x < x_1$

A homogeneous alloy containing more than 95% nickel. (These models are summarized in Figure 16.)

Range	x^a	Number of phases	Phase	x in outer phase
1	$x_2 < x < 1$	one		$x \geq x_2$
2	$x_1 < x < x_2$	two		x_2
2a	$x_1 < x < x_1 + \Delta x$	two		x_1 and x_2
3	$0 < x < x_1$	one	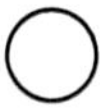	$x \leq x_1$

$^a x$ = atomic fraction of copper in alloy; x_1, x_2 minima in phase diagram

FIG. 16. Models for location of phases in copper-nickel alloys. (Reproduced with permission from Sachtler and Jongepier (1965c). *J. Catal.* **4**, 665.)

There appears to be some doubt as to the general validity of Sachtler and Jongepier's conclusion that all alloys between 3% and about 80% copper will, if equilibrated at 200 °C, produce solids in which the copper rich phase of a given composition is exposed on the surface of the crystallite (Campbell and Emmett, 1967). This is the Range 2 referred to above. Evidence for the views of Campbell and Emmett is presented in the following section dealing with catalytic activity and surface properties of alloy films.

IV. Surface Properties

A. work function

Sachtler and Dorgelo (1965b) have studied the photoelectric work functions of copper-nickel alloy films which had been prepared as previously described. Full details of the sliding-cathode photo-tube developed for the measurements are given in the original paper. The work function was derived from the spectral distribution of the photoelectric yields, which is the ratio of the number of electrons emitted to the number of photons adsorbed. The values were calculated from the photoelectric yield distributions from Fowler's theory (1931) using the approximations of Anderson et al. (1959).

The results obtained in this investigation showed that while sintering at 200 °C for several hours increased the work functions of pure copper and pure nickel, it was found to reduce those of the alloy systems. Work functions were also determined after the films had been allowed to adsorb carbon monoxide. The complete results are shown in Table 5 and Figures 17 and 18. Two sets of experiments were undertaken and it may be seen that while the work function of a freshly deposited film decreases as the copper content increases, after sintering at 200 °C there is a wide range of compositions for which the work function has a constant value of approximately 4.61 eV. Sachtler and Dorgelo estimated that the temperature of deposition of their films was 100 °C. Thus there

TABLE 5

Work functions (ϕ) and emission constants (α) of Ni–Cu alloy films.

Segment	Expt No.	ϕ after evaporations (eV)	α^a	ϕ after sintering (eV)	α^a	ϕ after CO admission (eV)	α^a
1 Pure Ni	I	5.03	6.0	>5.20	—	—	—
	II	5.04	7.1	>5.20	—	—	—
2 Ni-rich	I	4.86	3.0	4.61	3.2	4.71	4.0
	II	4.93	4.8	4.64	4.3	4.68	4.8
3 ~50/50	I	4.81	4.2	4.61	3.1	4.72	4.0
	II	4.84	7.4	4.61	4.3	4.67	4.8
4 Cu-rich	I	4.75	5.1	4.60	3.0	—	—
	II	4.80	11.2	4.62	4.2	4.66	3.8
5 Pure Cu	I	4.61	5.6	4.67	3.2	4.67	3.2
	II	4.60	4.7	4.65	3.7	4.66	3.6

[a] Emission constant [Electrons photon^{-1} degree $^{-2} \times 10^{12}$].

(Reproduced with permission from Sachtler and Dorgelo (1965b). J. Catal. 4, 654.)

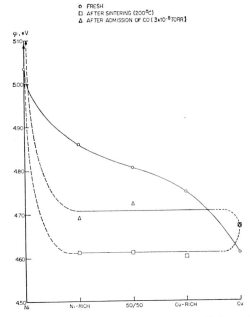

FIG. 17. Work functions of copper-nickel films, experiment **I**. Reproduced with permission from Sachtler and Dorgelo (1965b).

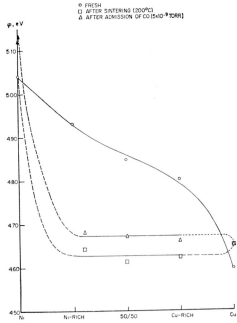

FIG. 18. Work functions of copper nickel films, experiment **II**. (Reproduced with permission from Sachtler and Dorgelo (1965b). *J. Catal.* **4**, 654.)

appears to be a wide range of overall nickel-copper compositions for which sintering tends to produce a surface phase of approximately the same composition, with a work function of 4.61 eV. This implies that the outer phase formed upon sintering has a work function which is lower than that of either pure nickel or pure copper, thereby showing that alloying has caused an increase in the Fermi level. This is in contrast to the generally held view that the work functions of alloys should be intermediate between those of the pure components. Work on alloys with extremely low or extremely high copper:nickel ratios showed that the work function was intermediate between that of the pure metal in the largest quantity and that of the 10% nickel–90% copper alloy presumed to be present in the surface of sintered films of medium composition.

It may also be seen from Table 5 and Figures 17 and 18 that chemisorption of carbon monoxide leads to an increase in the work function, except in the case of pure copper. Sachtler and his co-workers use these results to support the view that if the composition of the surface phase of alloys of medium composition is 10% nickel–90% copper, then either one must assume that these copper-rich alloys still possess incompletely filled d bands, or that some other basis for discussing chemisorption phenomena is required. These authors suggest that the orbitals of individual atoms would serve as a more suitable basis for accounting for chemisorption and catalytic phenomena.

Further evidence that the surface composition, and not a bulk property such as the average d band composition calculated from the overall composition of an alloy, is a determining factor in catalytic activity has recently been presented by van der Plank and Sachtler (1967). In this work the ratio of the amount of chemisorption of hydrogen to the amount of physical adsorption of xenon was determined for a series of nickel-copper films. Nickel will strongly chemisorb hydrogen at room temperature, whereas the chemisorption of hydrogen by copper, though finite, is small. Therefore, the values of the ratio

$$\alpha = \frac{\text{No. of H atoms adsorbed}}{\text{No. of Xe atoms adsorbed}}$$

should be indicative of the nickel content of the alloy surface. It was found that the value of this ratio was approximately constant (with some scatter of results) in the composition range of 25%–95% Ni, rising rapidly in the range 95%–100% Ni. α was also found to be independent of the order in which the original metals were evaporated to form the alloy film. The activity of these films for the hydrogenation of benzene at 150 °C was found to follow the same pattern with respect to film composition as did the ratio α. These results indicate that the catalytic activity of these films is approximately proportional to the nickel content of the alloy surface, which can be very different from the average nickel content of the whole film.

B. CATALYTIC ACTIVITY

One of the main reasons for the study of the catalytic activity of evaporated alloy films is providing information concerning various models for catalytic mechanisms. These films can be made reproducibly clean and of large surface area compared to filaments. They are also suitable for study by electron microscopy so that the effect of any structural defects in such films on their catalytic activity may also be studied.

The alloy films which have been studied in most detail, for catalytic work, are those in the palladium-silver and nickel-copper series. Both of these allow the d band structure to be varied continuously, although in the nickel-copper systems phase separation may be expected, which is not likely to occur in the palladium-silver series. In the following section a resumé will be given of work on several well-known catalytic "test" reactions which has been carried out on evaporated alloy films.

1. *Parahydrogen Conversion*

The simplest reaction for the study of catalytic activity, the para-ortho hydrogen conversion, has been investigated by many workers on a wide variety of substrates. Rossington and Runk (1967) studied the conversion of a series of palladium-silver alloy films ranging in composition from approximately 50% silver to 100% silver. Palladium-rich alloys were not studied in order to avoid any possible complications due to the absorption of hydrogen in palladium. Activation energies were obtained in the usual way from plots of the Arrhenius equation:

$$k_e = B \exp(-E/RT)$$

where k_e is the experimental rate constant obtained from the first order rate equation for the parahydrogen conversion reaction, and E is the apparent activation energy. Activation energies are given in Table 6, together with the pre-exponential factor $B°$, corrected for film surface area and reaction volume

$$B° = \frac{BV}{A} \qquad\qquad (IV)-(1)$$

TABLE 6

Parahydrogen conversion on Pd–Ag alloy films.

Weight % Ag in alloy	Temperature range of reactions (°K)	E (kcal mole^{-1})	log B^0	ΔH
48	273–413	0.47	0.44	—
50	273–413	0.37	0.27	—
68	296–393	1.91	1.00	−1.29
82	333–393	4.29	2.01	−2.11
100	363–413	5.90	2.64	−2.89
100	363–413	5.40	2.20	−2.54

Rossington and Runk (1967), Holden and Rossington (1964).

where   B = experimental pre-exponential factor, obtained from the intercept of the Arrhenius plot

V = reaction volume

A = film surface area

The corrected pre-exponential factor may be expressed as a linear function of the activation energy, as shown in Figure 19. This type of compensation effect is not at all unusual in hetereogeneous catalytic

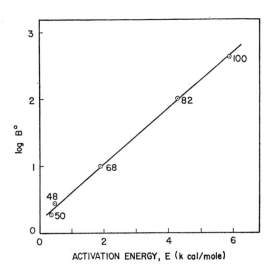

FIG. 19. Compensation effect for parahydrogen conversion on palladium-silver films (% Ag indicated). (Reproduced with permission from Rossington and Runk (1967). *J. Catal.* **7**, 365.)

reactions (Cremer, 1955). The results in Figure 19 can be expressed by the equation

$$\log B^{\circ} = 0 \cdot 42E + 0 \cdot 19 \qquad \text{(IV)--(2)}$$

where E is the experimental activation energy.

Several interpretations of the compensation effect have been presented in the literature (Cremer, 1955; Kemball, 1950, 1953). Bond (1962) has shown that when the compensation effect can be expressed in the form of equation (IV)--(2) there must be some characteristic temperature T_c, at which the reaction rates are equal for a given catalyst series. That is, the Arrhenius plots will converge and cross over at this temperature. Bond also shows that the relationship between the

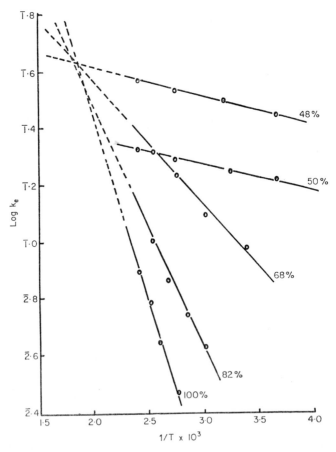

FIG. 20. Arrhenius plots for parahydrogen conversion on palladium-silver films (% Ag. indicated).

slope of the compensation effect line, m, and this characteristic temperature is

$$m = (2 \cdot 303 R T_C)^{-1} \qquad \text{(IV)–(3)}$$

In Figure 20 the Arrhenius plots for the parahydrogen conversion on palladium-silver films are presented. These lines have been extrapolated and it may be seen that they do, indeed, converge and cross over at a given point. The marked difference in activity between the 50% silver and 48% silver films could be due to some contamination of the former. The convergence point in Figure 20 is represented by the co-ordinates $\log k_e = \overline{1}.6 \pm 0.1$ and $1/T \times 10^3 = 1.85 \pm 0.01$, corresponding to a characteristic temperature of 540 ± 25 °K. The value of m from equation (IV)–(3) may be calculated as 0.41 ± 0.02, compared with the experimental value of 0.42 from equation (IV)–(2).

The relationship between activation energy and film composition is shown in Figure 21. In bulk palladium-silver alloys the d band of the palladium becomes filled at a composition of approximately 60% silver (Pugh and Ryan, 1958; Bond, 1962), whereas in Figure 21 the rise in

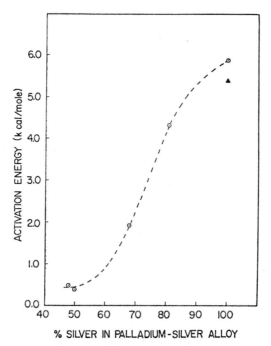

FIG. 21. Activation energy as a function of alloy composition. (Reproduced with permission from Rossington and Runk (1967). *J. Catal.* 7, 365.) ○, Rossington and Runk, 1967; ▲, Holden and Rossington, 1964.

activation energy occurs at a silver content somewhat higher than 60% silver. There are uncertainties as to the exact composition at which the palladium d band is filled and there is also the question of whether the surface d orbitals are all paired at the same composition at which the d band in the bulk alloy is filled. The change of total electron Fermi energy, arising from the transfer of electrons from the s band of silver to the d band of palladium, shows a minimum at 70% silver (Pratt, 1960).

Also presented in Table 6 are values for the heat of adsorption of hydrogen on these alloy films. It has been shown previously that a study of the pressure dependency of the parahydrogen conversion at different temperatures may be used to calculate heats of adsorption (Holden and Rossington, 1964).

It is interesting to note that Rienacker and Engels (1963, 1965) have studied the parahydrogen conversion on palladium-silver alloys in the form of foils, in a somewhat higher temperature range than that used by Rossington and Runk. Rienacker and his co-workers found a wide variation in activity over the whole range of composition from pure palladium to pure silver. They showed that the activation energy rose from approximately 6 kcal mole^{-1} with an alloy of 50% silver to 11 kcal mole^{-1} for pure silver. For the palladium-rich alloys the activation energy was found to fluctuate, rising from 6 kcal mole^{-1} to 9 kcal mole^{-1} in the range of 50% silver to 30% silver, then falling from 9 kcal mole^{-1} to 4 kcal mole^{-1} as the palladium content increased to 100%.

2. Oxidation reactions

Palladium-silver alloy films have also been used to catalyse the oxidation of carbon monoxide and ethylene. For the carbon monoxide oxidation Moss and Thomas (1964b) found activation energies varying from 20 kcal mole^{-1} on a pure silver film to 36 kcal mole^{-1} on pure palladium, with a minimum of 10 kcal mole^{-1} at a film composition of 20% palladium –80% silver. In general, these authors found that when films were formed by slow simultaneous evaporation followed by annealing, the general effects of the filling of d band vacancies on catalytic behaviour formed by using bulk alloy catalysts could be reproduced. It was found that the activity of silver-rich films (where there is little probability of hydrogen absorption) could be repeatedly restored to their original values by a hydrogen reduction technique of heating in 50 torr of hydrogen at 250 °C for one hour. This method had been used by Gharpurey and Emmett (1961) to restore the activity of copper-nickel films which had been used for ethylene hydrogenation.

The oxidation of ethylene provides another useful tool for studying electronic and other factors involved in catalysis. This reaction has also been investigated by Moss and Thomas (1967b) over palladium-silver alloy films. Their reactions were carried out at 240 °C in a static system over a film which had been annealed previously at 400 °C. The reaction products were analysed by a mass spectrometer, less than 2% of the reaction material being removed from the reaction system per hour for analytical purposes.

The rates of formation of ethylene oxide and carbon dioxide, expressed as a function of alloy composition, are shown in Figure 22 together with the magnetic susceptibility data, indicating d band

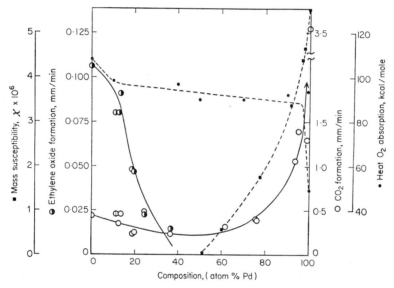

FIG. 22. Rates of formation, torr (0 °C) per minute at 240 °C of ethylene oxide ◖ and CO_2, ○. Rate of formation of CO_2 over film annealed in H_2, ◐. Magnetic susceptibility data from Hoare et al. (1953). Heats of oxygen chemisorption from Bortner and Parravano (1957). (Reproduced with permission from Moss and Thomas (1967b). J. Catal, 8, 162.)

vacancies, and heats of chemisorption of oxygen on these alloys. The heats of chemisorption data were obtained on co-precipitated bulk alloys by Bortner and Parravano (1957).

The yields of ethylene oxide were found to be highest on pure silver, decreasing to zero on alloy films with more than 40% palladium. The rate of carbon dioxide formation remained reasonably constant with alloy composition from pure silver to approximately 80% palladium and then increased rapidly, by a factor of seven, over pure palladium. The apparent activation energies and pre-exponential factors are given in Table 7, the figures in parenthesis being less certain due to a change in activity of the films after the reactivation procedure involving reduction in hydrogen.

TABLE 7

Activation energies and frequency factors
for CO_2 formation over Pd–Ag alloy films.

Film composition	Ag	11% Pd	26% Pd	43.5% Pd	62% Pd	100% Pd
Activation energy (kcal/mole)	12.5	16.4	(18.5)	(13.9)	30.2	13.6
Log_{10} (frequency factor)*	2.64	4.19	(5.29)	(3.21)	10.01	4.03

*frequency factor in torr CO_2 min^{-1} per cm^2 of apparent area.
(Reproduced with permission from Moss and Thomas (1967b). *J. Catal.* 8, 162.)

The high activity of pure palladium for the complete oxidation of ethylene to carbon dioxide is discussed by Moss and Thomas (1967b) in terms of the moderate heat of chemisorption of oxygen on palladium. The relatively low heat may not support a full oxygen coverage and it is possible that the ethylene is chemisorbed on the remaining bare portions of the palladium surface. With the addition of a few percent silver to palladium the heat of adsorption of oxygen rises abruptly and remains reasonably constant as the silver content of the alloy increases to pure silver; at the same time the activity decreases sharply, then remaining approximately constant as the silver content increases.

This variation in activity is not due to any simple variation in activation energy or frequency factor, as shown in Table 7. From pure silver to 62% palladium there is the usual compensation effect operative between activation energy and frequency factor, resulting in reasonably constant activity for films in this composition range. The sharp rise in activation energy between pure palladium and an alloy with a small percentage of silver is only partially compensated by a corresponding change in frequency factor, and hence the activity falls markedly in this range.

Apart from any simple correlations between activity and electronic properties of these alloys Moss and Thomas have also shown that there are other factors to be considered. The heat treatment of palladium-rich alloys in pure oxygen at 250 °C results in a degree of inhomogeneity in the films due to the diffusion of silver atoms from the bulk to the surface of the alloy (Moss and Thomas, 1967a). This is referred to by the authors as "de-alloying". It is possible that heat treatment with an oxygen/ethylene mixture could result in a similar modification of the amount of silver in the film surface. Also, as ethylene undergoes oxidation the desorption of hydrogen as water molecules competes with the absorption of hydrogen into the bulk, in the case of palladium-rich alloys. Thus, the possible silver enrichment of the alloy surface would result in less d band vacancies than indicated by bulk magnetic susceptibility measurements and d band vacancies in the palladium-rich alloys may also be filled by electrons from absorbed hydrogen which decreases the activity of palladium (Couper and Eley, 1950). These effects could be factors dominant over the change in d band composition, causing the rapid change in activity between pure palladium and alloys containing a small percentage of silver.

It is seen from Figure 22 that the rate of ethylene oxide formation decreases with increasing palladium content, nearing zero at approximately the composition at which d band vacancies start to be present. Moss and Thomas relate this change in activity to a change in the mode of oxygen chemisorption with a decrease in the Fermi level rather than a direct correlation with d band vacancies. They suggest that there are two states of binding of oxygen on the films, a weakly bound and a strongly bound state. It is postulated that adsorbed fragments of ethylene combine with the strongly adsorbed oxygen to form carbon dioxide, and that the less stongly bound oxygen dissociates and oxidizes the ethylene to ethylene oxide.

3. Ethylene Hydrogenation

Alexander and Russell (1965) studied this reaction on nickel-copper, nickel-iron and nickel-palladium films formed by the induction evaporation of homogeneous alloy pellets. The hydrogenation reaction was carried out in the temperature range 240 °K to 345 °K. For the alloys studied the activation energy for the reaction was found to fall into two general classes, a low activation energy of approximately 6.8 kcal/mole and a higher value of 9.1 kcal mole^{-1}. Two series of runs were carried out in the absence and in the presence of a pre-adsorbed hydrogen layer. The activation energies determined, together with the relative activities of the films, may be summarized as shown in Table 8.

The relative activity was defined as the ratio of the number of millimoles of ethylene reacted per square centimetre of alloy film surface compared to that for pure nickel. For the case of pre-adsorbed hydrogen present the relative activity is that relative to the same film in the absence of pre-adsorbed hydrogen.

It may be seen from Table 8 that, in the absence of a pre-adsorbed hydrogen layer, alloying with diamagnetic copper or ferromagnetic iron produces only a very small change in activity, whilst alloying with paramagnetic palladium increases the activity greatly. Pre-adsorbed hydrogen increases the activity of the nickel-copper and nickel-iron films but

TABLE 8

Activity data for C_2H_4 hydrogenation over alloy films.

Film Composition	Activity*	Activation Energy kcal mole^{-1}	
100% Ni	1.0	7.0	—
54.6 atom % Cu	1.04	—	8.9
10.8 atom % Fe	1.19	—	8.9
5.5 atom % Pd	31.7	7.0	—
16.0 atom % Pd†	(0.002)	—	(9.8)
100% Ni	1.23	7.0	—
54.6 atom % Cu	8.8	—	8.9
10.8 atom % Fe	3.0	6.5	—
16.0 atom % Pd†	(1.14)	—	(9.8)
Average		6.8	9.1

H_2 pre-adsorbed (for rows 100% Ni, 54.6 atom % Cu, 10.8 atom % Fe, 16.0 atom % Pd†)

*The activity of the films in the absence of pre-adsorbed hydrogen is relative to that of 100% Ni. In the presence of pre-adsorbed hydrogen the activity is relative to that of the same film in the absence of pre-adsorbed hydrogen.

† Poisoned in preparation.

(Reproduced (in part) with permission from Alexander and Russell (1965). *J. Catal.* **4,** 184.)

not that of the nickel-palladium films. Alexander and Russell interpret the effects of pre-adsorbed hydrogen in terms of a strongly adsorbed hydrogen film (type A chemisorption) believed to be located in pits around surface atoms (Sachtler and Dorgelo, 1960), and that this strongly held hydrogen was already present in the mixed-palladium catalysts. As this type A chemisorption increased, the amount of type C chemisorption (weakly or reversibly adsorbed hydrogen) would increase also, as its transition to a type A chemisorption is slowed down due to an increase in the activation energy for such a transition, (Bond, 1962). In terms of such a model the type C chemisorption was regarded as being largely responsible for the ethylene hydrogenation and the authors associated the lower activation energy of 6.8 kcal mole^{-1} with an Eley-Rideal mechanism involving reaction between a strongly held chemisorbed layer and a weakly adsorbed layer. The higher activation energy of 9.1 kcal mole^{-1} was associated with a Langmiur-Hinshelwood mechanism involving reaction within a strongly held chemisorbed layer.

Alexander and Russell (1965) conclude that, because of the metals studied for alloying with nickel, only palladium does not appreciably change the saturation magnetization of nickel and only palladium greatly changes the catalytic activity of nickel, the indications are that the electronic factor does not appear to be directly involved in the hydrogenation process investigated.

One of the first instances of the use of evaporated alloy films in catalytic studies was that of Gharpurey and Emmett (1961) who made a brief investigation of the hydrogenation of ethylene over nickel-copper films. This work was continued and extended by Campbell and Emmett (1967). These authors prepared films of nickel-copper alloys by the method already described followed by sintering in 50 torr of hydrogen at 300 °C or at 500 °C. Some work was also carried out on nickel-gold films. X-ray diffraction measurements showed that films homogenized by heating in 50 torr hydrogen formed an alloy containing 40% nickel, the excess element being present as pure metal. Films homogenized at 300 °C did not give any evidence of this alloying effect.

The results obtained by Campbell and Emmett for ethylene hydrogenation on copper-nickel films are shown in Figure 23. The activity per

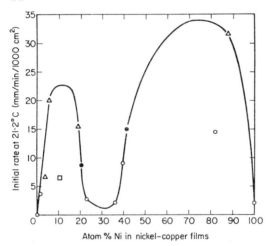

FIG. 23. Initial rates of reaction for ethylene hydrogenation on Cu–Ni films. △, films prepared by depositing Cu on Ni. ○, ● films prepared by depositing Ni on Cu. Open symbols: films homogenized at 300 °C in 50 torr H_2. Closed symbols: films homogenized at 500 °C in 50 torr H_2. □ Cu and Ni deposited simultaneously. (Reproduced with permission from Campbell and Emmett (1967). *J. Catal.* **7**, 257.)

unit area increased greatly for a catalyst containing 10 to 20% copper compared to that for pure nickel. Films prepared by depositing copper on top of nickel appeared to be more active than those prepared by depositing nickel on top of copper. The activities of the films sintered in hydrogen at 500 °C lay on the smooth curve drawn through the other experimental points.

Campbell and Emmet consider that the findings of Sachtler *et al.*

(1965b, 1965c) on the composition of copper-nickel alloy films surfaces may not be generally valid. They point out that the results of Best and Russell (1954) showed that a 90% Cu–10% Ni film was ten times as active as pure nickel for ethylene hydrogenation and one containing 64% Cu was almost a hundredfold as active. Alexander and Russell (1965) found that a 54.6% Cu–43.4% Ni film was about eight times as active as pure nickel, in agreement with results shown in Figure 23.

Sachtler and Dorgelo (1965b) carried out their work function experiments on an alloy the composition of which varied as the thickness of copper deposited on a strip of nickel increased. Campbell and Emmett suggest that, due to the rapid diffusion of copper on the surface of nickel, the entire surface area for the work function measurements may have been, in fact, of uniform composition, and not varying between 3% and 80% copper as assumed by Sachtler *et al*. The results of Campbell and Emmett may be generally summed up by saying that small amounts of nickel deposited on copper or small amounts of copper deposited on nickel, result in film surfaces which are many times more active for ethylene hydrogenation than pure nickel. It is well known that the amount of adsorption of molecular hydrogen by copper at room temperature is very small and Campbell and Emmett propose that the adsorption of hydrogen on copper-rich alloys takes place on adjacent copper-nickel atoms. The heat of adsorption of hydrogen on nickel is sufficient to dissociate a hydrogen molecule on adjacent copper-nickel sites. It is suggested that the adsorbed hydrogen atoms thus produced might be free to migrate over the copper surface and react with ethylene adsorbed on the copper surface. Holden and Rossington (1965) concluded that the film of hydrogen on a sparsely populated copper surface was mobile. The ethylene molecules adsorbed on the copper surface may be assumed to be more susceptible to hydrogenation that those on a nickel surface because of the lower heat of bonding on copper compared to nickel.

Figure 24 shows the variation of the pre-exponential factor with film composition. It is seen that this factor increased greatly upon the addition of a few percent of nickel to copper, consistent with an increase in the number of catalytically active sites as the number of copper-nickel site pairs increased.

The nickel-rich films containing 82% and 88% nickel have approximately the same pre-exponential factors as pure nickel and in this composition region the increase in activity appears to be due to a lowering of the activation energy for the reaction.

The results obtained on the copper-nickel films followed the usual compensation relationship as shown in Figure 25. Also in this figure are

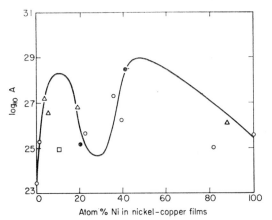

FIG. 24. Pre-exponential factors as a function of film composition. Symbols are the same as for figure 23. (Reproduced with permission from Campbell and Emmett (1967). *J. Catal.* **7,** 257.)

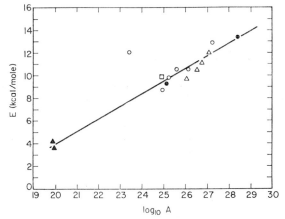

FIG. 25. Compensation effect for C_2H_4 hydrogenation on alloy films. Open circles, triangles and square as for figure 23. ▲ Au–Ni alloy. (Reproduced with permission from Campbell and Emmett (1967). *J. Catal.* **7,** 257.)

given the results for ethylene hydrogenation on nickel-gold alloy films, containing 45.7% nickel and 83.7% nickel. The work of the various authors cited for the hydrogenation of ethylene on alloy films indicates that there is a great deal of information yet to be obtained in order to completely elucidate reaction processes. Campbell and Emmett found great differences between the behaviour of nickel-copper and nickel-gold films for which there is as yet no satisfactory explanation and the reasons for the large variation in pre-exponential factors and activation energies across the nickel-copper system is not yet fully understood.

REFERENCES

Alexander, E. G. and Russell, W. W. (1965). *J. Catal.* **4**, 184.

Anderson, J. S., Faulkner, E. A. and Klemperer, D. F. (1959). *Aust. J. Phys.* **12**, 469.

Barrett, C. S. (1952). "Structure of Metals." McGraw-Hill, New York.

Belser, R. B. (1960). *J. Appl. Phys.* **31**, 562.

Best, R. J. and Russell, W. W. (1954). *J. Am. Chem. Soc.* **76**, 838.

Bond, G. C. (1962). "Catalysis by Metals." Academic Press, London.

Bortner, M. H. and Parravano, G. (1957). *Adv. Catalysis* **9**, 424.

Burr, A. A., Coleman, H. S. and Davey, W. P. (1944). *Trans. Am. Soc. Metals* **33**, 73.

Campbell, J. S. and Emmett, P. H. (1967). *J. Catal.* **7**, 257.

Coles, B. R. (1952). *Proc. Phys. Soc.* **65B**, 221.

Coles, B. R. (1956). *J. Inst. Metals* **84**, 346.

Couper, A. and Eley, D. D. (1950). *Discuss. Faraday Soc.* **8**, 172.

Couper, A. and Metcalfe, A. (1966). *J. Phys. Chem.* **70**, 1850.

Cremer, E. (1955). *Adv. Catalysis* **7**, 75.

Dowden, D. A. (1950). *J. Chem. Soc.* **242**.

Dowden, D. A. and Reynolds, P. W. (1950). *Discuss. Faraday Soc.* **8**, 184.

Dushman, S. (1951). "Scientific Foundations of Vacuum Technique." Wiley, New York.

Edwards, H. W. (1933). *Phys. Rev.* **43**, 205.

Fowler, R. H. (1931). *Phys. Rev.* **38**, 45.

Fujiki, Y. (1959). *J. Phys. Soc. Japan* **14**, 913.

Gharpurey, M. K. and Emmett, P. H. (1961). *J. Phys. Chem.* **65**, 1182.

Goldsztaub, S. and Michel, P. (1951). *C. R.* **232**, 1843.

Harris, L. and Siegel, B. M. (1948). *J. Appl. Phys.* **19**, 739.

Hoare, F. E., Matthews, J. C. and Walling J. C. (1953). *Proc. R. Soc.* **A216**, 502.

Holden, S. J. and Rossington, D. R. (1964). *J. Phys. Chem.* **68**, 1061.

Holden, S. J. and Rossington, D. R. (1965). *J. Catal.* **4**, 403.

Holland, L. (1956). "Vacuum Deposition of Thin Films." pp. 182–195, Wiley, New York.

Jones, F. W. (1938). *Proc. R. Soc.* **A166**, 16.

Jost, W. (1933). *Z. Phys. Chem.* **B21**, 158.

Kemball, C. (1950). *Adv. Catalysis* **2**, 233.

Kemball, C. (1953). *Proc. R. Soc.* **A217**, 376.

Konig, H. (1944). *Reichsber Phys.* **1**, 7.

Mader, S., Nowick, A. S. and Widmer, H. (1967). *Acta Metall.* **15**, 203.

Moss, R. L., Duell, M. and Thomas, D. H. (1963). *Trans. Faraday Soc.* **59**, 216.

Moss, R. L. and Thomas, D. H. (1964). *Trans. Faraday Soc.* **60**, 1110.

Moss, R. L. and Thomas, D. H. (1967a). *J. Catal.* **8**, 151.

Moss, R. L. and Thomas, D. H. (1967b). *J. Catal.* **8**, 162.

Mott, N. F. and Jones, H. (1936). "The Theory of the Properties of Metals and Alloys." Oxford University Press, London.

Nachtrieb, N. H., Petit, J. and Wehrenberg, J. (1957). *J. Chem. Phys.* **26**, 106.

van der Plank, P. and Sachtler, W. M. H. (1967). *J. Catal.* **7**, 300.

Pratt, J. N. (1960). *Trans. Faraday Soc.* **56**, 975.

Pugh, E. W. and Ryan, F. M. (1958). *Phys. Rev.* **111**, 1038.

Rapp, R. A. and Maak, F. (1962). *Acta Metall.* **10**, 62.

Reynolds, P. W. (1950). *J. Chem. Soc.* **265**.

Rienacker, G. and Engels, S. (1963). *Mber. Dt. Akad. Wiss. Berl.* **5**, 706.
Rienacker, G. and Engels, S. (1965). *Z. Anorg. Allg. Chem.* **336**, 259.
Rienacker, G. and Vormun, G. (1956). *Z. Anorg. Allg. Chem.* **283**, 287.
Rossington, D. R. and Runk, R. B. (1967). *J. Catal.* **7**, 365.
Ruer, R. (1906). *Z. Anorg. Allg. Chem.* **51**, 315.
Sachtler, W. M. H. and Dorgelo, G. J. H. (1960). "Proc. 4th Int. Conf. on Electron
    Microscopy," (E. Möllintead, H. Nhiers and E. Ruska, eds.) Springer, Berlin,
    p. 801.
Sachtler, W. M. H., Dorgelo, G. J. H. and Jongepier, R. (1965a). *J. Catal.* **4**,
    100.
Sachtler, W. M. H. and Dorgelo, G. J. H. (1965b). *J. Catal.* **4**, 654.
Sachtler, W. M. H. and Jongepier, R. (1965c). *J. Catal.* **4**, 665.
Vecher, A. A. and Gerasimov, Y. I. (1963). *Zh. Fiz. Khim.* (Eng. trans.) **37**, 254.
Vogt, E. (1932). *Annln. Phys.* **14**, 1.

Chapter 10

The Oxidation of Evaporated Metal Films

I. M. RITCHIE

Department of Physical Chemistry,
University of Melbourne,
Melbourne, Australia

I. INTRODUCTION

In this chapter, the oxidation of thin metal films will be considered. The most common type of metal oxidation is that in which oxygen reacts with a metal to form a solid oxide on the metal surface. But the generalized chemical meaning of an oxidation reaction is one which involves the transfer of electrons, and so the reacting gas need not necessarily be oxygen. It could be chlorine, for example. This chapter will be concerned with this more general class of reaction. We will take as our brief any gas/metal reaction which results in the formation and growth of a discrete product phase on the metal, since the possible reaction mechanisms will be the same in each case. Of course, by far the largest amount of experimental work in this field has been carried out using oxygen as the reacting gas. This is partly because oxygen/metal systems are relatively accessible to experimentation, and partly because much of this work has a direct or indirect bearing on corrosion problems in industry. Thus it is convenient to couch the discussion entirely in terms of oxygen oxidation. The extension of the principles involved to other gas/metal systems is quite straightforward. It is also convenient to limit the discussion to systems in which the reacting materials are pure or contain at most a trace of impurity. Systems involving gas mixtures or metal alloys will not be considered.

The free energy of formation of most oxides from their elements is negative. The metal is therefore thermodynamically unstable in oxygen, and in principle will be converted to the oxide. In practice, of course, this is often not so. Considering the vast use to which metals are put, it is fortunate that while the thermodynamics of oxide formation are favourable, the kinetics are not. Consider the equilibrium between the metal M, its oxide M_xO_y and oxygen

$$O_2 + \frac{2x}{y}M \rightleftharpoons \frac{2}{y}M_xO_y \qquad \text{(I)–(1)}$$

The dissociation pressure of the oxide is the reciprocal of the equilibrium constant for the reaction as written above, and is the pressure of oxygen above which the oxide will start to form from the metal. Thus it can be seen that the dissociation pressure p is given by,

$$p = \exp(\Delta G_T^0/RT) \qquad \text{(I)–(2)}$$

where ΔG_T^0 is the free energy of formation of the oxide per mole of oxygen consumed, and R and T have their usual meanings. Typically, free energies of formation of metal oxides are of the order of tens of kilocalories per mole. For example, ΔG_{1000}^0 for nickel oxide is -59.8 kcal mole^{-1} (Richardson and Jeffes, 1948) which corresponds to a dissociation pressure of 1.2×10^{-13} torr. This pressure is barely accessible even with the best ultrahigh vacuum techniques.

When a clean metal surface is exposed to oxygen, the first step in the oxidation process is the formation of an adsorbed layer of oxygen on the surface. Chapter 4 discusses the adsorption of gases on thin metal films in some detail, and this topic will only be considered briefly here. As more gas is taken up by the metal, an oxide phase first nucleates, then separates out, and finally grows according to one of several possible growth laws. Relatively little is known of the very first stages of the oxidation process. Only in recent years has some progress been made in this area by low energy electron diffraction. On the other hand, oxidation rates have been measured for some forty years. It is generally conceded that the first quantitative kinetic study of an oxidation reaction was that of Tammann (1920) who investigated the iodination of copper, silver and lead in iodine-containing air. To determine the thickness of the iodide layer, he used an optical technique. Not surprisingly, this particular piece of work and all the other early studies were carried out using metal foils, wires or specimens fabricated in some other way from the bulk metal. So far as we have been able to establish, the first use that was made of an evaporated film in an oxidation study was that of Steinheil (1934). He observed how an evaporated aluminium film became increasingly transparent as the metal oxidized, and deduced that the oxide grew to a limiting thickness of about 40 Å. Since then, despite some serious drawbacks which will be discussed later, increasing use of evaporated metal films has been made by those investigating oxidation behaviour. For, not only have the unique advantages conferred by the use of evaporated films enabled significant advances to be made in this field, but these studies have also stimulated interest in the properties of vacuum deposited metal layers.

This chapter will consist of two main sections. The first section will comprise a review of the theory of the oxidation of metals. The second section will consider the advantages and disadvantages of using thin metal films in oxidation work, and finally discuss their use in kinetic and mechanistic studies.

For further detailed accounts of the topic of metal oxidation, the books by Evans (1960), Kubaschewski and Hopkins (1967) and Hauffe (1965) should be consulted. Recently Kofstad (1966) has published a book dealing with high temperature gas/metal reactions. The bias in this review of oxidation will be towards low temperature oxidation rather than high, since most evaporated film studies are conducted at low temperatures.

II. A Review of the Oxidation of Metals

A. the early stages of oxidation

1. *Adsorption and Place-Exchange*

The first stage in the oxidation process must be the adsorption of molecular oxygen, although dissociation to atoms may follow immediately. If sufficient adsorption takes place, the adsorbed species will often form a regular two dimensional array on the exposed planes of the metal lattice. These two dimensional structures differ from plane to plane, and also vary with the exposure, defined as the product of the oxygen pressure and the time of gas/metal contact. Initially the surface structure may only grow in patches, but with increasing exposure, the patches unite to cover the whole surface. Further exposure still, may lead to the formation of a different type of surface array. Some of the two dimensional structures involve surface rearrangement in which the metal atoms are displaced from their original positions. One such surface rearrangement is aptly called place-exchange, and involves the interchange of a metal and an oxygen atom. An oxygen atom lying inside the

M

metal surface after surface reconstruction has a choice of two courses of action. It can either diffuse into the bulk of the metal (atom burial), or with its fellows, it can nucleate and form the ordered structure of a sub-oxide or oxide at the oxygen/metal interface. A suboxide is a reaction intermediate which will be converted to the oxide itself as the reaction proceeds. Very likely both diffusion into the metal, and oxide formation will take place simultaneously. Which process will predominate, will depend on a number of factors such as the diffusion coefficient of oxygen in the metal, the stability of the oxide, the gas pressure and the temperature. The sequence of events is shown schematically in Figure 1.

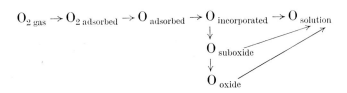

FIG. 1. The incorporation of oxygen on a clean metal surface.

While the interaction of oxygen with most metals seems to follow the general pattern of behaviour outlined above, it is only to be expected that the details differ from metal to metal. Moreover, certain features of the pattern are still not clear. For example, it is not known in general, to what extent surface steps, edges and other defects are involved in adsorption and the subsequent processes.

The picture given here has been constructed from several ex-perimental sources, chiefly low energy electron diffraction (LEED) studies and work function measurements. Before going on to the results, certain inherent deficiencies of these techniques should be noted. From LEED patterns, the unit cells of the surface structure can be deduced. However, the theory of back scattering from a low energy electron beam impinging on a crystal surface has not been sufficiently developed for the atoms in the unit cell to be identified positively. Furthermore, only those structures with sufficient scattering centres on the crystal surface will be identified. It is therefore possible for oxide nucleation and growth to take place without being detected in the LEED camera. This was found to be so for copper (Simmons *et al.*, 1967). Most LEED experiments are conducted at very low pressures, and to extrapolate the conclusions to higher pressures may be unsound.

As Huber and Kirk (1966) have shown, any work function measure-ments which have not been made under the most rigorous experimental conditions must be treated with circumspection. They established that the large increase in work function, previously observed during the initial exposure of aluminium to oxygen, was due to water contamination. The change brought about by oxygen alone was extremely small. In addition, the interpretation of data from which work functions are calculated is not always straightforward (see for example, the field emission study of Bell *et al.*, 1968).

Potentially the field ion microscope is the most powerful of all tools for probing surface processes, since with this technique, individual atoms on the surface may be imaged. But because the electric fields across the emitter tip, whose surface is being magnified, are so large (of the order of a volt per Å), field evaporation of the metal is possible and surface distortion inevitable. For this reason, few metals other than tungsten have been studied so far. During imaging, field desorption of any gas adsorbed on the emitter tip is also possible. Consequently, very little which is relevant to this chapter has been achieved using field ion microscopy. The work of Anderson (1967) is one notable exception.

Adsorbed oxygen may be either molecular or atomic. According to Simmons *et al.* (1967), when sufficient oxygen is adsorbed on a copper (111) face, a close packed layer of O_2^- ions is formed. On the other hand, it would appear (Farnsworth, 1964) that the adsorbed species on the germanium (100) plane is almost exclusively atomic. The presence of randomly oriented oxygen molecules adsorbed on the (100) nickel crystal surface was inferred by Farnsworth and Madden (1961) from their LEED studies. They found that the intensity of the nickel diffraction pattern diminished when the surface was first exposed to oxygen. Further experiments suggested that the oxygen molecules diffused to lattice imperfections where they dissociated into atoms. Diffusion of the atoms and ordering into a face centred surface array about the imperfection ensued. LEED studies have shown that most metals exhibit a wide variety of ordered surface structures (e.g. tungsten, Chang and Germer, 1967). Some apparently comprise an adsorbed species only, arranged in a pattern compatible with the metal substrate (Simmons *et al.*, 1967). Others involve surface reconstruction, and the formation of a suboxide. Oxygen on tantalum yields a monolayer of what is virtually TaO (Boggio and Farnsworth, 1964). In the case of nickel (Farnsworth and Madden, 1961), rearrangement of the surface leads to the formation of Ni_3O.

The chemisorption bond holding the oxygen atom to the surface might be expected to have an ionic component although in some cases such as germanium (Ernst, 1967) and nickel (Delchar and Tompkins, 1967), it is apparently very small. A partially charged oxygen atom chemisorbed to the metal comprises a dipole, the negative end of which points outwards from the metal. An array of such dipoles can reduce their total energy by reversing the polarity of one of their number. Lanyon and Trapnell (1955) have suggested that place exchange may occur in this way. This is shown in Figure 2.

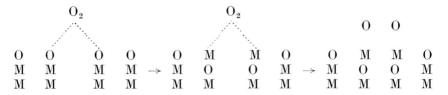

FIG. 2. One of Lanyon and Trapnell's place exchange mechanisms.

The place exchange derives its energy of activation from the heat liberated by the chemisorption of the oxygen molecule, whose presence is therefore essential to the process. This is in accord with their experimental observation that the half life of a dose of oxygen admitted to the metal surface was unaltered if the initially oxygenated surface was maintained in a vacuum for some hours before admission of the dose. In other systems, such as silicon (Law, 1958), the half life of the oxygen dose has decreased while standing in vacuum. Place exchange has here proceeded without the heat of chemisorption. It is clear from the work function study of Quinn and Roberts (1964) that the stability of chemisorbed layers with respect to place exchange varies enormously from one metal to another. The Lanyon and Trapnell mechanism has been extended to several layers by Eley and Wilkinson (1960). Other variations on the ionic place exchange theme have been proposed by Green (1957) and Huber and Kirk (1966). If the adsorbed atom is covalently bonded, the metal-oxygen and metal-metal bonds may be strained to accommodate it to the surface. This strain, according to Green and Liberman (1962), can also provide the driving force for a place exchange.

That surface reconstruction does take place is not doubted. Several exponents of LEED (e.g. Germer, 1966) have argued that the intensity of the diffraction patterns is evidence of this. Anderson (1967) observed a regular surface structure, different from that of the clean tungsten surface, in the field ion microscope. Using field evaporation to dissect this structure layer by layer, he found it to consist of slabs some three atomic layers in depth. Measurements of work function changes associated with oxygen exposure show that metal displacement does take place; in some cases even before a monolayer of oxygen has been adsorbed (Quinn and Roberts, 1962). Typically, oxygen adsorbed at low temperatures causes the work function to become larger (Quinn and Roberts, 1964). This is to be expected if adsorbed oxygen ions are formed. Allowing the metal to stand, or heating without further addition of oxygen, brings about a decrease in the work function, i.e. it approaches the clean metal value. The decrease is associated with place exchange and oxygen solution. More oxygen can then be taken up by the surface, confirming the presence of fresh metal atoms in the surface. Similarly, it is found that LEED patterns characteristic of a clean surface can sometimes be regenerated by heating after oxygenation. Ultimately, regeneration cannot be effected in this way since the metal becomes saturated with oxygen (Simmons *et al.*, 1967). The recent work function measurements of Roberts and Wells (1966) are interesting. It would appear that these results can be interpreted in terms of a competition for the oxygen atoms between different suboxides. One of the suboxides may be nothing more than a dilute solution of oxygen in nickel.

2. Nucleation

In oxidation literature, the term nucleus is used rather loosely, and usually means an oxide particle which is huge on the atomic scale. Nucleation is the formation of a viable oxide particle and presumably only involves a cluster of a few atoms. Particulate growth as such will be reviewed later. However a study of particulate growth may yield information about the nucleation process, just as the study of a man may tell us something of his infancy. The genesis of an oxide nucleus is not completely understood. It seems possible that two types of nucleation can occur; the one being nucleation at a surface defect, and the other, random nucleation on a flat surface. Nucleation having taken place, growth of the nucleus to an oxide particle follows.

We will consider nucleation at a surface defect first. This has been observed to occur on very clean metal substrates exposed to low partial pressures of oxidizing agent at or around room temperature. This type of nucleation may well be common to all metal oxidations, but will not be noticed unless special precautions are taken. The chance of nucleation taking place at any site goes up when the partial pressure of oxygen is raised, and unless special care is taken to minimize the amount of oxygen present, nucleation at random sites over the whole metal surface will take place simultaneously with the formation of a thin film of oxide. The reaction of oxygen with a metal is an activated process, and when the supply of oxygen is restricted, oxide growth at sites of high chemical potential will be favoured. Such sites occur at surface steps, edges and kinks, since the atoms there are less strongly bound to the bulk than the atoms in a low index plane. Preferred reaction at this type of site, and a correlation between the extent of reaction and the ease of removal of the metal atoms has been demonstrated by Allpress and Sanders (1964) in an elegant electron microscope study of the sulphide tarnishing of silver.

Other evidence for nucleation at defects is largely inferential. As mentioned previously, Farnsworth and his associates (Farnsworth and Madden, 1961; Lee and Farnsworth, 1965) believe that the surface structures formed on nickel and copper nucleate and grow about defects in the surface. The suboxide Ni_3O was shown to be an immediate precursor of the oxide NiO and it seems reasonable to suppose that NiO itself is nucleated about the defect site. In another LEED investigation of oxygen uptake on nickel, Germer and Macrae (1962) have inferred the presence of patches of NiO, about 25 Å across, on the nickel surface. Bloomer (1957a), investigating the sorption power of barium getters, found that an induction period preceded gettering. The induction period, which was observed only with barium films deposited slowly and under low residual gas pressures, can be ascribed to nucleation about surface defects. Cohen (1960) drew similar conclusions about the oxidation of magnesium. It is not conclusively established that all surface defects can act as nucleation centres. Perhaps only dislocations which have Burgers vectors with components perpendicular to the surface are suitable (Jaeger and Sanders, 1968). Whatever the precise mechanism involved, it is clear from Figure 1, that the formation of nuclei will be favoured on a metal containing extensive amounts of oxygen in solution. In fact it has been suggested (Kofstad, 1961) that oxide nucleation can occur if the metal becomes supersaturated with oxygen.

Nearly all observations of the second type of nucleation, that is at random positions on a flat surface, have been made at relatively high temperatures when oxide particles have been seen to nucleate and grow on a base film of oxide. A discussion of this phenomenon is therefore more appropriately left until the section on oxide morphology which follows.

B. OXIDE GROWTH

1. Oxide Morphology

To elucidate completely the reaction mechanism of any metal oxidation, one must know, inter alia, which oxide is formed, what its structure and texture are, and to what extent these features are dictated by the properties of the underlying metal.

Which oxide is formed is usually governed by the thermodynamics of the system, but these may be overruled by the kinetics of the process. Thus one finds that Cu_2O, not CuO, is formed at low temperatures

(Ronnquist and Fischmeister, 1960) although the latter is thermo-dynamically favoured. Again, sodium superoxide (NaO_2) grows on sodium at 273 °K in preference to the more stable Na_2O (Anderson and Clark, 1963). It is for kinetic reasons that multiple layers of oxides form on transition metals. On iron (Boggs *et al.*, 1967), for example, one finds in juxtaposition with the metal a layer of Fe_3O_4 capped by Fe_2O_3. To find the metal rich oxide adjacent to the metal is usual and to be expected, since metal is most readily available there and oxygen least. Corresponding variations in nonstoichiometry are found across some single phase oxides (Deal *et al.*, 1967).

The variation in oxide structure is considerable. On one hand there are the amorphous oxides such as that formed on aluminium below a temperature of approximately 425 °C (Beck *et al.*, 1967). These have little or no crystal structure and the nature of the bonding in these amorphous layers remains to be elucidated. On the other hand, there are highly ordered oxides which epitax onto their metal substrates. One of the finest examples of this is reported by Michell and Smith (1968). A single crystal film of CdO_2 is formed in air on a single crystal of cadmium, with the (111) plane of the oxide parallel to the (0001) plane of the metal, and the $[1\bar{1}0]$ parallel to $[11\bar{2}0]$. The oxide is laid down in such a way that the misfit between the two lattices is small, but when this condition cannot be met, the epitaxy between the oxide and the metal is usually less complete. The structure of an oxide, like its composition, depends to some extent on the mechanism by which the oxide was formed. Although aluminium oxide films are amorphous when formed below 425 °C, they contain crystallites of γ-alumina if grown at higher temperatures (Beck *et al.*, 1967). Anodically formed alumina can also be partly crystalline (Young, 1961).

Many of the early oxidation theories were developed on the assumption that the product layer was an isotropic film of uniform thickness. This was a gross oversimplification. Evans, in a series of papers which are summarized in his book (Evans, 1960), was probably the first to recognize that deviations from the coherent parallel-sided oxide film can occur in many ways, each capable of causing a modification of the rate constant if not the rate law. The importance of the deviations, some of which will now be considered, has been recently emphasized by Bénard (1967).

(a) *Particulate Growth*—It is a well established experimental fact that oxides do not thicken uniformly, but grow preferentially on certain sites giving a wide variety of surface textures of which Figures 3 and 4 are examples. Figure 3 shows nodular growth of tellurium oxide on a cleaved tellurium surface which has been exposed to air at room temperature for a few minutes. The oxide whiskers of Figure 4 were the result of heating a copper electron microscope grid for about 12 hours at 300 °C in one atmosphere of oxygen.

Fig. 3. Nodular oxide growth on a cleaved tellurium surface.

Preferential growth at some point on the surface can only occur when that location can acquire metal atoms or oxygen atoms, whichever are in short supply, more efficiently than other parts of the surface. Fischmeister (1960) has suggested that the particulate growth which is commonly observed at high oxidation temperatures and low gas pressures, results from a deficiency of oxygen at the surface. These particles, the so-called nuclei, are found to form on top of a thin relatively uniform oxide

FIG. 4. Oxide whiskers formed on a copper electron microscope grid.

film about 50–100 Å thick. They are all much of a size which depends on the duration of oxidation, the gas pressure, the temperature, and the orientation of the exposed face of the substrate. The particle density is also a function of these variables. There have been numerous attempts to correlate the particle location with the site of a metal dislocation.

These have met with little success and the work of Faust (1963) is the most convincing demonstration of this point. The lack of correlation is to be expected. Surface defects are far less likely to provide nucleation centres at high temperatures when the effects of differences in chemical potential between different spots on the surface are small, than they are at low temperatures. As a result the particles can form at random and will then grow by surface diffusion. The importance of surface diffusion in oxidation has been stressed by Anderson (1967). The growth process has been analysed by Rhead (1965), who pointed out that it was an example of Ostwald ripening, that is, the big particles expand at the expense of the small.

If the conditions are suitable, whisker and lamellar growth are possible. The ways in which whiskers can grow, and their importance in oxidation kinetics have been considered by Fischmeister (1960). He concluded that in most cases, the whiskers grow out from a base oxide film around a screw dislocation. Both diffusion along the dislocation and across the whisker surface are involved: lamellae result from adjacent dislocations. It is Fischmeister's opinion that whiskers, which comprise but a small fraction of the product layer, can exert only an indirect influence on the rate of reaction.

At lower temperatures, say a few hundred degrees centigrade at the most, surface diffusion is greatly reduced, and random oxide agglomeration curtailed. The thickness of the base film and the dimensions of the oxide nodules formed on it are much smaller. Possibly the supply of metal atoms to the oxide surface has been cut back. But the relative importance of oxidation at surface defects has been enhanced. Thus two or more types of surface structure can coexist. When Harris *et al.*, (1957) oxidized copper at 150 °C, they observed "nuclei", some 30 Å across, and structures which they called polyhedra, which were about 1000 Å across. The concentration of the nuclei per unit area varied with time, but that of the polyhedra did not. This suggests that the latter grew at defect sites, the density of which was fixed. In another low temperature study of copper oxidation, Bassett *et al.* (1959) used electron transmission microscopy to demonstrate that some oxide particles are formed at dislocations and others are not.

(b) *Blisters, Cracks, Cavities and Voids*—The oxidation product of most metals occupies a larger volume than the metal from which it is derived. From this fact has stemmed the empirical rule of Pilling and Bedworth (1923), which states that the oxide M_xO_y will form a protective coating on the metal if its volume ratio ϕ, defined as the quotient of the molar volume of $MO_{y/x}$ and the atomic volume of the metal itself, is greater than unity. To Pilling and Bedworth, protective oxidation meant any oxidation rate which diminished as the oxide thickened. They argued that if the volume of oxide formed was insufficient to cover the area of metal it replaced, direct access of the gas to the metal was possible and a constant reaction rate, independent of thickness, would be observed. Only if the oxide completely covered the metal would it be protected. Their results seemed to bear this out. But now it is known that all metals, even those whose oxides have volume ratios less than unity, can oxidize protectively. For example, MgO which has a volume ratio of 0.81 grows to a limiting thickness on magnesium in very pure oxygen at 550 °C (Castle *et al.*, 1962). Thus the rule has a very limited value although it is still occasionally invoked (Anderson and Clark, 1963). However the idea that the volume ratio plays a part in determining oxidation behaviour is a useful one; for when the volume ratio exceeds 1, the oxide can be in a state of compressive stress, whereas when ϕ is less than one, the product layer can be subject to tensile forces. This has been very neatly demonstrated by Dankov and Churaev (1950) who oxidized films of metal evaporated onto thin mica sheets, which, as the reaction proceeded, flexed in a convex or concave manner according to the value of ϕ for that oxide/metal system. The strains in the oxide are not as large as one might have expected from a simple calculation based on the relative volumes of oxide and metal. Vermilyea (1957) has suggested that when the oxidation product grows at the oxygen/oxide interface, it is expanding without restriction into the gas phase, and so no strain in the oxide should arise. The same cannot be true if fresh oxide is created at the oxide/metal interface. The fact remains that many oxides grow in a state of stress, and this stress is often relieved by rupture of some sort. Naturally, those oxide/metal systems which have values of ϕ very different from unity (e.g. niobium pentoxide/niobium; $\phi = 2.68$), are more susceptible to mechanical breakdown than most. In addition, there is an increasing body of evidence that impurities in the metal or gas can be built into the product layer, thereby aggravating its tendency to fail. The oxidation of magnesium in the presence of traces of an organic impurity (Castle *et al.*, 1962) is one instance of this.

At high temperatures, the stress can be alleviated by plastic flow of the oxide, which Moore (1953) deduced had taken place during the oxidation of copper at 1000 °C.

The various ways in which mechanical failure of the oxide can take place were first enumerated by Evans (1947). He listed blistering, void formation, cracking and flaking. The topic of noncoherent oxides is undoubtedly more complex than implied by this simple classification, and much remains to be understood concerning their formation.

When an oxide is under compression, it can sometimes relieve the stress by lifting from the metal in several places to form blisters. Blister formation is favoured when the cohesion of the oxide is greater than its adhesion to the metal, and so they are less likely to be found in oxygen/metal systems which have a wide solubility and nonstoichiometric range. Figure 5 shows some oxide blisters on the surface of a cadmium single crystal. They have been observed on several other metals as well.

Cavities can also form along the oxide metal interface without blistering. Evans (1955) has proposed that cavities can result from the aggregation of cation vacancies which have moved from the oxygen/oxide interface to the metal during the growth of the oxide. This mechanism is only applicable when the oxide formed is a metal deficient p-type oxide such as stannous oxide (SnO). Cavities of the type postulated by Evans have been observed by Boggs et al. (1961) who used transmission electron microscopy, as well as replication techniques, to examine stannous oxide films which had grown on tin foil oxidized at 190 °C. They found that the cavities grew in size as the reaction proceeded.

In general, blisters or cavities arrest the transport of matter across the oxide and the reaction in that area must cease. But if in some way the cavity became permeable to oxygen, as it would do if the oxide dome cracked slightly, oxidation would recommence at the seat of the cavity. Thus Boggs et al. (1961) were able to correlate an increase in the rate of oxidation of tin with the onset of cracking in the oxide film over the cavities. As a result of the further growth, the cavity would be left behind by the advancing oxide/metal interface as a void within the product layer. This would explain why the nickel oxide formed on various nickel alloys during oxidation at 800 °C contained pores in the vicinity of the metal (Kennedy et al., 1959). As the ingenious experiments of Evans (1946) have shown, it is possible for an oxide film, even one formed at low temperatures, to be continually cracking open. The cracks self-heal by further oxidation. If there is little adhesion to the base metal, oxidation at high temperatures can be accompanied by an extensive cracking and flaking of the oxide. Cooling an oxidizing metal can also cause the oxide to peel, since the stresses introduced between the metal and its oxide are high, and cooling the latter enhances its brittle nature. But if the oxide layer adheres strongly to the metal, it will prefer to fracture internally rather than along the junction with the metal. Continual breakup of the oxide leads to the growth of a loose porous layer on top of a thin adherent oxide film whose average thickness remains unchanged as the oxidation proceeds. One instance of this type of duplex layered structure has been reported by Wallwork

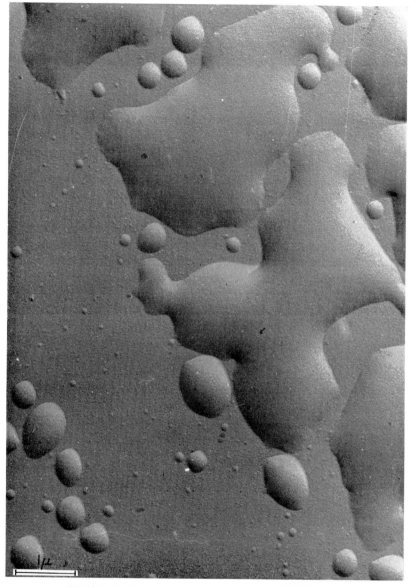

FIG. 5. Oxide blisters on cadmium (courtesy of D. Michell).

et al. (1965) in the oxidation of zirconium. Undoubtedly many other factors are involved in the formation of oxides which do not conform to the ideal of a plane parallel layer. For example, in the same study of zirconium oxidation, Wallwork *et al.* (1965) point out that in addition to the oxide, the metal can crack as a result of the surface stresses generated during the reaction. The freshly exposed metal of the cracks are immediately attacked by the oxygen and a repetition of this process can lead to disintegration of the metal surface.

However, as already mentioned, most oxidation theories have been developed on the assumption that the oxide is a uniform coherent film. Too much attention has been paid to the business of mathematically refining these theories when their basis is unsound. This effort is misplaced, in our opinion, and theories which take cognizance of the non-uniform nature of the oxide are a pressing need. Some of the few which have been developed already are described later.

2. Oxidation Kinetics

Experimentally, it is found that most oxides grow according to one or more of the following growth laws:

The linear law	$x = k_1 t$	(II)–(1)
The parabolic law	$x^2 = k_2 t$	(II)–(2)
The cubic law	$x^3 = k_3 t$	(II)–(3)
The direct logarithmic law	$x = k_4 \ln (k_5 t + t_0)$	(II)–(4)
The inverse logarithmic law	$1/x = k_6 - k_7 \ln t$	(II)–(5)

In these equations, x is the average oxide thickness, t is the time of the reaction and all the k's and t_0 are constants. It is often tacitly assumed that the several ways of measuring the average oxide thickness are equivalent. This need not be so (Evans, 1945). In addition to this, the equations imply a somewhat idealized relationship which is not realised in practice. Other growth laws are also observed. The linear, parabolic and cubic rate laws are all particular cases of the general equation

$$x^n = kt \qquad\qquad \text{(II)–(6)}$$

and non integral values of n (Smeltzer and Simnad, 1957) as well as values of n in excess of 3 (Hauffe et al., 1968) have been reported. Additive constants in all these equations are not unusual. The direct and inverse logarithmic rate laws are difficult to distinguish one from another, and from other rate laws which describe growth to a limiting thickness, such as the asymptotic equation of Evans (1945). Indeed, it has been shown (Evans, 1960) that when $k_7 \log t \ll k_6$, the inverse logarithmic law reduces to the direct logarithmic equation. The same reduction can be made with the asymptotic equation. It is also possible for the rate law to change during the course of a reaction: thus Rosenberg et al. (1960), when studying the oxidation of antimony, found that the kinetics changed from parabolic to linear as the oxidation proceeded.

For any given oxygen/metal system, it is found that both the rate and the rate law usually depend on the temperature, pressure and crystallographic orientation of the exposed metal face. In most cases one may say that as the temperature is increased, the rate increases, and the value of n in the equation $x^n = kt$ decreases. That is, at low temperatures logarithmic type laws are observed; as the temperature is raised, these give way successively to the cubic law and parabolic law; at a sufficiently high temperature all metals oxidize linearly. Not all metals follow this pattern and it is not uncommon for one rate law to be absent from the sequence; cubic oxidation is not always observed. In some instances, the kinetics pass directly from logarithmic to linear without the intermediate parabolic region (Anderson and Clark, 1963). However inversion of the sequence of rate laws is rare. The transition between the different types of oxidation behaviour naturally varies from one metal to another but for typical structural metals, for logarithmic growth to be replaced by parabolic it is a few hundred degrees centigrade, and for linear to take over from parabolic, about a thousand degrees centigrade. Clearly there exists a rough correlation between oxide thickness and rate law. Thin films formed at low temperatures grow logarithmic-wise; very thick films or scales thicken at a constant rate; and oxide films of an intermediate thickness grow according to the parabolic law. It is often found that for any given rate law, the rate constant is related to the temperature by the Arrhenius equation. Moore and Lee (1951) verified this for the oxidation of zinc.

In proposing a reaction mechanism, it is necessary for the mechanism to lead to the observed rate law, but the rate law itself cannot be regarded as conclusive evidence for a particular mechanism. There are two reasons for this. Firstly there is considerable difficulty in distinguishing one rate law from another, particularly when the experimental data are of limited accuracy and apply to only a few hours of oxidation. Secondly, as will be shown, the relationship between mechanism and rate law is never unique, and several widely dissimilar mechanisms may yield the same growth kinetics. Kinetic measurements alone are therefore insufficient to completely elucidate an oxidation mechanism. Other mechanistic experiments must be undertaken, and to measure the variation of reaction rate with gas pressure is one way in which more information about the reaction mechanism can be obtained.

Most oxidation rates vary with oxygen pressure, and it is usual to express the variation by the empirical expression

$$\frac{dx}{dt} = k_0 f(x) \, p^m \qquad \text{(II)–(7)}$$

where k_0 is the pressure independent rate constant, f(x) denotes the functional dependence of the rate on the thickness, and p is the partial pressure of the oxidizing gas. The values of m, which are not usually known with any great precision, lie between ± 1, and while they are approximately constant for any given set of experimental conditions, they can vary with the gas pressure itself, as well as the oxidation temperature.

There are numerous examples of the variation of oxidation rate with crystal face, but the most aesthetically pleasing demonstration of this is described by Young et al. (1956) who oxidized single crystal spheres of copper. When the sphere was illuminated in white light, beautiful patterns of interference colours could be seen corresponding to the change in oxide thickness over the various crystal planes on the metal surface. It would appear from this and other studies that the ease of oxidation of a particular crystal plane may be correlated with the atomic roughness of that crystal plane.

3. Oxidation Mechanisms and Possible Rate Determining Steps

Once the first few atomic layers have been formed, a barrier is interposed between the gas and the metal, and further growth of the oxide can only take place if oxygen is transported across the oxide to the metal, or conversely if metal is moved to the oxygen. Figure 6 is a schematic diagram of the reaction paths by which the oxide can continue to thicken.

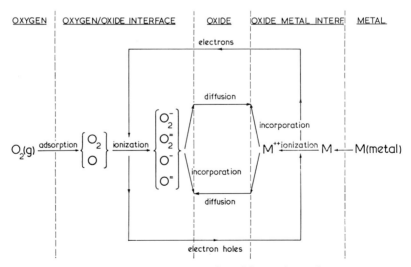

FIG. 6. Schematic diagram of possible reaction paths.

For simplicity, the charged species in the diagram are shown as ions with integral charges. From the diagram, it can be seen that the slow step in an oxidation reaction can be either mass or charge transport across the oxide, mass or charge transport across one of the interfaces, or one of the surface steps involving oxygen. By far the largest number of oxidation theories are based on the supposition that transport across the oxide, in one form or another, is the slowest step in the reaction sequence. This possibility will be considered first.

a. *Slow ion or electron transport across an ideal parallel sided oxide film*—One of the simplest and earliest theories of diffusion controlled oxidation is due to Pilling and Bedworth (1923).

They assumed that the oxidation rate was determined by the speed at which oxygen moved across the oxide. It is implicit in their derivation that the diffusing species is uncharged and we will take it to be atomic. According to Fick's first law, the oxygen diffusion current (j) is given by

$$j = -D \frac{d[O]_y}{dy} \tag{II)–(8}$$

where D is the diffusion coefficient, and $[O]_y$ the oxygen concentration at a distance y from the oxygen/oxide interface, where $0 \leqslant y \leqslant x$. If j and D are constants, and the oxygen concentration at each interface fixed, this equation may be integrated to give

$$j = -\frac{D([O]_x - [O]_0)}{x} \tag{II)–(9}$$

When an oxygen atom is incorporated, a volume Ω of fresh oxide is created. Thus

$$\frac{dx}{dt} = \Omega j = -\frac{\Omega D ([O]_x - [O]_0)}{x} \tag{II)–(10}$$

and the growth law is parabolic. The assumption that j is constant implies that the concentration gradient across the oxide is linear, which as Booth (1948) has pointed out cannot be correct. However, no serious error is introduced if the concentration of the diffusing species is small, a condition which will hold true in most practical cases. This condition is also sufficient to ensure that the known variation of D with concentration is small. The third assumption concerning the constancy of the interfacial concentrations is reasonable provided the rates of the surface processes are fast.

While the diffusion of a neutral entity, such as an oxygen atom or molecule is possible, it is now generally considered that at high temperatures diffusion in compact oxide layers will take place via charged

N

lattice defects (interstitial atoms, vacancies, etc.). The movement of the charged defects will be governed not only by their concentration gradients, but also by any electric fields which are present in the oxide. Such fields can arise in two ways. The first of these ways has its origin in the high electron affinity of oxygen. Oxygen atoms or molecules adsorbed at the oxygen/oxide surface trap electrons out of the underlying metal or oxide. If the trapped electrons come from the metal, there is a positive charge in the metal near the oxide/metal interface, but if the electrons are drawn from the oxide a region of positive space charge adjacent to the oxygen/oxide interface is formed. In either case an electric field results from the charge imbalance. The second way in which an electric field can occur is also due to a space charge within the oxide. The transport of charged ions or vacancies during oxidation requires an approximately equal countercurrent movement of charge carriers (electrons or electron holes) to keep the system electrically neutral as a whole. However, the mobilities of ions and charge carriers are usually quite different and they tend to separate with the consequent formation of a space charge.

The diffusion current of a charged particle in an electric field is given by a modified Fick's law equation

$$j = - D\frac{dC}{dx} - \mu zq \frac{dV}{dx} \tag{II)—(11}$$

where C is the concentration of the particles, μ their mobility, z the particle charge number, q the electronic charge and V the potential of the field. The diffusion co-efficient D and the mobility μ are related by the Einstein equation

$$Dzq = \mu kT \tag{II)–(12}$$

in which k and T have their usual meanings. The current j, and hence the oxidation rate (using equation (II)–(10)), can only be evaluated if V is known, but to calculate V exactly is not yet possible. Accordingly, solutions to the diffusion equation (II)–(11) can only be approximate and are obtained by making various simplifying assumptions about the way in which V changes across the oxide. In a mathematical sense, the most tractable of these approximations is to neglect all space charge effects, and to assume that the electric field is homogeneous and due only to the uncompensated charges at each oxide interface. The conditions under which this approximation is valid have been discussed by Cabrera and Mott (1949) and Fromhold (1963b). As a result of the affinity of oxygen for electrons, one would expect that the oxygen/oxide interface would always be negative with respect to the metal. This is certainly true for the formation of thin oxide films, as surface potential measurements have shown (e.g. Boggio and Plumb, 1966). One would also expect that the electric field would be a function of the oxide thickness, and would change during the course of oxidation. Grimley and Trapnell (1956) have considered the surface charges to be fixed by the equilibria between the various oxygen species on the oxide surface. This approach has been elaborated by Ritchie and Hunt (1969). The conclusions from this type of analysis are as follows: as the oxide grows, the field which was initially proportional to the oxide thickness (i.e. constant potential difference across the oxide), tends to become independent of thickness; for a given oxide thickness, the effect of increasing the oxidation temperature is to move from a region of constant potential difference to one of constant field. It should be noted that in order for the field to be constant, the number of field creating ions must be unchanged. Fromhold (1963a) and Anderson and Ritchie (1967a) have examined the situation in which the surface charge is determined by the difference in transport rates of the positively and negatively charged particles. When the transport of only one ionic and one electronic species is involved in the growth of the oxide, and the two species move by diffusion, the potential difference is independent of the oxide thickness. But when the electrons tunnel through the oxide, an approximately constant field is more likely to result.

It can be seen from these arguments that the two limiting cases, one of constant potential difference, and the other of constant field, are arrived at irrespective of whether the surface charge is fixed by surface or transport processes. In discussing the effect of surface charge on oxidation kinetics, it is usual to consider both possibilities.

There is general agreement that diffusion under the action of a constant potential difference leads to a parabolic growth law. Fromhold (1963b) has shown this to be true by integrating equation (II)–(11) and assuming that the diffusion current (j) is the same at all points within the oxide. A similar conclusion was reached by Anderson and Ritchie (1967a) who used random walk theory to describe the movement of atoms and electrons. Like most oxidation theories, there is little or no experimental evidence that can be quoted in its support. The existence of a constant potential difference, which seems reasonable, requires experimental verification. Engell and Hauffe (1952) have invoked an approximately constant potential difference to explain the pressure dependence of the rate constant for the parabolic oxidation of zinc between 375 and 400 °C. They postulated that the rate of zinc oxidation in this temperature range is determined by the speed at which cations move across the growing oxide with the aid of a field. If the oxide film is very thin, and the electric field is very large, it is no longer valid to use equation (II)–(11) to describe the ionic motion. The appropriate equation for ions travelling in a strong electric field was developed by Mott and Gurney (1940). The application of this equation to field assisted ion transport yields an inverse logarithmic law as demonstrated by Evans (1955).

When the electric field rather than the potential difference is fixed, it is found that a wide variety of rate laws are possible. These range from an approximately linear law which results when the transport of the slow moving species is aided by the electric field, to a quasi-logarithmic law which describes the kinetics when the field opposes transport (Fromhold, 1963b; Anderson and Ritchie, 1967a). Opposition of the field to electronic transport seems physically more likely than opposition to ionic transport. The case in which the oxide film is thin, and the electric field large, has been treated by Grimley and Trapnell (1956). While ions are restricted to diffusion through the product layer, electrons have the option of tunneling across the layer, perhaps in stages from electron trap to electron trap. Slow electron tunneling has sometimes been suggested as the rate determining step in the oxidation process. Mott (1939) was the first to make this suggestion and deduced that if the tunneling took place through a simple rectangular potential barrier (i.e. there was no electric field across the oxide), growth would proceed according to the direct logarithmic law. However the predicted growth rate is independent of the temperature of oxidation and the oxygen pressure, which is contrary to experiment. Fromhold (1963b) sought to remedy this defect in the Mott theory by supposing that electron tunneling was assisted by a constant electric field (i.e. through a trapezoidal potential barrier). He succeeded in deriving a growth law of the logarithmic type which was temperature dependent. Further work on this idea has been undertaken by Fromhold and Cook (1967). There is some evidence that the field across growing oxides can be constant, at least at low temperatures. In the oxidation of zinc at 300° C for example, Uhlig (1956) has shown that the surface potential becomes increasingly negative, the oxide/metal interface being positive with respect to the oxygen/oxide. Its magnitude was found to be approximately proportional to the logarithm of the time. The kinetics of oxidation are also logarithmic at this temperature (Vernon et al., 1939). The potential difference across the oxide is therefore proportional to the oxide thickness, and in the absence of a space charge, this indicates a constant field. A field of the reverse polarity, as required by Fromhold's theory, is far less likely. Possibly the only experimental evidence that can be cited in its support is the data of Hirschberg and

Lange (1952). They observed a positive potential on zinc samples which had been oxidized at low temperatures.

When the space charge within the oxide is large enough to affect materially the oxidation rate, the problem of deriving the growth law is extremely difficult. This is because it is necessary to couple diffusion equations, similar to equation (II)–(11) for each of the diffusing species, via Poisson's equation,

$$\frac{d^2V}{dx^2} = -q\,\frac{\sum z_i C_i}{\epsilon} \qquad \text{(II)–(13)}$$

where ϵ is the permittivity of the oxide. On occasions, it is possible to avoid the use of Poisson's equation by assuming some simple form for the charge distribution within the oxide. The best known and the most successful of these simplifying assumptions is due to Wagner (1933). He assumed that in each portion of the oxide thermal equilibrium prevails, and the system is in a state of quasi-neutrality. This requires that a virtual potential be set up in the oxide such that there is no net transfer of charge across any plane in the product layer throughout the oxidation. In most oxides this means that ion transport is increased and charge transport held back. The process is known as ambipolar diffusion (Hauffe, 1965). To obtain an expression for the oxidation rate, Wagner combined the appropriate diffusion equations (II)–(11) by equating the flux of positive charges to the flux of negative charges. The resulting equation can be used to eliminate the unknown quantity dV/dx from the expression for the growth rate which is proportional to the total flux of positive or negative charge. For the simple case in which diffusion is restricted to only one ionic and one electronic species, the growth rate reduces to an expression analogous to (II)–(8) and containing only one unknown chemical potential gradient. The integration of this equation is not straightforward, but can be achieved in certain cases. The growth law is found to be parabolic, and the rate constant can be expressed in terms of physical quantities such as transport numbers, free energies, etc., which can be measured independently. The same growth equation can be arrived at in another way by considering the system oxygen/oxide/metal as an electrochemical cell and load resistance in one (Hoar and Price, 1938): although this way of regarding the oxidation process is a useful one, no new results are obtained.

For some metal oxidations, mostly those taking place at high temperatures, the assumptions in the Wagner theory would seem to be reasonable. In these cases, it has been shown that there is good agreement between the values of the parabolic rate constants, as determined by kinetic studies, and those values estimated from the Wagner expression (Wagner, 1950). For the particular case of cobalt oxidation, it has been possible to demonstrate that the concentration gradient of cobalt in cobalt (II) oxide conforms closely to the theoretical distribution predicted from Wagner's equation (Carter and Richardson, 1955). Because ions are much less mobile than electrons, Wagner's theory requires that ion transport be rate controlling in most metal oxidations, and there is a lot of qualitative evidence to show that many high temperature scaling reactions which conform to the parabolic law are governed by slow ion transport. Jorgensen's field experiments (1962, 1963) are consistent with this supposition, although Raleigh (1966) has disputed this interpretation of the results. Both the variation of oxidation rate with oxygen pressure (Wagner and Grunewald, 1938) and the effect of dissolving small quantities of an altervalent element in the growing oxide film (Hauffe, 1953) have provided evidence of slow ion transport. However, the fact that ion diffusion determines the rate of growth of any given oxide, does not mean that the Wagner model is necessarily the correct one. The assumption that the oxide is in a state of quasi-neutrality may be incorrect, and the movement of ions and electrons may be considerably modified by electric fields within the oxide. This is known to be possible because Ilschner (1955) has shown that the diffusion potential, which arises from the tendency of the ions and charge carriers to separate, and the affinity potential, created by oxygen atoms trapping electrons at the oxide surface, can be of the same magnitude. The presence of potentials other than the diffusion potential may account for the failure of the Wagner theory at low and intermediate oxidation temperatures.

If no simplifying restrictions such as quasi-neutrality, are placed upon the charge distribution in the growing oxide, the problem of determining the oxidation rate becomes impossibly complex. To obtain an exact solution, it would be necessary to solve the following set of simultaneous equations (Fromhold, 1963a):

   (i) a diffusion equation (II)–(11) and an equation of continuity for each of the diffusing species;
  (ii) Poisson's equation (II)–(13) which includes a term for the charge distribution across the oxide, and a term for charges trapped at the interfaces;
 (iii) an equation expressing overall charge neutrality for the oxidizing system throughout the oxidation;
 (iv) equations expressing the rates of the several possible surface reactions;
  (v) a growth equation similar to (II)–(10).

The most exhaustive, but still incomplete treatment of this problem and of the topic of space charge in oxides is to be found in the series of papers by Fromhold. In his first paper (1963a), he simplifies the above series of equations by assuming that the flux of each diffusing particle is constant across the oxide at any instant of time; he also assumes that the interfacial concentrations of the diffusing particles are fixed. Using a computer, the growth laws were calculated for the cases of variable and fixed surface charge. When the surface charge was allowed to vary, it did so in a manner which was approximately proportional to the inverse of the oxide thickness; the corresponding growth law was parabolic. When the surface charge was fixed, the kinetics varied from linear to approximately logarithmic. So far as the rate law is concerned, these results are similar to those obtained when space charge is neglected. The only effect of an appreciable space charge appears to be a change in the growth rate. However, the calculated distribution of the diffusing species is very different from that computed using the Poisson-Boltzmann distribution law (Fromhold, 1963c). Fromhold (1964) has also discussed the effect of space charge on oxidation kinetics using a perturbation technique.

Of the theories described in this section, only the Wagner theory, which is of limited applicability, can be said to have adequate experimental verification. Probably most high temperature oxidation reactions are controlled by ion diffusion, but to obtain a satisfactory agreement between theory and experiment requires a knowledge of the electric field within the oxide. At low temperatures, it is not known whether the oxidation rate is diffusion controlled or not. In general, the diffusion mechanism itself is not understood. By extrapolating high temperature diffusion data to lower temperatures, it can be shown that the diffusion process which sustains high temperature oxidation reactions is insufficient to maintain low temperature tarnishing processes. Clearly some other low activation energy diffusion process is operative. If the diffusing species is neutral, the transport of mass should not be affected by any electric field within the oxide.

b. *Slow reaction steps at the surfaces of an ideal parallel-sided oxide film*—Mott (1947), in a paper which was subsequently elaborated by Cabrera and Mott (1949), proposed that ions could be transported across an oxide at low temperatures if they were pulled across by a sufficiently large electric field. By assuming a constant potential difference across the oxide, and a rate controlling movement of ions across the oxide/metal interface, he was able to show that an approximately logarithmic law would result. The original theory neglected space charge in the oxide, but has since been extended by Boggio and Plumb

(1966) to include this factor. Grimley and Trapnell (1956) have shown
that if a constant field rather than a constant potential difference is
assumed, a linear rather than a logarithmic law obtains.

The Mott-Cabrera theory is the most successful of the low temperature oxidation theories, and has been used by a number of authors to
interpret their kinetic data. We may cite as examples the work of
Boggio and Plumb (1966) on the oxidation of aluminium at room
temperature, and the study of the chlorination of sodium carried out by
Dignam and Huggins (1967). In the aluminium oxidation study, surface
potential as well as oxygen uptake measurements were made, and when
the combined data were analysed, the results were found to agree with
theory.

Slow electron transport across the oxide/metal interface has also been
suggested as the rate determining step which leads to the direct logarithmic growth law observed at low temperatures (Uhlig, 1956). It was
proposed that the growth rate diminished because the space charge,
assumed to be uniform, grew more negative as the reaction proceeded.
However the concepts and mathematical development of this model
have received extensive criticism (Fromhold, 1968).

The possibility that reactions at the oxide surface can control the
overall reaction rate has received little attention, mainly because it
has been generally believed that such reactions must lead to linear
kinetics (see for example, Fromhold 1963a). The papers of Moore (1952)
and Ritchie and Hunt (1969) show that this need not be so, and that all
the major growth laws, linear, parabolic, cubic and logarithmic can be
derived by assuming that one or other of the surface steps is slow. For
the oxidation of a metal such as zinc, whose oxidation product is an n-
type oxide, Moore (1952) proposed that the slow step in the reaction
was the ionization-incorporation step.

$$Zn^+ \text{ (interstitial)} + O^- \text{ (adsorbed)} \rightarrow ZnO \text{ (crystal)}$$

By assuming that the concentration of interstitial sites was inversely
proportional to the oxide thickness, he was able to show that the
growth law would be parabolic. Ritchie and Hunt (1969) considered all
the various pathways following the adsorption of molecular oxygen, by
which the oxygen can be ionized, dissociated or incorporated at the
oxygen/oxide interface. In principle, any one step in these pathways
can be rate determining, in which case a rate law can be deduced by
assuming that all steps prior to the slow step are at equilibrium, and
that the rate is given by simple collision theory. Electron transport
across the oxide, being a step prior to the rate controlling surface step,
must also come to equilibrium, and it is assumed that the concentration

of electrons at the oxygen/oxide interface is given by the Boltzmann distribution. If a charged species, such as O_2^-, is resident upon the surface, then the electron concentration is appropriately modified by the electric field across the oxide. In setting up the equations, it is also necessary to assume that the total number of sites, both occupied and vacant, is a constant. Proceeding in this way, rate laws corresponding to each of the several possible slow steps could be deduced. For example, if the ionization of a charged species, such as O_2^-, governs the reaction rate, either cubic or logarithmic kinetics will be observed. There is some evidence (Ritchie et al., 1970) that the logarithmic oxidation of nickel below 350 °C is controlled by a surface reaction.

Noting the similarity between the logarithmic rate laws governing chemisorption and oxidation, Landsberg (1955) has suggested that the act of chemisorption is the rate determining step in each case. The idea has been extended by Hooke and Brody (1965). However its application to tarnishing reactions has been criticized by several authors, including Hurlen (1960) and Cope (1964), and the theory is probably restricted to the uptake of only a monolayer or two.

c. *Slow reaction steps involving non-isotropic and mechanically imperfect oxide films*—As already mentioned, real oxide films seldom approach the uniform, parallel-sided ideal. At the best, all contain such structural defects as dislocations, and usually the imperfections are much more gross.

Structural defects are often important as paths of easy diffusion. Cathcart and Petersen (1968), using several different techniques to study the structural characteristics of the oxide formed on the (111) face of a copper crystal, came to the conclusion that there was a correlation between the number of structural defects they observed and the rate of oxidation of the copper surface. If diffusion along dislocations and other defects provides the most important means of mass transport in low temperature oxidations, the blocking of these defects by continued product growth at their base can lead to cessation of the reaction. This "blocked pore" theory of oxidation was first suggested by Evans and his co-workers (Davies et al., 1954), who derived logarithmic type laws for the model. Further development of the theory by Harrison (1965) has shown that a linear law is possible when the rate of production of pores is balanced by the rate at which they are blocked. While quite a feasible mechanism, pore blocking has not been observed experimentally. The diffusion necessary for continued growth may be curtailed in other ways. The mathematics of the situation in which cavities or blisters act as a barrier to diffusion has been developed by

Evans (1960), and applied by Boggs *et al.* (1961) to the oxidation of tin. It is also possible that after an initial period of oxide growth, a second crystalline form of the oxide will develop which slows down the overall movement of ions. Boggs *et al.* (1965) ascribed the rather curious effect that oxygen pressure had on the oxidation rate of iron to the nucleation of Fe_2O_3 on the initially formed Fe_3O_4. On the other hand, Beck *et al.* (1967), in a very careful investigation of the oxidation of aluminium in the temperature range 450–575 °C, interpreted the rather complex kinetics they observed by assuming that the crystals of γ-Al_2O_3, which started to grow at the junction between the amorphous oxide and the metal, did not interfere with the growth of the amorphous material. Recrystallization of the existing oxide can also lead to a change in its diffusion characteristics, and a modification of the growth rate. Meijering and Verheijke (1959) have proposed that the cubic region of copper oxidation is caused by ageing of the oxide.

Defects can assist as well as retard oxidation. When the oxide contains macroscopic imperfections, such as cracks, the growth rate can be accelerated by the gas having easy access to the metal. Linear kinetics at high temperatures are often ascribed to this cause.

Some of the theories which have been described in this section have been gathered into Table 1 in which they are classified according to the rate law they predict. Not all theories have been included in Table 1; the particular selection given is meant to illustrate the wide variety of rate determining steps that have been postulated, and to show that several widely differing mechanisms can lead to the same rate law. Table 1 also illustrates that a large number of oxidation theories still await the test of experiment, which in many cases will be carried out using evaporated metal films as described in the next section.

III. The Oxidation of Evaporated Metal Films

A. KINETIC MEASUREMENTS

1. *Comparison of Film and Bulk Sample Measurements*

It is clear from the foregoing discussion of oxidation theories that a kinetic study per se is of very limited value to an investigation of the reaction mechanism. The rate law is of some use in elucidating the rate determining step, but with very few exceptions, the measured magnitude of the rate constant has very little significance. This statement is especially applicable to kinetics determined using evaporated metal films, for these are complicated by the nature of the films, which have often been laid down on glass and are usually particulate as a result.

TABLE 1

Growth Law Postulated	Rate Determining Step	Originator of Theory	Example of System on which Theory has been used
Linear	Direct contact between gas and metal following repeated cracking of the oxide	Pilling and Bedworth (1923)	Ta + O₂ (800–950 °C); Stringer (1967)
	Slow surface reaction	Several authors e.g. Grimley and Trapnell (1956)	Fe + O₂ (20 °C); Kruger and Yolken (1964)
	Ion or electron transport assisted by a constant electric field	Fromhold (1963b)	—
Parabolic	Ambipolar diffusion	Wagner (1933)	Co + O₂ (1000–1350 °C); Carter and Richardson (1955)
	Ion transport against a constant potential difference	Fromhold (1963b)	—
	Incorporation of oxygen at the oxygen/oxide interface	Moore (1952)	—
Cubic	Diffusion characteristics altering as the oxide ages	Meijering and Verheijke (1959)	—
Logarithmic	Ionization of surface oxygen species	Ritchie and Hunt (1969)	Ni + O₂ (200 °C); Hauffe and Ilschner (1954)
	Slow electron tunneling	Mott (1939)	
	Ion movement across the oxide/metal interface assisted by a constant potential	Mott (1947)	Al + O₂ (20 °C); Boggio and Plumb (1966)
	Electron transport retarded by a constant electric field	Anderson and Ritchie (1967a)	Zn + O₂ (300–375 °C); Anderson and Ritchie (1967b)
	Ion diffusion blocked by cavities in the oxide	Evans (1960)	Sn + O₂ (190–220 °C); Boggs et al. (1961)
	Ion diffusion along "pores" which are blocked as the reaction proceeds	Davies et al. (1954)	—
	Ionization of surface oxygen species	Ritchie and Hunt (1969)	—

Our knowledge of the surfaces of such films is often limited: how defective they are, which crystal planes are exposed, and what fraction of the surface is taken up by any one crystal plane.

There are in the literature various mathematical treatments of the oxidation of non-uniform metal surfaces which could perhaps be adapted to an irregular, vacuum evaporated film. For example, Irving (1965) has discussed the effect of paths of easy diffusion, such as grain boundaries, on the rate of oxygen uptake. An apparently accelerated oxidation rate is observed because oxide growth along the surfaces adjacent to the grain boundaries takes place, and in extreme cases, it is possible for parabolic kinetics to appear to be quartic. Irving's work has been criticized by Gibbs (1967). If the attacking gas has free access to most of the sides of an evaporated crystallite, the size and the shape of the crystallite may be important variables in determining the overall oxidation rate. Jander (1927) has discussed the simple case of a cubic crystallite whose size diminishes as the reaction proceeds. More complicated geometries have been considered by Romanski (1968).

Even a film which is initially uniform may become non-uniform during oxidation. An account has already been given of the work of Dankov and Churaev (1950) in which the strains generated during the oxidation of metal films bent the thin mica substrates on which the films had been evaporated. When the substrate is too stiff to flex, the film itself may crack or blister, exposing fresh metal to the oxygen. Figure 7 shows cracking in an evaporated film of nickel deposited on glass, the oxidation temperature being 350 °C (Ritchie and Scott, 1968).

For these reasons it is not to be expected that the oxidation rate of an evaporated film will necessarily be the same as that of a bulk sample such as a foil. Naively, one would anticipate that the oxygen uptake of an evaporated metal per unit area of apparent surface, would be much greater than the corresponding weight increase for a metal foil, since the film would have a larger roughness factor and hence higher true surface area. To compare satisfactorily the oxidation rates of films and foils requires the use of the same starting material for the samples, and the same technique for measuring the amount of oxygen incorporated into the oxide. So far as we are aware, this has not been done. Because the results of different workers can vary considerably (see the review of the oxidation of copper by Ronnquist and Fischmeister, 1960), a comparison between one study, using films, and a second, using foils, cannot be made with any degree of confidence. There are not all that many cases in which a comparison can be made anyway. The oxidation of copper at a temperature of approximately 200 °C is one example. According to the data of Weissmantel (1962) and Campbell and Thomas (1947), the

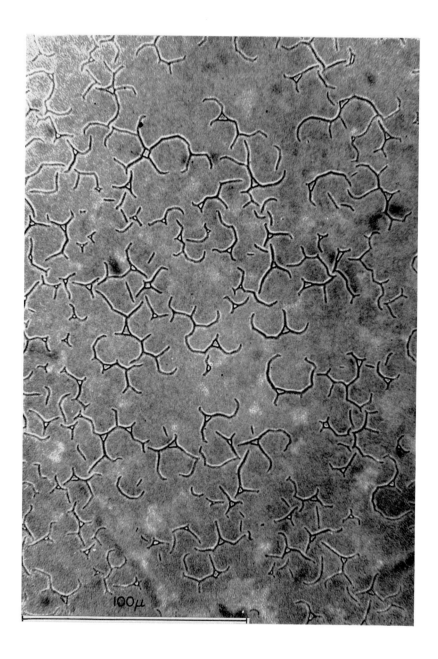

FIG. 7. Cracking in a nickel film deposited on glass: oxidized at 350 °C.

oxidation rate of an evaporated copper film at 200 °C is slightly less
than the oxidation rate of a copper foil at 194 °C. This is somewhat
curious since the roughness factor of the evaporated film has not been
taken into account.

The rate constants reported for the oxidation of evaporated metal
layers are often corrected for the roughness factor, which can be sometimes estimated from the kinetic data (e.g. Weissmantel et al., 1961).
When the freshly prepared metal surface is exposed to oxygen, there
is initially a rapid uptake of the gas, which is followed by a slower
sorption process. It is assumed that the fast process is chemisorption;
the true film area, and hence roughness factor, can then be estimated if
the area occupied by a chemisorbed atom is known. Given the weight of
the metal film, it is also possible to make some estimate of the size of the
particles comprising the film (Law, 1958). According to Roykh et al.
(1966), the film area of zinc, cadmium and magnesium condensate is unaltered during oxidation. The same is not true for germanium (Bennett
and Tompkins, 1960). Reduction of the BET area occurred upon exposure to oxygen, and the film was said to have sintered. Possibly the
sintering was caused by local heating in the film. The heat liberated
during oxidation might not be adequately dissipated in a film consisting
of minute crystallites, and it is well known that finely divided metals
can be pyrophoric.

Films and foils have been used by Addiss (1963) in a gravimetric
study of the oxidation of magnesium, but the films were only used at
room temperature, and the foils only at 400 °C and above. The almost
exclusive use of films at low temperatures (< 200 °C) and foils at high
temperatures (> 400 °C) is usual and is the explanation of why there
are relatively few film and foil kinetic studies which can be compared.
It is not hard to understand the reasons for each type of metal sample
having its own restricted range of use. Films tend not to be used at high
temperatures because (i) they are thin and can be oxidized rapidly to
completion; (ii) they are unstable and sinter rapidly; (iii) the chance of
contamination by the substrate is increased. Furthermore, vacuum
deposited metal layers have a large surface area/weight ratio. Thus,
although the total oxide thickness formed at low temperatures is small,
the oxygen uptake can sometimes be measured by using a film of sufficiently large surface area. Foils, on the other hand, are always covered
by an air-formed oxide coating which, being difficult to remove, would
interfere with low temperature measurements. The removal of the coating cannot always be accomplished prior to measurement, and there is
always the risk of further contamination during cleaning. In general,
this coating is considered to have little effect on the growth of thick

oxides at high temperatures. For the very reactive metals such as
sodium, the air formed layer is thick and grows rapidly. In these cases,
the oxidation studies have been carried out exclusively with vacuum
deposited films, which have been deposited under conditions of very
high or ultra high vacuum (e.g. Anderson and Clark, 1963).

One further advantage possessed by evaporated films is that of flexibility, in method of preparation, film structure and sample shape.
Normally, an evaporated film, as used in an oxidation reaction, is prepared by evaporation from a hot hairpin or spiral of the element (e.g.
tungsten), or by evaporation of the element from a hot inert filament or
crucible (e.g. zinc from tungsten spiral). Sputtering is less common, but
has been used by van Itterbeek *et al.* (1950) in the preparation of nickel,
iron and lead films. Thus the actual deposition of the film can usually be
considered as the last stage in the purification of the metal. Dignam and
Huggins (1967) distilled sodium from a sodium still onto a quartz slide
prior to its reaction with chlorine or bromine, and in the sodium/
oxygen study of Anderson and Clark (1963), the sodium was prepared by
diffusion through pyrex: the sodium ions reaching the interior of the
diffusion vessel, combined with electrons from a hot filament before
condensing on the cooler portions of the vessel. The film as formed was
reacted in situ with oxygen gas admitted after the film preparation had
been completed. Film preparation methods are discussed in detail in
Chapter 2.

The structure of an evaporated film depends enormously upon the
conditions of evaporation and upon the substrate used. This fact,
although common knowledge, has not been acted upon by many of those
engaged on research into the oxidation of metals. Thus, in even the
most recent papers, it is not unusual to find no mention of the film
structure (e.g. the oxidation of vacuum deposited films of lead by
Hapase *et al.*, 1968a). Most authors are content to know that a metal
surface prepared under good vacuum conditions is clean, and that the
film has been sintered sufficiently to ensure that the film structure will
not change appreciably during the course of oxidation. Eley and
Wilkinson (1960) found that the kinetics of oxide growth on aluminium
films deposited on glass at 0 °C conformed to the direct logarithmic law.
However, films deposited at -183 °C showed deviations from this law
which were ascribed to sintering in the film concurrent with its oxidation.
It is small wonder that the control of film structure which is possible
with vacuum deposited metals has not been capitalized upon in the way
that it might have been. The oxidation of aluminium (Boggio and
Plumb, 1966) is one of the very few cases in which the conditions of
evaporation were chosen so that the metal film approached the parallel

sided, pore-free ideal. The dependence of the structure of the film on
the evaporation conditions had been previously evaluated by Swaine
and Plumb (1962). In another study, Bloomer (1957b) evaporated
barium onto various substrates to test whether the number of defects
in the metal changed its gettering power. There have been few attempts
to make use of the phenomenon of epitaxy, by which films of known
and preferred orientation can be prepared, although their utility was
stressed by Rhodin (1953) some years ago. One recent attempt was that
of Jackson *et al.* (1968) who investigated the interaction of oxygen with
an aluminium film grown on a Ta (110) face.

For all these criticisms of the way in which vacuum deposited metals
have been utilized in oxidation studies, their inherent advantages of
cleanliness of surface, flexibility of preparation and structure, and high
surface area/weight ratio are undeniable, and have led to their widespread use. Table 2 shows a selection of oxidation reactions which have
been investigated using evaporated metal films. Some of the techniques
for following the rate of an oxidation reaction are described next.

2. *Methods of Measurement*

Nearly all of the methods which can be used to measure the oxidation
rate of a metal are equally applicable to foil and film samples. The
various possible experimental techniques have been described by
Kubaschewski and Hopkins (1967) and will therefore not be given here,
except for those new features which arise when the metal is in the form
of an evaporated layer. Thus the use of a microbalance, a manometric
method (change of pressure in a constant volume), or a gas burette
(change of volume at constant pressure) will only be mentioned briefly.
The microbalance, whose use in low temperature gas absorption studies
has been described by Rhodin (1953), is much less popular as an experimental method than manometry. There are two reasons for this.
Firstly, the manometric technique is the more sensitive and is therefore
better adapted to low temperature studies. The quantity of oxygen
necessary for the formation of 1 cm² of nickel oxide 25 Å thick would
exert a pressure of 3×10^{-4} torr in a dead volume of 1000 ml, but the
corresponding weight change would be only 0.5 μg. Secondly, the area of
metal film used in a constant volume system would be much larger than
that used on a microbalance. In a microbalance, the area of evaporated
metal is restricted by the balance capacity, and the dimensions of the
vacuum envelope. An area of only 2.5 cm² (Hapase *et al.*, 1968a) is not
unusual. To offset these advantages, there is the disadvantage that the
oxygen pressure varies during a kinetic run in a constant volume apparatus. A possible change in oxidation mechanism, brought about by

the changing gas pressure, could pass unnoticed. The work of Kruger and Yolken (1964) indicates that such changes are possible.

a. *Wagener Flow Method*—Related to the gas burette technique is the Wagener (1956) flow method which has been used only in conjunction with evaporated films. It is a very sensitive technique capable of measuring uptakes corresponding to a fraction of a monolayer and requires the perfectly clean surface that a freshly evaporated metal film provides. As a consequence of the sensitivity, the technique is valuable for studying the initial period of oxidation including nucleation and the results are often expressed in terms of the sticking coefficient rather than the average oxide thickness. The method is based on the Knudsen flow of gas in a capillary at low pressures. The apparatus consists of a capillary of known gas conductance (B), at the ends of which are mounted low pressure gauges. One end of the capillary is also connected to a vacuum line and gas injection system; the other is connected to the absorption vessel. The rate of gas absorption is equal to the rate of flow of gas through the capillary (G) which can be shown to be given by

$$G = B\,(p_1 - p_2)/p_1 \qquad\qquad \text{(III)–(1)}$$

where p_1 and p_2 are the pressures recorded by the gauges on the high and low pressure sides respectively. Since the gas pressures involved are very low, being 10^{-5} torr or less, it is necessary to use ultrahigh vacuum techniques and a low pressure gauge such as an ionization gauge. A sensitivity of $1 \times 10^{-5}\,\mu g\ cm^{-2}$ of metal surface has been claimed for the technique by Wagener (1956). Bloomer (1957a) has pointed out that the ionization gauges can lead to an enhancement of the oxidation rate, since ionized gases can be more reactive than their nonionized counterparts. It is also possible for the oxygen to be depleted by reaction with the hot filament.

b. *Transparency Method*—It is simple enough to evaporate metal films which are sufficiently thin to be nearly transparent. During oxidation, the film will be converted to oxide, and become less opaque. By making measurements of the optical absorption of the film at a particular wavelength, and knowing the optical properties of both metal and oxide, it is possible to estimate the metal thickness at any instant, and so determine the oxidation rate. The method was originated by Steinheil (1934) who used it to study the oxidation of aluminium, as did Cabrera *et al.* (1947). It is a particularly convenient technique to employ when the oxidizing gas is highly reactive since the reaction can be carried out in an all glass apparatus. Dignam and Huggins (1967) followed the chlorination and bromination of sodium films in this way,

o

TABLE 2

Metal	Film Preparation Method	Substrate	Gas[d]	Pressure (torr)	Oxidation Temperature (°K)	Method of Following Kinetics	Reference
Al	W[a]	Lucite		760	293	Ellipsometry	Boggio and Plumb (1966)
Al	W	P[c]		10^{-3}–10^{-2}	293	Manometry	Eley and Wilkinson (1960)
Ba	Evaporation from commercial getter	P and Others		$<10^{-5}$	253–343	Wagener Flow	Bloomer (1957b)
Bi	W	Glass		30	448–523	Microbalance	Hapase et al. (1967)
Ca	Evaporation from molybdenum	P	N_2	10^{-2}	296–473	Gas burette	Roberts and Tompkins (1959)
Cu	W	P		30	423–473	Microbalance	Hapase et al. (1968b)
Fe	E[b]	P		10^{-6}–10^{-1} / 10^{-2}	78–393	Manometry and Gas burette	Roberts (1961)
Ge	Evaporation from alumina coated tungsten	P		10^{-1}–1	78–493	Manometry	Bennett and Tompkins (1960)
Mg	Evaporation from tantalum	P		$<10^{-5}$	197–409	Wagener Flow	Addiss (1963)
Mg	Evaporation from tantalum	P		10^{-7}–10^{-6}	293	Microbalance	Cohen (1960)
Na	Distillation	Quartz	Cl_2 Br_2	3–50	238–323	Optical Density	Dignam and Huggins (1967)
Na	Electrolysis through glass	P		10^{-3}–10^{-1}	90–273	Manometry	Anderson and Clark (1963)
Ni	E	Jena glass		<1	298–673	Manometry	Weissmantel et al. (1961)

Metal		Substrate	Gas	Pressure	Temperature	Method	Reference
Pt	E	Jena glass		<1	298–673	Manometry	Weissmantel et al. (1961)
Pb	W	P		10^{-5}–10^{-2}	90–548	Manometry	Anderson and Tare (1964b)
Pb	W	P		30	500–580	Microbalance	Hapase et al. (1968a)
Sb	Not specified	Glass	Cs	Not specified	Not specified	Resistance of metal film	Miyake (1961)
Si	E	P		10^{-4}–10^{-3}	273–323	Manometry	Law (1958)
W	E	P		$<10^{-3}$	195–273	Manometry	Lanyon and Trapnell (1955)

(a) W – evaporation from tungsten: (b) E – evaporation from elements: (c) P – pyrex glass: (d) Oxygen unless stated otherwise

o*

and found that the light used to measure the optical absorption could affect the reaction rate. Evans (1945) has objected to the technique on the grounds that the measurements do not give the average oxide thickness, but yield a result which is closer to the modal thickness.

c. *Metal Resistance Method*—Another technique for following the kinetics of an oxidation reaction is the metal resistance method which, like the transparency technique, depends on the fact that very thin evaporated layers can be prepared. The resistance along the length of a thin evaporated film is a measure of its thickness. For a thin isotropic metal sheet of length l, thickness d, width w and resistivity ρ, the resistance (R) is given by

$$R = \rho \, l/wd \, . \qquad\qquad \text{(III)–(2)}$$

When the metal is oxidized, it is replaced by an oxide whose resistivity is usually many orders of magnitude larger than the metal. As a result, the resistance of the film increases during oxidation, and if the oxide (of volume ratio ϕ) grows to a thickness x, the resistance becomes

$$R = \rho \, l/w(d - x/\phi)$$

or

$$d - x/\phi = \rho \, l/wR \, . \qquad\qquad \text{(III)–(3)}$$

In principle, the growth law and rate constant can be obtained by plotting the reciprocal of the film resistance against a suitable function of time by combining equation (III)–(3) and one of the rate equations (II)–(1) to (6).

The principle of the method was originally established by Pilling and Bedworth (1923) who measured the increase in resistance of a thin copper wire which was heated in air. It was first applied to sputtered films by van Itterbeek *et al.* (1950) who reported the sigmoid resistance–time plots which are typical of the technique. No attempt was made to determine the growth law or kinetic parameters from the data however, and it was left to Miyake (1961) to be the first to deduce a growth law from a set of resistance versus time measurements for the reaction between evaporated antimony films and caesium vapour.

The chief advantage of the metal resistance method is that it is quite a sensitive technique; this can be illustrated by considering the oxidation of an evaporated nickel film having the dimensions of 2 cm × 1 cm × 100 Å. If the film were uniform, and scattering effects neglected, its resistance would be 4 ohm. The oxidation of 25 Å of the metal would increase the resistance by $33\frac{1}{3}\%$. Resistance can of course be measured quite accurately. Ruhl (1963) has studied the oxidation of several metals in the temperature range 1.5–100 °K in this way. However, it should be noted that changes in resistance can also occur when a gas is chemisorbed onto the metal and when the film recrystallizes (Smith, 1964).

When the vacuum deposited layer does not conform to the plane parallel sided ideal, deductions about the growth law and rate constant can be in error. This has been shown to be so by Ritchie and Scott (1968) who simulated the evaporated metal film by a network of resistances similar to that depicted in figure 8.

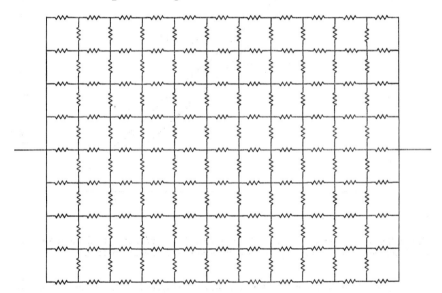

FIG. 8. Network of resistances representing the resistance of a thin evaporated film composed of small crystallites or grains.

Each element in the network was supposed to correspond to a particle in the metal film. The nonuniformity of a metal film deposited on an amorphous substrate such as glass could then be represented by assigning different values to each of the elements. Using a computer, it was possible to calculate the total resistance of the array as a function of time by assuming that the decrease in conductance caused by oxidation of the metal was a constant over the whole network. The results of one computation are shown in figure 9.

It can be seen from the graph that the oxidation rate of a nonuniform film is similar to that of the uniform film for about the first 25% of oxidation, but that serious differences develop at longer oxidation times. The calculations can be extended to include cracking in the film.

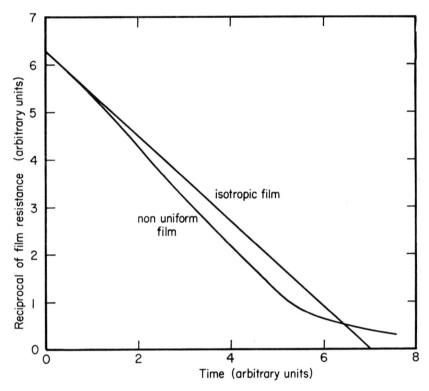

FIG. 9. The computed increase in resistance with time of an isotropic film and an aniso-
tropic film assuming that each particle oxidizes at the same constant rate.

A related technique to the one given above has been described by
Fehlner (1968). It involves measuring the d.c. resistance of a discon-
tinuous evaporated metal film during oxidation and is restricted to the
low pressure, low temperature region.

B. MECHANISTIC EXPERIMENTS

To characterise fully an oxidation reaction, a large amount of information is needed in addition to a knowledge of the growth law and
the appropriate rate constants. It is necessary to know the oxide composition, structure and morphology; the charge and mass transport
mechanisms must be understood. Suitably designed experiments which
make use of the unique properties of evaporated metal films have
yielded much information on these points. One experiment which has
already been described is that of Dankov and Churaev (1950) who
studied stresses in oxides using vacuum deposited metal layers; other

experiments will be given later. It is also necessary to know the sequence of reaction steps which make up the growth mechanism, and which one of these controls the oxidation rate.

Some details of the growth mechanism can be inferred from the properties of the oxide which forms on the metal and from the growth law, but this information is insufficient in itself to specify the rate determining step in the reaction. Further experiments in which the oxidizing system is perturbed, and its response to the perturbation studied, are essential if the rate controlling step is to be identified. One type of perturbation that can be applied is the application of an electric field across the growing oxide. A second is the introduction of a small quantity of impurity or dopant into the metal or gas. Both are described in this section on mechanistic experiments, which is not meant to be exhaustive in its coverage, but only to illustrate the varied use to which evaporated metal films can be put in oxidation studies.

1. *Determination of the Ultimate Oxide Composition*

In any oxidation study, it is clearly necessary to determine the stoichiometry of the oxide formed. Thick oxides can often be stripped from the base metal and analysed, but this is not always possible with the thin oxide layers formed at low temperatures. Wieder and Czanderna (1962) have used evaporated metal films in a rather cunning way to overcome this difficulty. Knowing the weight of metal film they started with, and the amount of oxygen taken up by the film when oxidized to completion, they were able to determine the stoichiometry of the oxide directly. Using a variety of film thicknesses and oxidation conditions, they came to the unexpected conclusion that the oxide formed on copper in the temperature range 110–200 °C at an oxygen pressure of 100 torr has the composition of $CuO_{0.67}$ and is a grossly defect form of Cu_2O. Their work has since been confirmed by Hapase *et al.* (1968a). It is perhaps surprising that this type of experiment has not been more widely employed.

2. *Determination of the Most Mobile Species by Resistance Marker Measurements*

The ionic components of an oxide tend to diffuse through the oxide at totally dissimilar rates. As a result, most oxides grow almost exclusively by either cation diffusion or by anion diffusion. Which species is primarily responsible for mass transport is usually ascertained by means of a marker experiment whose aim is to discover the interface at which growth takes place. Growth at the oxygen/oxide interface will occur when metal transport through the oxide lattice is more rapid than oxygen transport, but if the converse is true, the oxide will grow into the metal. Wagner (1933) identified the growing interface by directly measuring the extent of reaction at each interface, and found that silver migration is more rapid than sulphur migration in silver sulphide. More usually, the movement of the interfaces relative to a fixed point or marker is determined. In the experiment introduced by Pfeil (1929), an inert marker is placed on the metal surface. After some oxidation has taken place, the marker's position relative to the metal surface is observed optically or by radiochemical techniques. Jorgensen (1962, 1963) has shown that the position of an inert metal marker relative to the oxidizing metal can be determined by measuring the resistance between the marker and the base metal. An increasing resistance signifies growth at the oxide/metal interface, due to oxygen movement across the oxide. The inert metal marker of Jorgensen's experiments was ingeniously prepared by sputtering platinum onto the oxidizing metal crystal in an argon atmosphere, giving a platinum film which is porous to oxygen. The resistance marker method is extremely sensitive since most oxides have very high resistivities, and a slight increase in oxide thickness is accompanied by a large change in resistance. It is therefore particularly useful for studying thin film tarnishing reactions. Anderson and Ritchie (1967b) adapted Jorgensen's technique to thin oxide films by evaporating the base metal, the porous platinum layer and the necessary electrical contacts in such a way that the chance of breaking through the thin oxide film and shorting the platinum and base metal during the setting up of the experiment was minimized. They were able to show that the oxidation of zinc and tungsten at low temperatures proceeds principally by oxygen transport. Using the same apparatus, they were also able to identify the conductivity type of the oxide.

The ambiguity of marker experiments has been discussed by Mrowec and Werber (1960). Sachs (1956) has used markers of different metals and sizes and shown that the position of the marker is a function of the marker size and material. It would appear from the work of Fiegna and Weisgerber (1968) that the deposition of some metals on a growing oxide surface can accelerate the oxidation rate. They found that several of the noble metals, when sputtered onto oxidizing zirconium, affected its reactivity in this way. It may be concluded that the choice of marker material must be made with care.

3. *Measurement of Surface Potentials*

The measurement of surface potentials, or changes in work function, is of great importance in studying the early stages of oxidation. Vacuum deposited metal films have been used extensively in surface potential measurements because they can be prepared with clean surfaces, free of contaminants such as water which are known (Huber and Kirk, 1966) to give rise to spurious values. By the same token, much of the earlier work in which the metals studied were prepared under vacua of only 10^{-6} torr must be regarded as suspect.

Changes in the work function can be measured in a variety of ways which have been reviewed by Culver and Tompkins (1959). The vibrating condenser, the diode, and the photoelectric method have all been used for surface potential measurements on evaporated films. The data obtained from a vacuum deposited layer is subject to the same sort of limitations that apply to kinetic data. The film must be sintered sufficiently to be stable. The surface potential is structure sensitive, and the measured value is some sort of average value fixed by the size and orientation of the crystallites comprising the evaporated film. Despite these disadvantages, the results have been qualitatively useful in unravelling what happens when a metal surface is first exposed to an oxidizing gas, as previously discussed in section IIA. It is convenient at this stage to consider the type of results obtained and their implications in somewhat more detail. The most exhaustive study of the way in which the work function of a clean metal surface changes when the surface interacts with oxygen has been carried out by Roberts and his co-workers on nickel (Quinn and Roberts, 1962, 1964, and 1965; Roberts and Wells, 1966). Their results will be described here to illustrate the power of the technique. Both capacitor measurements and photoelectric measurements were made, the information from the two types of experiment being complementary. For example, the capacitor method is limited to measurements of the surface charge, and this need not yield information about the band structure of the oxide formed. However measurements of the photoelectric current at a particular wavelength, both as a function of time and as a function of "stopping potential" (i.e. the potential applied between the metal and an inert collector electrode) can provide such information (Quinn and Roberts, 1965).

Several variables were found to be important in the study of work function changes. The change in work function was not only a function of the time of exposure, the temperature of the metal, and the oxygen pressure, but it also depended upon the way in which the surface had already been exposed to oxygen; whether the gas was admitted slowly or rapidly, and whether the surface was heat treated between gas doses or not.

From the capacitor measurements it was found that when a clean nickel surface was exposed to a dose of oxygen at $-195\ ^\circ C$, the work function increased rapidly to attain a value which did not change with time. The magnitude of the change in work function depended on the size of the oxygen dose, but reached a maximum value when the amount of oxygen was sufficient to form a monolayer on the surface. These characteristics led Quinn and Roberts (1964) to suggest that the oxygen was present on the nickel surface as a chemisorbed layer, the oxygen being somewhat negatively charged with respect to the nickel. If a nickel surface which had been partially covered with oxygen at -195 °C was subsequently heated in vacuum, the work function gradually decreased, and the change in work function with respect to the clean metal value could even become negative (Roberts and Wells, 1966). The change was an activated process, having an activation energy of 1–2 kcal mole^{-1} (Quinn and Roberts, 1964), and the surface become regenerated in some way, for further oxygen (approximately 0.3 monolayer) could be chemisorbed at -195 °C, with a corresponding increase in work function. Roberts and Wells have interpreted these results in terms of the formation of a low work function suboxide of nickel.

When a clean nickel surface was exposed to oxygen at temperatures above -195 °C both chemisorption and suboxide formation occurred. Thus immediately oxygen was admitted to the nickel, a chemisorbed layer was formed, and the work function increased rapidly. Incorporation of the oxygen as the low work function suboxide followed, and the work function gradually decreased. If the oxygen pressure was sufficiently high to maintain a chemisorbed layer on the suboxide throughout the incorporation process, the decrease in work function due to the formation of the suboxide was masked by the large increase in work function caused by the chemisorbed layer (Quinn and Roberts, 1964). Conversely, at very low oxygen dosage rates, the formation of the suboxide could be favoured. Paralleling the changes:—oxygen chemisorbed on metal → suboxide → oxygen chemisorbed on suboxide, there was an increase, then a decrease and finally an increase in the work function (Roberts and Wells, 1966). Analogous results have been obtained from photoelectric measurements.

Only a few surface potential measurements have been made on oxides having an average thickness in excess of about 20 Å (e.g. Hirschberg and Lange, 1952) but these could improve our knowledge of the electric fields which exist in a growing oxide layer.

4. Field Experiments

The importance of field experiments, that is, studying the effect of an electric field on the oxidation rate of a metal, has already been mentioned. The electric field can be conveniently applied between the base metal and a porous platinum layer of the type used in resistance measurements. This arrangement was used by both Jorgensen (1962, 1963) and Anderson and Ritchie (1967b). On the other hand, Bradhurst et al. (1965) found it convenient to use doped nickel oxide powder as the electrode at the oxygen/oxide interface. Various techniques are used by these authors to monitor the oxidation rate during the application of the field. Jorgensen (1962, 1963) used an optical method for the oxidation of silicon and measured the resistance between the porous platinum and the metal for the oxidation of zinc. Bradhurst et al. (1965), who were following the fairly large oxygen uptakes associated with the high temperature oxidation of zirconium, adapted the gravimetric method to their purpose. But for the low temperature oxidation of zinc and tungsten where the reaction rates are small, Anderson and Ritchie (1967b) used their evaporated film version of the resistance marker method for determining the most mobile species. With this technique, the rate of change of resistance with time under fixed oxidation conditions is not particularly reproducible. This being so, it was found convenient to study the effect of the electric field by repeatedly reversing its polarity during the same kinetic run, rather than by comparing the results of two different runs, one with, and one without the applied electric field.

There are thus two resistance methods for following the kinetics of an oxidation reaction. The one used by Anderson and Ritchie (1967b) depends on the resistance between the porous platinum and the base metal being a measure of the oxide thickness. It is only applicable when the oxide grows at the oxide/metal interface, and cannot be used for those oxides which grow at the oxygen/oxide interface. The latter class includes all the metal deficient p-type oxides and certain n-type oxides, such as aluminium oxide, which contain interstitial cations. For metals which form this type of oxide, the other resistance method in which the conductance of a thin metal layer is a measure of the amount of unoxidized metal, and hence of the oxide thickness, can be used to monitor the oxidation rate during a field experiment. Ritchie et al. (1970) employed this procedure to carry out field experiments on oxidizing nickel which gives p-type nickel oxide. Again it was found to be convenient to alternate the polarity of the applied electric field and observe the consequent rate changes. A typical plot is shown in Figure 10.

As Fromhold (1963d) has pointed out, the interpretation of field experiments is not always unambiguous; they do however serve to reduce the number of possible rate determining steps.

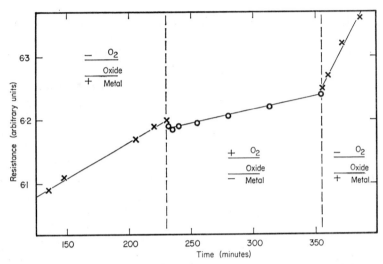

FIG. 10. The effect of an applied electric field on the oxidation rate of nickel at 371 °C. Reproduced with permission from Ritchie, Scott and Fensham (1970). *Surface Sci.* **19**, 230.

5. *Doping Experiments*

In a doping experiment, the oxidizing system is perturbed by the introduction of a small quantity of impurity into the metal or gas, and the changes in oxidation rate which have been effected by the dopant are studied. Doping of the metal is a common procedure with bulk metal samples, but almost unknown with evaporated film specimens. When interpreting the changes in rate, it is usually assumed that the dopant is dissolved in the metal and also in the oxide which is formed. Hauffe's doping rule (1953) depends on this assumption. One cannot be sure in evaporating a doped metal that the impurity will be uniformly distributed in the resultant film since the evaporation rates of two elements are in general totally dissimilar. Nor can one be sure, without experimental verification, that the oxide itself has been doped, for Boggs *et al.* (1963) have shown that in the oxidation of tin, certain dopants will oxidize in preference to the solvent metal. Neither of these uncertainties would appear to have been completely eliminated in the evaporated film study of the oxidation of lead doped with bismuth and thallium carried out by Anderson and Tare (1964a).

When evaporated films are used, doping of the gas phase would seem to be a more satisfactory procedure since doubts concerning the composition of the film would be disposed of, although it would still be necessary to check the composition of the product layer. Figure 11 shows the effect on the oxidation rate of zinc in oxygen at 1 atm

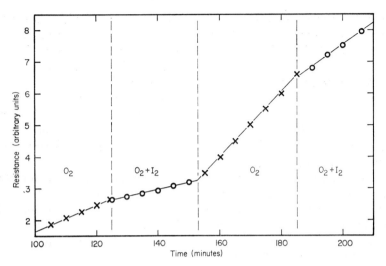

Fig. 11. The effect of iodine doping on the oxidation rate of zinc at 393 °C.

pressure of introducing a small quantity ($\sim 0 \cdot 1$ mole %) of iodine into the oxygen (Ritchie and Tandon, 1970).

Doping experiments, like field experiments, do not uniquely determine the rate determining step. They do curtail the number of possibilities, and by performing a number of such mechanistic experiments, it should be possible to specify completely the reaction mechanism.

IV. CONCLUSIONS

In this chapter, we have attempted to outline the current state of knowledge in the field of metallic oxidation, and to show the part in developing the topic that has been played by experiments using vacuum deposited metal films. While the way in which thick oxides grow at high temperatures is at least partially understood, the same cannot be said about the early stages of oxide formation, and the growth of thin layers at low temperatures. It is in the investigation of the initial oxide/low temperature region that vacuum deposited films have found their greatest use.

Most of the published work with films is restricted to kinetic studies on films of uncertain structure. No new major contribution can be expected from studies of this type which tell us little about the make-up of the film and about the reaction mechanism. Their chief value has been in providing information about the very first stages of oxide growth, for the study of which a clean uncontaminated metal surface is essential. If further kinetic studies are to be carried out with evaporated metal films, they should not only be prepared under ultra-high vacuum conditions, but also epitaxed onto suitable substrates to ensure that the surface is well characterised. This would be valuable for there is little information on the oxidation of clean oriented surfaces.

The flexibility of specimen preparation afforded by the process of vacuum deposition has often been exploited to good purpose. This has been valuable not so much in measurements of the oxidation rate as such, but in the so-called mechanistic experiments in which other important variables have been measured or controlled.

In order to obtain a better understanding of the way in which metals oxidize, more precise kinetic data for well characterised surfaces are required, but particularly we require information which can only be derived from mechanistic experiments. In view of this it seems likely that evaporated metal films will play an increasingly important part in the study of the oxidation of metals.

REFERENCES

Addiss, R. R. (1963). *Acta Metall.* **11**, 129.
Allpress, J. G. and Sanders, J. V. (1964). *Phil. Mag.* **10**, 827.
Anderson, J. R. and Clark, N. J. (1963). *J. Phys. Chem.* **67**, 2135.
Anderson, J. R. and Ritchie, I. M. (1967a). *Proc. R. Soc.*, **A299**, 354.
Anderson, J. R. and Ritchie, I. M. (1967b). *Proc. R. Soc.*, **A299**, 371.
Anderson, J. R. and Tare, V. B. (1964a). *In* "Electrochemistry" (J. A. Friend, F. Gutmann and J. W. Hayes, eds.). Pergamon, London, p. 813.
Anderson, J. R. and Tare, V. B. (1964b). *J. Phys. Chem.* **68**, 1482.
Anderson, J. S. (1967). *Rec. Aust. Acad. Sci.* **1**, 109.
Bassett, G. A., Menter, J. W. and Pashley, D. W. (1959). *Discuss. Faraday Soc.* **28**, 7.
Beck, A. F., Heine, M. A., Caule, E. J. and Pryor, M. J. (1967). *Corros. Sci.* **7**, 1.
Bell, A. E., Swanson, L. W. and Crouser, L. C. (1968). *Surface Sci.* **10**, 254.
Bénard, J. (1967). *J. Electrochem. Soc.* **114**, 139c.
Bennett, M. J. and Tompkins, F. C. (1960). *Proc. R. Soc.* **A259**, 28.
Bloomer, R. N. (1957a). *Br. J. Appl. Phys.* **8**, 40.
Bloomer, R. N. (1957b). *Br. J. Appl. Phys.* **8**, 321.
Boggio, J. E. and Farnsworth, H. E. (1964). *Surface Sci.* **3**, 62.
Boggio, J. E. and Plumb, R. C. (1966). *J. Chem. Phys.* **44**, 1081.
Boggs, W. E., Trozzo, P. S. and Pellissier, G. E. (1961). *J. Electrochem. Soc.* **108**, 13.
Boggs, W. E., Kachik, R. H. and Pellissier, G. E. (1963). *J. Electrochem. Soc.* **110**, 4.
Boggs, W. E., Kachik, R. H. and Pellissier, G. E. (1965). *J. Electrochem. Soc.* **112**, 539.
Boggs, W. E., Kachik, R. H. and Pellissier, G. E. (1967). *J. Electrochem. Soc.* **114**, 32.

Booth, F. (1948). *Trans. Faraday Soc.* **44**, 796.

Bradhurst, D. H., Draley, J. E. and van Drunen, C. J. (1965). *J. Electrochem. Soc.* **112**, 1171.

Cabrera, N. and Mott, N. F. (1949). *Rep. Progr. Metal Phys.* **12**, 163.

Cabrera, N., Terrien, J. and Hamon, J. (1947). *C.R.* **224**, 1558.

Campbell, W. E. and Thomas, U. B. (1947). *Trans. Electrochem. Soc.* **91**, 623.

Carter, R. E. and Richardson, F. D. (1955). *Trans. Am. Inst. Min. Metall. Engrs.* **203**, 336.

Castle, J. E., Gregg, S. J. and Jepson, W. B. (1962). *J. Electrochem. Soc.* **109**, 1018.

Cathcart, J. V. and Petersen, G. F. (1968). *J. Electrochem. Soc.* **115**, 595.

Chang, G. C. and Germer, L. H. (1967). *Surface Sci.* **8**, 115.

Cohen, M. S. (1960). *Acta Metall.* **8**, 356.

Cope, F. W. (1964). *J. Chem. Phys.* **40**, 2653.

Culver, R. V. and Tompkins, F. C. (1959). *Adv. Catalysis* **11**, 67.

Dankov, P. D. and Churaev, P. V. (1950). *Dokl. Akad. Nauk S.S.S.R.* **73**, 1221.

Davies, D. E., Evans, U. R. and Agar, J. N. (1954). *Proc. R. Soc.* **A225**, 443.

Deal, B. E., Sklar, M., Grove, A. S. and Snow, E. H. (1967). *J. Electrochem. Soc.* **114**, 266.

Delchar, T. and Tompkins, F. C. (1967). *Surface Sci.* **8**, 165.

Dignam, M. J. and Huggins, D. A. (1967). *J. Electrochem. Soc.* **114**, 117.

Eley, D. D. and Wilkinson, P. R. (1960). *Proc. R. Soc.* **A254**, 327.

Engell, H. J. and Hauffe, K. (1952). *Metall* **6**, 285.

Ernst, L. (1967). *Surface Sci.* **6**, 487.

Evans, U. R. (1945). *Trans. Faraday Soc.* **41**, 365.

Evans, U. R. (1946). *Nature, Lond.* **157**, 732.

Evans, U. R. (1947). *Trans. Electrochem. Soc.* **91**, 547.

Evans, U. R. (1955). *Rev. Pure Appl. Chem.* **5**, 1.

Evans, U. R. (1960). "The Corrosion and Oxidation of Metals". Edward Arnold, London.

Farnsworth, H. E. (1964). *Adv. Catalysis* **15**, 31.

Farnsworth, H. E. and Madden, H. H. (1961). *J. Appl. Phys.* **32**, 1933.

Faust, J. W. (1963). *Acta Metall.* **11**, 1077.

Fehlner, F. P. (1968). *J. Electrochem. Soc.* **115**, 726.

Fiegna, A. and Weisgerber, P. (1968). *J. Electrochem. Soc.* **115**, 369.

Fischmeister, H. F. (1960). *In* "Reactivity of Solids" (J. H. de Boer *et al.*, eds.). Elsevier, Amsterdam, p. 195.

Fromhold, A. T. (1963a). *J. Phys. Chem. Solids* **24**, 1081.

Fromhold, A. T. (1963b). *J. Phys. Chem. Solids* **24**, 1309.

Fromhold, A. T. (1963c). *J. Chem. Phys.* **39**, 2278.

Fromhold, A. T. (1963d). *Nature, Lond.* **200**, 559.

Fromhold, A. T. (1964). *J. Phys. Chem. Solids* **25**, 1129.

Fromhold, A. T. (1968). *J. Electrochem. Soc.* **115**, 882.

Fromhold, A. T. and Cook, E. A. (1967). *Phys. Rev.* **158**, 600.

Germer, L. H. (1966). *Surface Sci.* **5**, 147.

Germer, L. H. and Macrae, H. U. (1962). *J. Appl. Phys.* **33**, 2923.

Gibbs, G. B. (1967). *Corros. Sci.* **7**, 165.

Green, M. (1957). *In* "Semiconductor Surface Physics" (R. H. Kingston ed.), University of Pennsylvania Press, Philadelphia, p. 362.

Green, M. and Liberman, A. (1962). *J. Phys. Chem. Solids* **23**, 1407.

Grimley, T. B. and Trapnell, B. M. W. (1956). *Proc. R. Soc.* **A234**, 405.

Hapase, M. G., Tare, V. B. and Biswas, A. B. (1967). *Acta Metall.* **15**, 131.

Hapase, M. G., Gharpurey, M. K. and Biswas, A. B. (1968a). *Surface Sci.* **9**, 87.

Hapase, M. G., Gharpurey, M. K. and Biswas, A. B. (1968b). *Surface Sci.* **12**, 85.

Harris, W. W., Ball, F. L. and Gwathmey, A. T. (1957). *Acta Metall.* **5**, 574.

Harrison, P. L. (1965). *J. Electrochem. Soc.* **112**, 235.

Hauffe, K. (1953). *Prog. Metal Phys.* **4**, 71.

Hauffe, K. (1965). "Oxidation of Metals and Alloys". Plenum Press, New York.

Hauffe, K. and Ilschner, B. (1954). *Z. Elektrochem.* **58**, 382.

Hauffe, K., Pethe, L., Schmidt, R. and Morrison, S. R. (1968). *J. Electrochem. Soc.* **115**, 456.

Hirschberg, R. and Lange, E. (1952). *Naturwissenschaften* **39**, 187.

Hoar, T. P. and Price, L. E. (1938). *Trans. Faraday Soc.* **34**, 867.

Hooke, R. and Brody, T. P. (1965). *J. Chem. Phys.* **42**, 4310.

Huber, E. E. and Kirk, C. T. (1966). *Surface Sci.* **5**, 447.

Hurlen, T. (1960). *J. Inst. Metals* **89**, 128.

Ilschner, B. (1955). *Z. Elektrochem.* **59**, 542.

Irving, B. A. (1965). *Corros. Sci.* **5**, 471.

Jackson, A. G., Hooker, M. P. and Haas, T. W. (1968). *Surface Sci.* **10**, 308.

Jaeger, H. and Sanders, J. V. (1968). *J. Res. Inst. Catalysis Hokkaido Univ.* **16**, 287.

Jander, W. (1927). *Z. Anorg. Allg. Chem.* **163**, 1.

Jorgensen, P. J. (1962). *J. Chem. Phys.* **37**, 874.

Jorgensen, P. J. (1963). *J. Electrochem. Soc.* **110**, 461.

Kennedy, S. W., Calvert, L. D. and Cohen, M. (1959). *Trans. Metall. Soc. A.I.M.E.* **215**, 64.

Kofstad, P. (1961). *J. Inst. Metals* **90**, 253.

Kofstad, P. (1966). "High Temperature Oxidation of Metals". Wiley, New York.

Kruger, J. and Yolken, H. T. (1964). *Corrosion* **20**, 29t.

Kubaschewski, O. and Hopkins, B. E. (1967). "Oxidation of Metals and Alloys". Butterworth, London.

Landsberg, P. T. (1955). *J. Chem. Phys.* **23**, 1079.

Lanyon, M. A. H. and Trapnell, B. M. W. (1955). *Proc. R. Soc.* **A227**, 387.

Law, J. T. (1958). *J. Phys. Chem. Solids* **4**, 91.

Lee, R. N. and Farnsworth, H. E. (1965). *Surface Sci.* **3**, 461.

Meijering, J. L. and Verheijke, J. L. (1959). *Acta Metall.* **7**, 331.

Michell, D. and Smith, A. P. (1968). *Phys. Status Solidi* **27**, 291.

Miyake, K. (1961). *J. Phys. Soc. Japan* **16**, 1473.

Moore, W. J. (1952). *Phil. Mag.* **43**, 688.

Moore, W. J. (1953). *J. Chem. Phys.* **21**, 1117.

Moore, W. J. and Lee, J. K. (1951). *Trans. Faraday Soc.* **48**, 501.

Mott, N. F. (1939). *Trans. Faraday Soc.* **35**, 1175.

Mott, N. F. (1947). *Trans. Faraday Soc.* **43**, 429.

Mott, N. F. and Gurney, R. W. (1940). "Electronic Processes in Ionic Crystals". Oxford University Press, Oxford.

Mrowec, S. and Werber, T. (1960). *Acta Metall.* **8**, 819.

Pfeil, L. B. (1929). *J. Iron Steel Inst.* **119**, 501.

Pilling, N. B. and Bedworth, R. E. (1923). *J. Inst. Metals* **29**, 529.

Quinn, C. M. and Roberts, M. W. (1962). *Proc. Chem. Soc.* 246.

Quinn, C. M. and Roberts, M. W. (1964). *Trans. Faraday Soc.* **60**, 899.
Quinn, C. M. and Roberts, M. W. (1965). *Trans. Faraday Soc.* **61**, 1775.
Raleigh, D. O. (1966). *J. Electrochem. Soc.* **113**, 782.
Rhead, G. E. (1965). *Trans. Faraday Soc.* **61**, 797.
Rhodin, T. N. (1953). *Adv. Catalysis* **5**, 39.
Richardson, F. D. and Jeffes, J. H. E. (1948). *J. Iron Steel Inst.* **160**, 261.
Ritchie, I. M. and Hunt, G. L. (1969). *Surface Sci.* **15**, 524.
Ritchie, I. M. and Scott, G. H. (1968). *In preparation.*
Ritchie, I. M. and Tandon, R. K. (1970). *Surface Sci.* **22**, 199.
Ritchie, I. M., Scott, G. H. and Fensham, P. J. (1970). *Surface Sci.* **19**, 230.
Roberts, M. W. (1961). *Trans. Faraday Soc.* **57**, 99.
Roberts, M. W. and Tompkins, F. C. (1959). *Proc. R. Soc.* **A251**, 369.
Roberts, M. W. and Wells, B. R. (1966). *Trans. Faraday Soc.* **62**, 1608.
Romanski, J. (1968). *Corros. Sci.* **8**, 67.
Ronnquist, A. and Fischmeister, H. F. (1960). *J. Inst. Metals* **89**, 65.
Rosenberg, A. J., Menna, A. A. and Turnbull, T. P. (1960). *J. Electrochem. Soc.* **107**, 196.
Roykh, I. L., Tolkachev, V. Ye., Pustotina, S. R. and Rafalavich, D. M. (1966). *Fizika Metall.* **21**, 131.
Ruhl, W. (1963). *Z. Phys.* **176**, 409.
Sachs, K. (1956). *Metallurgia* **54**, 11.
Simmons, G. W., Mitchell, D. F. and Lawless, K. R. (1967). *Surface Sci.* **8**, 130.
Smeltzer, W. W. and Simnad, M. T. (1957). *Acta Metall.* **5**, 328.
Smith, A. W. (1964). *J. Phys. Chem.* **68**, 1465.
Steinheil, A. (1934). *Annln. Phys.* **19**, 465.
Stringer, J. (1967). *J. Electrochem. Soc.* **114**, 428.
Swaine, J. W. and Plumb, R. C. (1962). *J. Appl. Phys.* **33**, 2378.
Tammann, G. (1920). *Z. Anorg. Allg. Chem.* **111**, 78.
Uhlig, H. H. (1956). *Acta Metall.* **4**, 541.
van Itterbeek, A., de Greve, L. and Francois, M. (1950). *Meded. K. Vlaam. Acad.* **22**, 4.
Vermilyea, D. A. (1957). *Acta Metall.* **5**, 492.
Vernon, W. H. J., Akeroyd, E. I. and Stroud, E. G. (1939). *J. Inst. Metals* **65**, 301.
Wagener, S. (1956). *J. Phys. Chem.* **60**, 567.
Wagner, C. (1933). *Z. Phys. Chem.* **B21**, 25.
Wagner, C. (1950). *In* "Atom Movements" Am. Soc. Metals, Cleveland, p. 153.
Wagner, C. and Grunewald, K. (1938). *Z. Phys. Chem.* **B40**, 455.
Wallwork, G. R., Rosa, C. J. and Smeltzer, W. W. (1965). *Corros. Sci.* **5**, 113.
Weissmantel, Ch. (1962). *Werkstoffe Korros., Mannheim* **13**, 682.
Weissmantel, Ch., Schwabe, K. and Hecht, G. (1961). *Werkstoffe Korros., Mannheim* **12**, 353.
Wieder, H. and Czanderna, A. W. (1962). *J. Phys. Chem.* **66**, 816.
Young, F. W., Cathcart, J. V. and Gwathmey, A. T. (1956). *Acta Metall.* **4**, 145.
Young, L. (1961). "Anodic Oxide Films". Academic Press, New York.

Author Index (Volume 2)

The numbers in *italics* refer to pages in the References at the end of each Chapter.

F

Fahrenfort, J., 85, 200, *208*
Farkas, A., 126, 127, 128, 131, 133, 153, 193, *205*
Farkas, L., 131, 133, 153, 193, *205*
Farnsworth, H. E., 261, 264, *304, 305, 306*
Faulkner, E. A., 239, *254*
Faust, J. W., 268, *305*
Fedak, D. G., 15, *61*
Fehlner, F. P., 296, *305*
Fejes, P., 22, *62*
Fensham, P. J., 283, 301, 302, *307*
Fiegna, A., 298, *305*
Fischmeister, H. F., 265, 267, 268, 286, *305, 307*
Fisher, M. E., 28, *61*
Flanagan, T. B., 95, *205*
Foss, T. G., 137, 138, *206*
Fowler, R. H., 23, 26, *61*, 239, *254*
Fox, P. G., 17, 35, 38, *60*
Francois, M., 289, 294, *306*
Frankenburger, W., 75, 79, *206*
Freel, J., 185, 186, *206*
Frennet, A., 88, *205*
Fromhold, A. T., 276, 277, 278, 280, 281, 282, 285, 301, *305*
Fujiki, Y., 223, *254*
Fukuda, K., 200, *206*

G

Galwey, A. K., 65, 70, 87, 92, 128, 129, 144, 145, 147, 158, 185, 186, *205, 206, 209*
Garnett, J. L., 137, 154, 155, *206*
Gault, F. G., 43, *61*, 94, 97, 104, 105, 111, 116, 118, 123, 162, 163, 169, 170, 173, 174, 176, 177, 178, 184, 185, *204, 205, 206, 207, 208*
Gaunt, D. S., 28, *61*
Geller, B., 114, *208*
Gerasimov, Y. I., 237, *255*
Gerdes, R., 2, 3, 4, 6, 16, *62*
Germer, L. H., 261, 262, 264, *305*
Gharpurey, M. K., 221, 246, 251, *254*, 289, 290, 292, 293, 297, *306*
Gibbs, G. B., 286, *305*
Gillam, E., 200, *208*
Gjosten, N. A., 15, *61*

Gleiser, M., 89, *206*
Goldsztaub, S., 221, *254*
Good, R. H., 80, *206*
Gosselain, P., 88, *205*
Green, M., 262, *305*
Greenhalgh, R. K., 151, 153, *206*
Gregg, S. J., 269, *305*
Greve L. de, 289, 294, *306*
Grimley, T. B., 277, 278, 282, 285, *305*
Grove, A. S., 265, *305*
Grover, S. S., 41, *61*
Grunewald, K., 280, *307*
Guggenheim, E. A., 23, 26, *61*
Gundry, P. M., 8, 17, *61*, 76, 80, *206*
Gurney, R. W., 278, *306*
Gutmann, J. R., 134, *206*
Gwathmey, A. T., 268, 274, *306, 307*

H

Haas, T. W., 290, *306*
Haber, J., 76, 80, *206*
Hall, W. K., 138, *206*
Halsey, G. D., 22, 34, *61, 62*, 139, *206*
Hamon, J., 291, *305*
Hansen, R. S., 75, 79, 82, *204, 205*
Hapase, M. G., 289, 290, 292, 293, 297, *306*
Hara, J., 143, *209*
Harper, R. J., 158, *206*
Harris, G. M., 42, *61*
Harris, L., 217, *254*
Harris, L. B., 35, *61*
Harris, W. W., 268, *306*
Harrison, P. L., 283, *306*
Hauffe, K., 259, 272, 277, 279, 280, 285, 302, *305, 306*
Hayward, D. O., 11, *60*, 83, 92, *206*
Head, A. K., 35, *60*
Hecht, G., 288, 292, 293, *307*
Heerden, R. van, 77, *204*
Heine, M. A., 265, 284, *304*
Heinemann, H., 163, *207*
Henning, C. A. O., 6, *61*
Heras, J. M., 81, *209*
Heras, L. V., 81, *209*
Heyne, H., 17, *62*
Hill, T. L., 21, *61*
Hillaire, L., 123, 145, *206*
Hironaka, Y., 143, *206*

P

Subject Index (Volume 2)

F